大智小技 Ⅵ

数据库生产实战漫笔

爱可生开源社区 著

中国市场出版社
·北京·

图书在版编目（CIP）数据

大智小技：数据库生产实战漫笔 / 爱可生开源社区著. -- 6版. -- 北京：中国市场出版社有限公司，2024. 12. -- ISBN 978-7-5092-2631-5

Ⅰ. TP311.132.3

中国国家版本馆 CIP 数据核字第 2024WK6421 号

大智小技Ⅵ：数据库生产实战漫笔
DAZHIXIAOJI VI: SHUJUKU SHENGCHAN SHIZHAN MANBI

作　　者：爱可生开源社区
责任编辑：刘佳禾

出版发行：中国市场出版社
社　　址：北京市西城区月坛北小街 2 号院 3 号楼（100837）
电　　话：（010）68040722/68034118/68021338
网　　址：http://www.scpress.cn
印　　刷：北京捷迅佳彩印刷有限公司
规　　格：185mm×260mm　　16 开本
印　　张：27　　　　　　　　字　　数：525 千字
版　　次：2024 年 12 月第 1 版　印　　次：2024 年 12 月第 1 次印刷
书　　号：ISBN 978-7-5092-2631-5
定　　价：99.00 元

版权所有　侵权必究　　　印装差错　负责调换

大智者，宏观把握，
对技术产生的背景以及相应的原理应用了然于胸。
小技者，具体细微，
无论什么问题一定要剖析入骨，捋清脉络。
拥有了大智小技，面对问题轻松自如，
一目了然，一锤定音。

推荐序

爱可生开源社区非常赞，一直坚持输出高质量内容，这本《大智小技 VI：数据库生产实战漫笔》沉淀了社区大量的实战内容，近 100 篇文章涵盖了 MySQL 8.0 的特效、MySQL 故障分析、OceanBase 知识库。其中，"MySQL 故障分析篇"讨论了大量用户遇到过的疑难杂症，如主从不一致问题、全文索引问题、Online DDL 问题等，"OceanBase 篇"分享了很多 OceanBase 实战经验，比如全局索引和局部索引如何选择、性能优化实战，还有更有意思的 ActionDB GIS 拓展。此外，还增加了"一问一实验（ChatDBA）篇"。强烈建议热爱技术的同学一起学习并参与社区建设。

——封仲淹　OceanBase 开源生态技术部总经理

作为一名技术管理者，我深知数据库在企业 IT 架构中的核心地位，我们一直在开源与商业、创新与稳定之间寻找平衡。这本《大智小技 VI：数据库生产实战漫笔》不仅涵盖 MySQL 实战精华，更收录了 OceanBase 实践经验，完美契合国产化转型需求，是一本难得的数据库运维实战指南。

——王健　中国电信全渠道运营中心研发部

在过去的一年，数据库行业飞速变化，使用的数据库产品跟随着大环境的变化而变化，爱可生持续跟随市场变化——我在本书中看到了头部国产数据库的内容，恰好我也在学习这方面的知识。感谢爱可生一直帮助大家成长，祝愿爱可生持续高速发展，助力国产数据库事业发展，共赢互助。

——刘华阳　"AustinDatabases"公众号作者

六载光阴，深耕技术，《大智小技VI：数据库生产实战漫笔》本书不仅详细介绍了MySQL 8.0数据库系统新特性，还引入OceanBase分布式数据库技术。同时，书中穿插了丰富的故障分析案例和一问一实验（ChatDBA）的互动环节，旨在帮助读者全面了解和掌握数据库技术的精髓。融大智于小技之间，令人受益匪浅。

——胡捷　华夏银行

在这个快速发展的技术时代，持续学习和不断探索是我们前进的动力。《大智小技VI：数据库生产实战漫笔》的出版，为广大数据库爱好者和从业者提供了一个难得的学习机会。它不仅有助于提升技术水平，更能激发我们对开源技术的热情和创造力。让我们一起走进《大智小技VI：数据库生产实战漫笔》的世界，探索开源数据库的无限可能。

——张珍源　中移智家数据库负责人

每当翻开《大智小技》，仿佛打开了一扇通往数据库世界的大门。这本书源自爱可生开源社区，凝聚着大量DBA的智慧与经验。它贴近工作、接地气，无论是初入开源数据库领域的新手，还是经验丰富的技术专家，都能从中汲取宝贵的知识和实践经验。《大智小技》不仅仅是一套书，更是一个社区成员共同创作的结晶，可以感受到社区成员的无私奉献。

——徐良　中移智家数据库专家

《大智小技》历经六版迭代，始终与时代同行。最新版本不仅融合了前沿技术，更首次将LLM引入数据库诊断领域，开启了智能运维的新篇章。同时，书中详述了分布式数据库OceanBase的实战案例，为读者提供了宝贵的经验。此书无疑是数据库工程师不可多得的实战宝典。

——蔡鹏　货拉拉数据库负责人

"大智小技"系列图书自推出以来，一直深受广大数据库从业者的喜爱。作为数据库实战领域的经典之作，该系列不仅提供了最新的技术解读，还涵盖了丰富的生产环境案例，使每一位读者都能够从中获得真正有价值的知识与技能。这些年，越来越多的DBA、开发者和数据库爱好者通过"大智小技"系列提升自我，不仅掌握了更为先进的数据库操作技巧，也在实际工作中游刃有余地解决了诸多棘手的问题。

《大智小技Ⅵ：数据库生产实战漫笔》延续了前几版的风格，以实战为基础，内容深入浅出且不乏创新。无论是MySQL 8.0的新特性详解，还是OceanBase数据库的最佳实践，书中的每一个章节都通过细致的讲解和深入的分析，帮助读者更好地理解数据库的原理和最佳操作手段。在这一版中，故障排查与性能优化案例更为丰富，涉及日常运维中常见却复杂的问题，为大家提供了宝贵的解决思路。

可以说，这本书不仅是一部技术指南，更是一部充满智慧与经验的实战宝典。每一个案例都凝聚了作者团队在数据库领域多年的心血。无论你是刚刚入行的初学者，还是已有多年经验的资深从业者，这本书都将成为你工作中的重要参考，帮助你在数据库的世界中更加从容自信地应对每一个挑战。

——尚雷　TechTalk技术交流社区创办者

本年度的《大智小技Ⅵ：数据库生产实战漫笔》依然干货满满，诚意满满。不仅有MySQL 8.0新特性实践总结，还有故障案例分析，对于OceanBase的经验总结也是如数家珍。值得一提的是，从书中ChatDBA对问题的解读，可以看到很多值得DBA深思的地方。未来已来，值得推荐。

——杨建荣　dbaplus社群发起人，腾讯云TVP，
《Oracle/MySQL工作笔记》图书作者

一年一度的《大智小技》来了，到今年已是第六版。

近年来，数据库行业整体非常艰难，然而越是艰难越要用心做好产品，"致广大而尽精微"才是成事之道。

"尽小者大，积微者著"，能把细节做好，且持之以恒，由衷为爱可生开源社区的这份坚持点赞。

——叶金荣　万里数据库开源生态负责人，腾讯云TVP，墨天轮MVP

一年一度的社区新书再次来到，我迫不及待地打开，被其中大量的原理介绍所吸引。总的来讲，本次社区新书依旧紧扣开源和信创两个热门话题，大多从经典案例出发，深入分析其背后的原理，不仅能帮助读者了解如何处理故障，还能使其更深刻理解背后的技术逻辑，真正让读者做到了然于心。

——高鹏（八怪）　《MySQL主从原理》图书作者

很欣喜地看到《大智小技》已经出到第六版了。

这本书不仅是一本技术手册，更是一部引领读者深入理解MySQL精髓的宝典。书中不仅涵盖了最新的MySQL版本特性，还提供了大量实际案例和最佳实践，帮助读者将理论知识应用于实际工作。

本书延续以往风格，注重实践和理论的结合，尤其是"一问一实验（ChatDBA）篇"——基于ChatDBA智能辅助方式整理了8篇文章，彰显了一种知识积累、沉淀和转移的新形式，体现了令人深思时代的走向。无论是数据库新手，还是希望进一步提升技能的专业人士，这本书都能成为你的良师益友，使你从中获得宝贵的启示和实用的知识。

——杨磊　中信建投证券

爱可生服务的客户众多，涉及不同的服务器硬件、不同的操作系统、不同的数据库版本、不同的业务场景，总是会遇到各种奇怪的问题，以及常见和不常见的Bug，《大智小技Ⅵ：数据库生产实战漫笔》汇集了这一年来比较有代表性的各种问题和Bug的处置方案，相信对各位读者有帮助。

——操盛春　"一树一溪"公众号作者，"MySQL核心模块揭秘"专栏作者

在数据库行业高速发展的今天，技术积累与经验分享已成为推动行业进步的关键。《大智小技Ⅵ：数据库生产实战漫笔》延续了爱可生开源社区对内容一贯秉持的专业性与实用性，精选了过去一年中精彩的技术内容，不仅涵盖了MySQL集中式数据库的经典案例和实战经验，还紧跟潮流，拓展了分布式数据库OceanBase的实践探索，并深入探讨了AI技术在数据库领域的创新应用。

全新专栏"一问一实验（ChatDBA）"开创性地将AI与数据库技术结合，打造了一款面向数据库从业者的智能助手，填补了市场空白，为提升效率与优化工作流程提供了新的可能。本书不仅是一部技术实战的指导手册，更是记录行业前沿动态与技术发展的力作。希望每一位读者都能从中汲取灵感，拓展对数据库技术的全新认知。

——陈俊聪（芬达） "芬达的数据库学习笔记"公众号作者

时光荏苒，转眼《大智小技》来到第六版。作为参与者和读者，我见证了《大智小技》的成长。本书从分享MySQL技术，拓展到国产数据库的OceanBase的学习与探索。随着AI技术的兴起，爱可生推出基于大模型的ChatDBA。本书通过实际案例展示了ChatDBA的专业能力——既能对数据库问题进行鞭辟入里的分析，又能提供完善的解决方案。阅读本书，相信各位技术爱好者都能从中受益良多。

——杨奇龙 "yangyidba"公众号作者

《大智小技》系列图书不知不觉已经陪伴我们六年了。《大智小技 VI：数据库生产实战漫笔》介绍了MySQL 8.0新特性及故障分析，并新增了目前流行度比较高的国产开源分布式数据库OceanBase的相关内容。

此外，《大智小技 VI：数据库生产实战漫笔》还引入了AI相关的内容，为广大DBA开阔了视野。作为MySQL和OceanBase相关的丛书，《大智小技》从技术广度和深度上来说都是不错的图书。

——田帅萌 同程旅行分布式数据库负责人

在这个技术飞速发展的时代，数据库作为信息技术的基石，其重要性不言而喻。《大智小技》系列图书，作为爱可生开源社区的瑰宝，一直深受数据库从业者的喜爱。如今，《大智小技 VI：数据库生产实战漫笔》即将面世，令人期待。

本书不仅延续了前五册的经典内容，如MySQL故障案例剖析和技术分享，更紧跟技术前沿，对日益流行的OceanBase进行了深入解读。同时，本书大胆创新，引入了数据库根因分析智能助手——ChatDBA,通过《一问一实验(ChatDBA)》专栏,展现了 AI 如何助力数据库运维,为行业带来了新的视角。

作为与爱可生长期合作的伙伴,我们深知其技术实力和对用户需求的精准把握。本书不仅是我们技术团队期待已久的宝贵资料,更是对整个数据库行业的一份贡献。

在此,祝贺《大智小技 VI:数据库生产实战漫笔》的出版,期待它能继续引领数据库技术的潮流,为更多从业者带来启发与收获。

——谢聪 上汽大众汽车首席数据库技术专家

目录 contents

01 MySQL 篇
——技术分享

1 如何缩短 MySQL 物理备份恢复时间？ / 2

2 MySQL 数据导入方案推荐 / 8

3 详解 MySQL 三表 JOIN / 15

4 如何通过 binlog 定位大事务？ / 22

5 MySQL 8.0 字段信息统计机制 / 26

6 InnoDB 全表扫描和全主键扫描一样吗？ / 31

7 MySQL 权限变更何时生效？ / 40

8 盘点 MySQL 创建内部临时表的所有场景 / 45

9 MySQL MGR 滚动升级可行吗？ / 50

10 你知道 MySQL 函数 sysdate() 与 now() 的区别吗？ / 56

11 细说 MySQL 用户安全加固策略 / 62

12 MySQL 创建表后神秘消失？揭秘零宽字符陷阱 / 74

13 MySQL VARCHAR 最佳长度评估实践 / 76

14 什么情况下 MySQL 连查询都能被阻塞？ / 83

15 MySQL 生产环境 GROUP BY 优化实践 / 90

16 MySQL 从库可以设置 sync_binlog 非 1 吗？ / 100

17 从实现原理来看为什么 Clone 插件比 Xtrabackup 更好用？ / 103

18 基于 MySQL 的机房容灾建设和切换演练 / 114

02 MySQL 篇
——故障分析

1 生产环境遇到 MySQL 数据页损坏问题如何解决？ / 128

2 一则 MySQL 从节点无响应问题分析 / 132

3 如何通过 blktrace 排查磁盘异常？ / 144

4 MySQL 主从复制遇到 1590 报错 / 151

5 MySQL 的隐式转换导致诡异现象的案例一则 / 157

6 主从数据不一致竟然不报错？ / 160

7 从一个死锁问题分析优化器特性 / 165

8 MySQL 全文索引触发 OOM 一例 / 174

9 MySQL 扩展 VARCHAR 长度遭遇问题的总结 / 180

10 MySQL 执行 Online DDL 操作报错空间不足？ / 188

11 server_id 引发的级联复制同步异常 / 192

12 TCP 缓存超负荷导致的 MySQL 连接中断 / 196

13 MySQL 通过 systemd 启动时无响应 / 200

14 ERROR 1709: Index column size too large 引发的思考 / 206

15 MySQL 5.7 连续 Crash 引发 GTID 丢失 / 210

16 为什么你的 show slave status 会卡住？ / 216

17 MySQL 含有下画线的数据库名在特殊情况下导致权限丢失 / 224

18 企业如何做好 SQL 质量管理？ / 231

03 OceanBase 篇

1 OceanBase 安全审计之传输加密 / 240

2 OceanBase 安全审计之透明加密 / 245

3 OceanBase 建表分区数超限报错 / 254

4 MySQL 迁移至 OceanBase 场景中的自增主键实践 / 258

5 如何通过日志观测冻结转储流程？/ 269

6 从 Oracle 迁移至 OceanBase 后存储过程语法报错问题的诊断 / 275

7 Join 估行不准选错执行计划该如何优化？/ 284

8 Oracle 中部分不兼容对象迁移到 OceanBase 的处理方式 / 289

9 一个关于 NOT IN 子查询的 SQL 优化案例 / 294

10 OceanBase 是如何关闭主备线程的？/ 305

11 如何有效使用 outline 功能？/ 308

12 1000s → 10s OceanBase 标量子查询改写案例 / 312

13 日志盘过小也会导致创建租户失败？/ 318

14 Oracle 模式竟然可以使用 Repeatable Read？/ 323

15 ActionDB 扩展 OceanBase GIS 能力：新增 ST_PointN 函数 / 329

16 ActionOMS 具备的延迟智能诊断功能 / 334

04 一问一实验（ChatDBA）篇

1 MySQL 日志报错定位 / 343

2 MySQL 半同步复制频繁报错 / 351

3 MySQL 创建了用户却无法登录 / 359

4 MySQL Too many open files 报错 / 367

5 MySQL 频繁 Crash 怎么办？ / 382

6 为什么不建议关闭 MySQL 严格模式？ / 391

7 MySQL 清理 undo log 居然用了 10 个小时？ / 396

8 MySQL Slave 异常掉电后主从同步出现异常怎么排查？ / 403

01 MySQL 篇 —— 技术分享

在数据库的世界里，MySQL 以其强大的功能和广泛的应用，成为世界上最流行的数据库。本篇我们将深入探讨 MySQL 的核心功能，并提供一系列的技术解读和实践指导。我们精心挑选了 18 篇文章，内容涵盖了 MySQL 的性能优化、备份恢复、数据迁移、权限管理等多个方面，旨在帮助读者全面理解 MySQL 的运行机制，并在遇到实际问题时，能够迅速找到解决问题的钥匙。

今年的 MySQL 篇，我们在往年的基础上进行了内容的整合与扩充，每一篇都基于实际案例，结合作者的深入分析，为读者提供一系列高效运用 MySQL 的技能。我们相信，通过这些精选的技术分享，读者不仅能够理解 MySQL 的内部运作原理，还能够在实际工作中游刃有余。

1 如何缩短 MySQL 物理备份恢复时间？

作者：李彬

1.1 背景

作为一名 DBA（数据库管理员），数据库的备份与恢复是异常重要的。日常我们也许关注的仅仅是提升备份效率，但在真实的运维场景下，数据恢复的时间成本考量更为重要，过长的恢复时间可能满足不了 RTO 的要求。本文以 Xtrabackup 工具为例，分别基于以下三个场景，来探讨如何加快数据的恢复速度。

(1) 场景一：全备之后，数据库故障，需要恢复全备+Binlog 的所有数据。

(2) 场景二：全备之后，误删除了某个库，需要恢复该库的所有数据。

(3) 场景三：全备之后，误删除了某个表，需要恢复该表的所有数据。

> 前置条件：你已经拥有了完整的 Xtrabackup 全量备份和 Binlog。

1.2 场景一

基于全备+Binlog 的恢复流程，实现恢复加速的妙招在于使用 SQL 线程进行 Binlog 回放，这样做有以下几点好处：

(1) 可以用到并行复制特性，速度更快。

(2) 可以使用复制过滤功能，只回放相应库表的 Binlog（单库或单表恢复场景）。

假设你已经恢复了完整的 Xtrabackup 全量备份到临时实例，使用 SQL 线程回放 Binlog 的操作过程如下文所述。

1.2.1 生成 index 文件

将全备后的所有 Binlog 均复制到临时实例的 relay log 目录中并重命名，然后生成 index 文件。

```
[root@localhost relaylog]$ rename mysql-bin mysql-relay mysql-bin*
[root@localhost relaylog]$ ls ./mysql-relay.0* > mysql-relay.index
[root@localhost relaylog]$ chown -R mysql.mysql mysql-relay.*
```

1.2.2 修改参数

修改 MySQL 参数（server_id 不能与原实例相同，relay_log_recovery 必须配置为 0，其余参数可以提升回放效率），重启临时实例。

```
[root@localhost relaylog]$ vim ../my.cnf.3306
[root@localhost relaylog]$ less ../my.cnf.3306 | grep -Ei "server_id|relay_
log_recovery|slave-para|flush_log_at|sync_binlog"
server_id                         = 4674
slave-parallel-type               = LOGICAL_CLOCK
slave-parallel-workers            = 8
sync_binlog                       = 0
innodb_flush_log_at_trx_commit    = 0
relay_log_recovery                = 0

[root@localhost relaylog]$ systemctl restart mysql_3306
[root@localhost relaylog]$ ps aux | grep 3306
```

1.2.3 建立复制通道并开启复制线程

```
[root@localhost relaylog]$ cat /data/mybackup/recovery/186-60-42/xtrabackup_
binlog_info
mysql-bin.000002  195862214  5af74703-a85e-11ed-a34e-02000aba3c2a:1-205
[root@localhost relaylog]$ mysql -S /data/mysql/3306/data/mysqld.
sock -uroot -p
mysql> CHANGE MASTER TO MASTER_HOST='1.1.1.1',RELAY_LOG_FILE='mysql-
relay.000002',RELAY_LOG_POS=195862214;
mysql> SELECT * FROM MYSQL.SLAVE_RELAY_LOG_INFO\G
mysql> START SLAVE SQL_THREAD;
```

看到这里，我们总结一下用到的加速技巧：

(1) 使用 SQL 线程回放 Binlog，并配置并行复制。

(2) 修改双一参数为双 0，进行复制加速。

1.3 场景二

针对从全备中恢复单库的场景，又该如何加速呢？除了 SQL 线程回放 Binlog，还需要用到我们第二个加速恢复的妙招——可传输表空间。

老规矩，先贴出官方文档的说明：https://dev.mysql.com/doc/refman/5.7/en/innodb-table-import.html

> 注意：使用可传输表空间的方式是有限制的，官方提出了 6 点使用前提，大家可以自行研究。

对于大表，使用表空间传输来进行表迁移比使用 SQL 恢复在效率上有很大的提升，且 Xtrabackup 也提供了 --export 参数支持，让我们在 Xtrabackup 恢复的 prepare 阶段就可以获取到 .cfg 等需要的文件。

以恢复 test 库为例（源库 3310，临时库 3311）：

1.3.1 准备表结构

首先我们需要有对应表的表结构，这里使用 mysqldump 导出，并在目标端进行导入：

```
# 逻辑导出
[root@localhost 3310]$ /data/mysql/3310/base/bin/mysqldump -uroot -p -h127.0.0.1 -P3310 --set-gtid-purged=off --no-data --databases test > ./testdb_schema_bak.sql

# 目标端导入
[root@localhost 3311]$ /data/mysql/3311/base/bin/mysql -uroot -p -h127.0.0.1 -P3311 < /data/mysql/3310/testdb_schema_bak.sql
```

1.3.2 Prepare

在全备中使用 --export 进行 prepare，生成用于表空间传输的相关文件：

```
# 全备 prepare 之前的文件
[root@localhost test]$ ll
-rw-r----- 1 root root     8632 Dec 14 10:45 sbtest1.frm
-rw-r----- 1 root root 30408704 Dec 14 10:45 sbtest1.ibd
-rw-r----- 1 root root     8632 Dec 14 10:45 sbtest2.frm
-rw-r----- 1 root root 30408704 Dec 14 10:45 sbtest2.ibd
...

# --export 之后的文件，可以看到针对 MySQL 的 .cfg 文件已经自动生成，代替了 FLUSH TABLES ... FOR EXPORT
[root@localhost 3310]$ xtrabackup --prepare --export --use-memory=1024MB --target-dir=/data/mysql/3310/backup/3310_20231214_full_bak

[root@localhost 3310]$ ll /data/mysql/3310/backup/3310_20231214_full_bak/test/
-rw-r--r-- 1 root root      490 Dec 14 10:47 sbtest1.cfg
-rw-r----- 1 root root    16384 Dec 14 10:47 sbtest1.exp
-rw-r----- 1 root root     8632 Dec 14 10:45 sbtest1.frm
-rw-r----- 1 root root 30408704 Dec 14 10:45 sbtest1.ibd
-rw-r--r-- 1 root root      490 Dec 14 10:47 sbtest2.cfg
-rw-r----- 1 root root    16384 Dec 14 10:47 sbtest2.exp
-rw-r----- 1 root root     8632 Dec 14 10:45 sbtest2.frm
-rw-r----- 1 root root 30408704 Dec 14 10:45 sbtest2.ibd
...
```

1.3.3 准备 SQL

拼凑出多表 DISCARD TABLESPACE 和 IMPORT TABLESPACE 命令，当表存在时，可使用 SQL 配合 information_schema.tables 表进行语句拼接。这里以 Shell 实现进行举例：

```
[root@localhost tmp]$ DATABASE='test'
[root@localhost tmp]$ for table in sbtest1 sbtest2 sbtest3 sbtest4 sbtest5
> do
>     echo "ALTER TABLE ${DATABASE}.${table} DISCARD TABLESPACE;" >> discard_ts.sql
>     echo "ALTER TABLE ${DATABASE}.${table} IMPORT TABLESPACE;"  >> import_ts.sql
> done
```

```
[root@localhost tmp]$ cat discard_ts.sql
ALTER TABLE test.sbtest1 DISCARD TABLESPACE;
ALTER TABLE test.sbtest2 DISCARD TABLESPACE;
ALTER TABLE test.sbtest3 DISCARD TABLESPACE;
ALTER TABLE test.sbtest4 DISCARD TABLESPACE;
ALTER TABLE test.sbtest5 DISCARD TABLESPACE;
[root@localhost tmp]$ cat import_ts.sql
ALTER TABLE test.sbtest1 IMPORT TABLESPACE;
ALTER TABLE test.sbtest2 IMPORT TABLESPACE;
ALTER TABLE test.sbtest3 IMPORT TABLESPACE;
ALTER TABLE test.sbtest4 IMPORT TABLESPACE;
ALTER TABLE test.sbtest5 IMPORT TABLESPACE;
```

1.3.4 将全备中对应的表文件与 SQL 文件复制至目标库目录

```
[root@localhost test]$ cp sbtest*.{cfg,ibd} /data/mysql/3311/tmp/
[root@localhost tmp]$ ll
total 148508
-rw-r--r-- 1 root root       225 Dec 14 14:00 discard_ts.sql
-rw-r--r-- 1 root root       225 Dec 14 14:00 import_ts.sql
-rw-r--r-- 1 root root       490 Dec 14 13:59 sbtest1.cfg
-rw-r----- 1 root root  30408704 Dec 14 13:59 sbtest1.ibd
-rw-r--r-- 1 root root       490 Dec 14 13:59 sbtest2.cfg
-rw-r----- 1 root root  30408704 Dec 14 13:59 sbtest2.ibd
-rw-r--r-- 1 root root       490 Dec 14 13:59 sbtest3.cfg
-rw-r----- 1 root root  30408704 Dec 14 13:59 sbtest3.ibd
-rw-r--r-- 1 root root       490 Dec 14 13:59 sbtest4.cfg
-rw-r----- 1 root root  30408704 Dec 14 13:59 sbtest4.ibd
-rw-r--r-- 1 root root       490 Dec 14 13:59 sbtest5.cfg
-rw-r----- 1 root root  30408704 Dec 14 13:59 sbtest5.ibd
# 注意权限
[root@localhost tmp]$ chown mysql. ./*
```

1.3.5 恢复数据

```
# 1. 丢弃表空间
[root@localhost tmp]$ /data/mysql/3311/base/bin/mysql -uroot -p -h127.0.0.1 -P 3311 < discard_ts.sql
# 2. 复制 .cfg 和 .ibd 到目标端的 test 库目录下
[root@localhost tmp]$ cp -a sbtest*.{cfg,ibd} /data/mysql/3311/data/test/
# 确认权限
[root@localhost tmp]$ ll /data/mysql/3311/data/test/
# 3. 导入表空间（可通过查看 mysql-error.log 确认该过程是否有报错）
[root@localhost tmp]$ /data/mysql/3311/base/bin/mysql -uroot -p -h127.0.0.1 -P 3311 < import_ts.sql
```

1.3.6 数据验证

```
mysql> use test;
mysql> select count(*) from sbtest1;
```

```
+----------+
| count(*) |
+----------+
|   100000 |
+----------+
1 row in set (0.21 sec)
...
```

至此，我们已经恢复了全备中的表数据。那么 Binlog 中的数据如何恢复呢？

其实我们仅需在临时实例中配置 SQL 线程回放 + 过滤复制，即可完成对表数据的全量恢复。与场景一不同，我们需要找到 DROP 操作的 GTID 或者 POS，配置过滤复制，并使 SQL 线程回放到 DROP 之前停止。

解析 binlog/relaylog，得到 DROP 操作的 GTID 或者 POS。

```
[root@localhost relaylog]$
while read relaylogname
do
/data/mysql/3311/base/bin/mysqlbinlog --base64-output=decode-rows -vvv $relaylogname | grep -Ei "drop" && echo "RELAYLOG 位置：$relaylogname"
done</data/mysql/3311/relaylog/mysql-relay.index

# DROP DATABASE `test` /* generated by server */
# RELAYLOG 位置：./mysql-relay.000006

# 解析 BINLOG/RELAYLOG 日志确认位点或者 GTID 信息（POS 信息：20135899）
[root@localhost relaylog]$ /data/mysql/3311/base/bin/mysqlbinlog --base64-output=decode-rows -vvv mysql-relay.000006 | less

# at 20135872
#231213 17:53:07 server id 60423306  end_log_pos 20135899    Xid = 7982902
COMMIT/*!*/;
# at 20135899
#231213 17:53:27 server id 60423306  end_log_pos 20135960    GTID    last_committed=9207    sequence_number=9208    rbr_only=no
SET @@SESSION.GTID_NEXT= '5af74703-a85e-11ed-a34e-02000aba3c2a:399350'/*!*/;
# at 20135960
#231213 17:53:27 server id 60423306  end_log_pos 20136076    Query    thread_id=70    exec_time=0    error_code=0
use `test`/*!*/;
SET TIMESTAMP=1675936407/*!*/;
SET @@session.pseudo_thread_id=70/*!*/;
/*!\C utf8 *//*!*/;
SET @@session.character_set_client=33,@@session.collation_connection=33,@@session.collation_server=46/*!*/;
DROP DATABASE `test` /* generated by server */
/*!*/;
# at 20136076
```

配置复制过滤。

```
mysql> CHANGE REPLICATION FILTER REPLICATE_DO_DB = (test);
Query OK, 0 rows affected (0.01 sec)
```

启动复制线程，到误删除的那个事务停止。

```
# 启动复制线程，到误删除的那个事务停止
mysql> START SLAVE SQL_THREAD UNTIL SQL_BEFORE_GTIDS = '5af74703-a85e-11ed-a34e-02000aba3c2a:399350';
# 若为基于 POS 的复制，则使用下面的语句
mysql> START SLAVE SQL_THREAD UNTIL RELAY_LOG_FILE = 'mysql-relay.000006', RELAY_LOG_POS = 20135899;
```

至此，大家应该对于在全备中如何快速恢复误删除库表有了一定的思路，场景三实际上与场景二的思路一致。当然，有些小伙伴可能有一个疑问：如果是误删除操作，源端的库表已经不存在了，如何获取表结构呢？这里提供两个方法：

(1) 相关的表结构可以从测试或者性能环境中导出，当然你需要确保各个环境的表结构是一致的。

(2) MySQL 8.0 之前，可以解析备份中的 .frm 文件获取表结构，如利用 mysqlfrm 工具。MySQL 8.0 之后，ibd2sdi 工具配合一些第三方脚本可助你一臂之力。

同样，我们总结一下用到的加速技巧：

(1) 配合 Xtrabackup 的 --export 参数，通过表空间传输只恢复对应的表，而无需恢复整个全备数据。在全备很大，但需要恢复的表很小时，这能节省很多时间。

(2) 针对大表，可以直接使用表空间传输进行表迁移，对比逻辑恢复效率提升明显（注意限制）。

(3) 在场景一的基础上，使用过滤复制的功能，针对单库或单表选择性地进行回放，进一步缩短了恢复的时间。

1.4 其他技巧

除了以上两个妙招，其实在恢复数据的整个流程中，还有一些影响速度的因素，如：

(1) 工具及其版本的选择。以 Xtrabackup 为例，8.0.33-28 版本针对 prepare 阶段进行了优化，效率提升明显。

(2) 结合实际的机器资源，合理配置工具的性能参数。如 Xtrabackup 的 --parallel 可以配合 --decompress 和 --decrypt 选项来进行并行解压缩和解密操作，--use-memory 指定 Xtrabackup--prepare 或者 Xtrabackup--stats 时使用的内存大小，对恢复效率也有一定影响。

(3) 恢复流程控制。prepare 阶段是需要一定时间的，我们可以在备份完成后直接做

prepare，从而省掉大量时间。此外，用于恢复的临时机器如何快速拿到备份文件也是优化的方向之一。

(4) 机器性能因素，如 CPU、磁盘性能、网络带宽（传输备份相关文件）等。

2 MySQL 数据导入方案推荐

作者：陈伟

2.1 需求背景

应用侧的同学需要对数据进行导出和导入，于是跑来找 DBA 咨询问题：MySQL 如何导入大批量的数据？

应用侧目前采用的方式为：

(1)mysqldump 工具。

(2)select outfile 语句。

(3) 图形化管理工具（MySQL Workbench、Navicat、DBeaver）。

DBA 听了觉得挺好的。

DBA 想，我的数据库我做主。于是，通知应用侧，目前先使用之前熟悉的方式进行对比，测试之后给建议。

> 建议：防止导入时出现大事务，造成主从延迟。

2.2 方案准备

待测方案：mysqldump、mydumper、select outfile 语句、Util.dumpTables、Util.exportTable。

2.2.1 环境配置信息

环境配置信息见表 1。

表 1

配置项	说明
MySQL 版本	5.7.39
磁盘随机读写	100 MB/s
测试表名	test.t_order_info
行数	1000W
字段数	6

2.2.2 建表语句

```
CREATE TABLE `t_order_info` (
  `ID` bigint(20) unsigned NOT NULL AUTO_INCREMENT COMMENT '自增主键ID',
  `order_no` varchar(64) NOT NULL DEFAULT '0000' COMMENT '订单编号',
  `order_status` varchar(2) NOT NULL DEFAULT '01' COMMENT '订单状态：00-异常、01-待处理、02-进行中、03-已完成',
  `flag` tinyint(4) NOT NULL DEFAULT '1' COMMENT '删除标识：1-正常、0-逻辑删除',
  `create_time` datetime NOT NULL DEFAULT CURRENT_TIMESTAMP COMMENT '创建时间',
  `modify_time` datetime NOT NULL DEFAULT CURRENT_TIMESTAMP ON UPDATE CURRENT_TIMESTAMP COMMENT '更新时间',
  PRIMARY KEY (`ID`),
  UNIQUE KEY `IDX_ORDER_NO` (`order_no`)
) ENGINE=InnoDB AUTO_INCREMENT=1 DEFAULT CHARSET=utf8mb4 COMMENT='订单表'
```

2.2.3 导出文件

(1) 包含数据结构和数据的备份文件（mysqldump、mydumper、Util.dumpTables）。

(2) 只包含数据的数据文件（select outfile、Util.exportTable）。

2.2.4 导出/导入命令

导出/导入命令见表2。

表2

导出	导入
mysqldump	source 或 mysql<
mydumper	myloader
select outfile	load data
Util.dumpTables	Util.loadDump
Util.exportTable	Util.importTable

2.3 方案测试

测试时首先考虑的是提升导入效率，并新增 MySQL Shell 的使用。

2.3.1 mysqldump

2.3.1.1 单表导出（备份文件）

```
mysqldump --default-character-set=utf8mb4 --master-data=2 --single-transaction --set-gtid-purged=off --hex-blob  --tables test t_order_info
```

(1)--master-data=2 参数会在备份期间对所有表加锁 FLUSH TABLES WITH READ LOCK，并执行 SHOW MASTER STATUS 语句以获取二进制日志信息。因此，在备份期间可能会影响数据库的并发性能。如果不需要进行主从复制，则可以考虑不使用该参数。

(2) --single-transaction 参数用于在备份期间"使用事务来确保数据一致性",从而避免在备份期间锁定表(必须有)。

2.3.1.2 备份文件

```sql
-- Table stricture for table `t_order_info`
--

DROP TABLE IF EXISTS `t_order_info`;
/*!40101 SET @saved_cs_client= @@character_set_client */;
/*!49101 SET character_set_client = utf8 */;
CREATE TABLE `t_order_info` (
`ID` bigint(2) unsigned NOT NULL AUTO_INCREMENT COMMENT '自增主键ID',
`order_no` varchar(64) NOT NULL DEFAULT `0000` COMMENT '订单编号',
`order_status` varchar(2) NOT NULL DEFAULT '01' COMMENT '订单状态:80-异常、81-待处理、2-进行中、03-已完成',
`flag` tinyint(4) NOT NULL DEFAULT '1' COMMENT '删除标识:1-正常、0-逻辑删除',
`create_time` datetime NOT NULL DEFAULT CURRENT_TIMESTAMP COMMENT '创建时间',
`modify_time` datetime NOT NULL DEFAULT CURRENT_TIMESTAMP ON UPDATE CURRENT_TIMESTAMP COMMENT '更新时间',
PRIMARY KEY (`ID`),
UNIOUE KEY `IDX_ORDER_NO` (`order no`)
) ENGINE=InnODB AUTO_INCREMENT=10129913 DEFAULT CHARSET=utf8m COMMENT='订单表';
/*!40101 SET character_set_client = @saved_cs_client */;

--
-- Dumping data for table `t_order_info`
--

LOCK TABLES `t_order_info` WRITE;
/*!40000 ALTER TABLE `t_order_info` DISABLE KEYS */;
```

2.3.1.3 文件内容解释

(1) 没有建库语句,因为是单表备份。

(2) 有删除表、建立表的语句。导入目标库时,删除表的语句可能造成数据误删。

(3) INSERT 语句没有字段名称,导入时表结构要一致。

(4) 导入过程中有 lock table write 操作,导入过程中相关表不可写。

(5) ALTER TABLE t_order_info DISABLE KEYS 语句将禁用该表的所有非唯一索引,这可以提高插入大量数据时的性能。对应的文件末尾有 ALTER TABLEt_order_infoENABLE KEYS;。

2.3.1.4 用途

可以将备份文件中的数据导入自定义库,"文件内容解释"部分遇到的问题可以使用下面的参数解决。

(1) --no-create-info 不包含建表语句（可以手动创建，如 create table tablename like dbname.tablename;）。

(2) --skip-add-drop-database 不包含删库语句。

(3) --skip-add-drop-table 不包含删表语句。

(4) --skip-add-locksINSERT 语句前不包含 LOCK TABLES t_order_info WRITE;。

(5) --complete-insertINSERT 语句中包含列名称（新表的列有增加的时候）。

2.3.1.5 单表导出备份数据（只导出数据）

```
mysqldump --default-character-set=utf8mb4 --master-data=2 --single-transaction --set-gtid-purged=off --hex-blob --no-create-info --skip-add-drop-table --skip-add-locks --tables dbname tablename

// 部分数据导出追加参数
--where="create_time>'2023-01-02'"
```

2.3.1.6 导出单库中的某表为 CSV

```
// 可选不导出表结构
--no-create-info --skip-add-drop-database --skip-add-drop-table
/data/mysql/3306/base/bin/mysqldump -uadmin -p123456 -P3306 -h127.0.0.1 --default-character-set=utf8mb4 --single-transaction --set-gtid-purged=OFF --triggers --routines --events --hex-blob --fields-terminated-by=',' --fields-enclosed-by='"' --lines-terminated-by='\n'  -T /data/mysql/3306/tmp test

// 其中 test 后面也可以指定表名，不指定就是全库
test t_order_info t_order_info01
其中 --single-transaction --set-gtid-purged=OFF  --triggers --routines --events --hex-blob
```

为了防止提示，可选 1GB 的备份文件，测试结果如下：

(1) 使用 mysql< 备份文件导入，耗时 5 分钟。

(2) 使用 source 备份文件导入，耗时 10 分钟。

推荐使用第一种，都是单线程。

2.3.2 mydumper

mydumper 版本为 0.14.4。

多线程导出如下：

```
mydumper -u admin -p 123456 -P 3306 -h 127.0.0.1 -t 8 --trx-consistency-only -G -E -R --skip-tz-utc --verbose=3 --compress --no-schemas --rows=1000000  -T test.t_order_info  -o /backup

// 导出时支持部分导出追加参数
```

```
    --where="create_time>'2023-01-02'"

    // 文件输出
    test01.t_order_info.00000.dat # 包含 CSV 数据
    test01.t_order_info.00000.sql # 包含 LOAD DATA 语句

    // 导入命令
    LOAD DATA LOCAL INFILE '/data/mysql/3306/tmp/test01.t_order_info.00005.dat'
    REPLACE INTO TABLE `t_order_info` CHARACTER SET binary FIELDS TERMINATED BY ','
    ENCLOSED BY '"' ESCAPED BY '\\' LINES STARTING BY '' TERMINATED BY '\n' (`ID`,`order_
    no`,`order_status`,`flag`,`create_time`,`modify_time`);
```

多线程导入如下：

```
    myloader -u admin -p 123456 -P 3306 -h 127.0.0.1 --enable-binlog -t 8
    --verbose=3 -B test -d /backup

    // 导入主库时需要添加
    --enable-binlog

    // 库名可以自定义
    -B test
```

耗时 2 分钟，建议如下：

(1) 在数据量大于 50GB 的场景中，更推荐用 mydumper。

(2) 补充场景，支持导出 CSV，也支持 --where 过滤。

```
    mydumper -u admin -p 123456 -P 3306 -h 127.0.0.1 -t 8 --trx-consistency-only
    -G -E -R --skip-tz-utc --verbose=3 --where="create_time>'2023-01-02'" --no-schemas
    --rows=1000000 --load-data --fields-terminated-by ',' --fields-enclosed-by '"'
    --lines-terminated-by '\n' -T test.t_order_info  -o /backup
```

导入命令同上，且可以按需手动进行 LOAD DATA。

2.3.3 SELECT OUTFILE 语句

适合单表数据的导出，不支持多表。

导出命令，耗时 15 秒。

```
    SELECT * from test01.t_order_info INTO OUTFILE "/data/mysql/3306/tmp/t_order_
    info0630_full.csv" CHARACTER SET utf8mb4 FIELDS TERMINATED BY ',' OPTIONALLY
    ENCLOSED BY '\'' LINES TERMINATED BY '\n';

    // 带列名导出，导入时需添加 IGNORE 1 LINES;
    SELECT *  INTO OUTFILE "/data/mysql/3306/tmp/t_order_info0630_full.csv"
    CHARACTER SET utf8mb4 FIELDS TERMINATED BY ',' OPTIONALLY ENCLOSED BY '\'' LINES
    TERMINATED BY '\n'  from (select 'id','order_no','order_status','flag','create_
```

```
time','modify_time' union all select * from test01.t_order_info) b;
```

导入命令,耗时 3 分钟。

```
mysql -uadmin -P3306  -h127.0.0.1 -p123456  --local-infile
 load data local infile '/data/mysql/3306/tmp/t_order_info0630_full.csv'  into
table test.t_order_info CHARACTER SET utf8mb4 fields terminated by ',' OPTIONALLY
ENCLOSED BY '\'' lines terminated by '\n';
```

总结如下:

(1) 支持跨表导入。A 表的数据可以导入 B 表,因为备份文件中只有数据。

(2) 可自定义导出部分列,导出导入速度较快,最常用。

2.3.4 MySQL_Shell > dumpTables

单表导出,耗时 4 秒。

```
util.dumpTables("test", ["t_order_info"], "/backup")
```

部分导出。

```
util.dumpTables("test", ["t_order_info"], "/backup", {"where" : {"test.t_order_info": "create_time>'2023-01-02'"}})
```

导入,耗时 3 分钟。

```
util.loadDump("/backup")
```

注意:不支持部分导入,不支持跨数据库版本。

导入时最多支持 2 个参数,可以将导出的部分数据全部导入到新的库中。

导入命令:

```
util.loadDump("/backup",{schema: "test_new"})
```

总结如下:

(1) 支持跨库导入,A 库的数据可以导入 B 库。表名需要一致。不支持增量到已有数据的表中。

(2) 导出时,和 SELECT OUTFILE 同效;导入时,比 LOAD DATA 快(默认 4 线程)。

注意:部分导出功能需要较新的 MySQL Shell 版本,如 8.0.33。LOAD DATA 单线程导入耗时 1 小时 20 分钟。

2.3.5 MySQL_Shell > exportTable

单表导出,耗时 10 秒。

```
util.exportTable("test.t_order_info",        "/backup/t_order_info.csv",
```

```
{defaultCharacterSet: "utf8mb4", fieldsOptionallyEnclosed: true, fieldsTerminatedBy:
",", linesTerminatedBy: "\n", fieldsEnclosedBy: '"', defaultCharacterSet: "utf8mb4",
showProgress: true, dialect: "csv"})
```

部分导出。

```
util.exportTable("test.t_order_info", "/backup/t_order_info.csv", {
dialect: "csv", defaultCharacterSet: "utf8mb4", fieldsOptionallyEnclosed:
true, fieldsTerminatedBy: ",", linesTerminatedBy: "\n", fieldsEnclosedBy: '"',
showProgress: true, where: "create_time>'2023-01-02'" } )
```

导入，耗时 10 分钟。

```
util.importTable("/backup/t_order_info.csv", { "characterSet": "utf8mb4",
"dialect": "csv", "fieldsEnclosedBy": "\"", "fieldsOptionallyEnclosed":
true, "fieldsTerminatedBy": ",", "linesTerminatedBy": "\n", "schema": "test",
"table": "t_order_info" })
```

部分导入（不推荐使用）。

```
util.importTable("/backup/t_order_info.csv", { "characterSet": "utf8mb4",
"dialect": "csv", "fieldsEnclosedBy": "\"", "fieldsOptionallyEnclosed":
true, "fieldsTerminatedBy": ",", "linesTerminatedBy": "\n", "schema":
"test100", "table": "t_order_info" })util.importTable("/backup/t_order_info0630.
csv", { "characterSet": "utf8mb4", "dialect": "csv", "fieldsEnclosedBy": "\"",
"fieldsOptionallyEnclosed": true, "fieldsTerminatedBy": ",", "linesTerminatedBy":
"\n", "schema": "test", "table": "t_order_info" })
```

有报错：MySQL Error 1205 (HY000): Lock wait timeout exceeded; try restarting transaction @ file bytes range [450000493, 500000518)。需要重复执行一次，才能保证数据完整。

根据报错提示可以使用以下命令导入：

```
LOAD DATA LOCAL INFILE '/backup/t_order_info0630.csv' INTO TABLE `test`.`t_order_info` CHARACTER SET 'utf8mb4' FIELDS TERMINATED BY ',' OPTIONALLY ENCLOSED BY '\"' ESCAPED BY '\\' LINES STARTING BY '' TERMINATED BY '\n';
MySQL 5.7 也推荐直接使用 LOAD DATA。
```

总结如下：

(1) 支持跨库导入，A 库的数据可以导入 B 库，表名需要一致。

(2) 导出时和 SELECT OUTFILE 同效。导入时，比 LOAD DATA 快（默认 8 线程）。

2.4 总结

可以根据数据大小选用不同方案，如表 3 所示。

表 3

导出	导入	优点	推荐度（效率）
mysqldump	source 或 MySQL<	原生，可远程	★★★（数据量<10GB）
mydumper	myloader	多线程	★★★（数据量>50GB）
SELECT OUTFILE	LOAD DATA	最灵活	★★（数据量<20GB）
Util.dumpTables	Util.loadDump	原生，多线程	★★★（数据量<50GB）
Util.exportTable	Util.importTable	原生，单线程	★（数据量<20GB）

(1) 使用 MySQL< 导入时，需要避免数据丢失。

(2) 前 3 种都支持 WHERE 过滤，mydumper 最快，SELECT OUTFILE 最常用（因为支持自定义导出部分列）。

(3) 前 2 种因为是备份工具，所以有 FTWRL 锁。

(4)Util.dumpTables 不支持增量到已有数据的表中，因为包含了库表的元数据信息，像 mydumper。

(5) Util.exportTable 备份是单线程，导入是多线程，不推荐用它的原因是导入时容易出错（多次导入可解决）。

(6) 按照数据量选择，全表备份用 Util.dumpTables 最快，部分备份用 SELECT OUTFILE。

(7) 测试之后再使用，导出和导入均需要进行数据验证。

3 详解 MySQL 三表 JOIN

作者：胡呈清

常听说 MySQL 中三表 JOIN 的执行流程并不是前两张表 JOIN 得出结果，再与第三张表进行 JOIN，而是三表嵌套的循环连接。

那这个三表嵌套的循环连接具体又是什么流程呢？与前两张表 JOIN 得出结果再与第三张表进行 JOIN 的执行效率相比如何呢？下面通过一个例子来分析。

3.1 前提

在关联字段无索引的情况下强制使用索引嵌套循环连接算法，目的是更好地观察扫描行数。

```
set optimizer_switch='block_nested_loop=off';
```

表结构和数据如下：

```
CREATE TABLE `t2` (
`id` int(11) NOT NULL,
`a` int(11) DEFAULT NULL,
`b` int(11) DEFAULT NULL,
PRIMARY KEY (`id`),
KEY `a` (`a`)
) ENGINE=InnoDB;

drop procedure idata;
delimiter ;;
create procedure idata()
begin
declare i int;
set i=1;
while(i<=1000)do
insert into t2 values(i, i, i);
set i=i+1;
end while;
end;;
delimiter ;
call idata();

create table t1 like t2;
create table t3 like t2;
insert into t1 (select * from t2 where id<=100);
insert into t3 (select * from t2 where id<=200);
```

示例 SQL：

```
select * from t1 join t2 on t1.b=t2.b  join t3 on t1.b=t3.b where t1.a<21;
```

3.2 通过扫描行数分析 JOIN 过程

通过 slow log 得知一共扫描 24100 行：

```
# Query_time: 0.016162  Lock_time: 0.000249 Rows_sent: 20   Rows_examined: 24100
SET timestamp=1617348099;
select * from t1 join t2 on t1.b=t2.b  join t3 on t1.b=t3.b where t1.a<21;
```

执行计划显示用的索引嵌套循环连接算法：

```
mysql> explain select * from t1 join t2 on t1.b=t2.b  join t3 on t1.b=t3.b
```

```
where t1.a<21;
+----+-------------+-------+------------+------+---------------+------+-------
---+------+------+----------+-------------+
 | id | select_type | table | partitions | type | possible_keys | key  | key_
len  | ref  | rows | filtered | Extra       |
+----+-------------+-------+------------+------+---------------+------+-------
---+------+------+----------+-------------+
 |  1 | SIMPLE      | t1    | NULL       | ALL  | a             | NULL | NULL
 | NULL | 100  |  20.00   | Using where |
 |  1 | SIMPLE      | t3    | NULL       | ALL  | NULL          | NULL | NULL
 | NULL | 200  |  10.00   | Using where |
 |  1 | SIMPLE      | t2    | NULL       | ALL  | NULL          | NULL | NULL
 | NULL | 1000 |  10.00   | Using where |
+----+-------------+-------+------------+------+---------------+------+-------
---+------+------+----------+-------------+
```

扫描行数构成：

(1) t1 扫描 100 行；

(2) t3 扫描 20×200=4000 行；

(3) t2 扫描 20×1000=20000 行。

总行数 =100+4000+20000=24100。

从这个结果来看，JOIN 过程像是先由 t1 和 t3 表 JOIN 得出 20 行中间结果，再与 t2 表进行 JOIN 得出结果。这个结论与我们通常认为的三表 JOIN 实际上是三表嵌套的循环连接不一样。

3.3 通过执行成本分析 JOIN 过程

查看执行计划成本：

```
mysql> explain format=json select * from t1 join t2 on t1.b=t2.b join t3 on t1.b=t3.b where t1.a<21\G
```

其他信息：

(1) t1 表 100 行，只有 1 个数据页（可通过 mysql.innodb_table_stats 查询）；

(2) t2 表 1000 行，有 4 个数据页；

(3) t3 表 200 行，只有 1 个数据页；

(4) io_block_read_cost=1.0，成本常数（MySQL 5.7）表示读取一个页面花费的成本默认是 1.0；

(5) row_evaluate_cost=0.2，成本常数（MySQL 5.7）表示读取以及检测一条记录是否符合搜索条件的成本默认是 0.2。

3.3.1 t1 是驱动表，全表扫描

(1) 扫描 100 行；

(2) 预估满足条件的记录只有 20%，即 100×20%=20，这表示 t1 的扇出。

$$IO 成本 = 1 \times 1.0 = 1$$

$$CPU 成本 = 100 \times 0.2 = 20$$

$$扫描 t1 的总成本 = 21$$

代码如图 1 所示。

```
"nested_loop": [
  {
    "table": {
      "table_name": "t1",
      "access_type": "ALL",
      "possible_keys": [
        "a"
      ],
      "rows_examined_per_scan": 100,
      "rows_produced_per_join": 20,
      "filtered": "20.00",
      "cost_info": {
        "read_cost": "17.00",
        "eval_cost": "4.00",
        "prefix_cost": "21.00",
        "data_read_per_join": "320"
      },
      "used_columns": [
        "id",
        "a",
        "b"
      ],
      "attached_condition": "(`join_test`.`t1`.`a` < 21)"
```

图 1

3.3.2 t3 是被驱动表，全表扫描

(1) 每次扫描 200 行；

(2) 因为驱动表的扇出为 20，所以要查找 20 次 t3，总共扫描 20×200=4000 行；

(3) 预估满足条件的行只有扫描行数的 10%，即 4000×10%=400，即为 t1 join t3 后的扇出，即 rows_produced_per_json。

$$IO 成本 = 1 \times 1.0 = 1$$

$$CPU 成本 = 200 \times 0.2 = 40$$

$$扫描 t3 表的总成本 = 驱动表扇出 \times (IO 成本 + CPU 成本) = 20 \times (1+40) = 820$$

阶段性总成本 =21+820=841

此处 eval_cost=80，实际上是为驱动表扇出 × 被驱动每次扫描行数 *filtered* 成本常数，即 20×200×10%×0.2。

简化公式为：

$$eval\_cost = rows\_produced\_per\_json \times 成本常数$$

代码如图 2 所示。

```
"table": {
  "table_name": "t3",
  "access_type": "ALL",
  "rows_examined_per_scan": 200,
  "rows_produced_per_join": 400,
  "filtered": "10.00",
  "cost_info": {
    "read_cost": "20.00",
    "eval_cost": "80.00",
    "prefix_cost": "841.00",
    "data_read_per_join": "6K"
  },
  "used_columns": [
    "id",
    "a",
    "b"
  ],
  "attached_condition": "(`join_test`.`t3`.`b` = `join_test`.`t1`.`b`)"
}
```

图 2

3.3.3 t2 也是被驱动表，全表扫描

(1) 每次查找扫描 1000 行；

(2) 要查找 400 次，总共扫描 400×1000=400000 行；

(3) 预估满足条件的行只有扫描行数的 10%，即 400000×10%=40000，即为 t2 的扇出，即 rows_produced_per_json。

IO 成本 =4×1.0=4

CPU 成本 =1000×0.2=200

扫描 t2 表的总成本 = 前 2 表 JOIN 的扇出 ×(IO 成本 +CPU 成本)

=400×(4+200)=81600

阶段性总成本 =841+81600=82441

此处 eval_cost=8000，即 rows_produced_per_json* 成本常数，即 40000×0.2。

代码如图 3 所示。

```
"table": {
  "table_name": "t2",
  "access_type": "ALL",
  "rows_examined_per_scan": 1000,
  "rows_produced_per_join": 40000,
  "filtered": "10.00",
  "cost_info": {
    "read_cost": "1600.00",
    "eval_cost": "8000.00",
    "prefix_cost": "82441.00",
    "data_read_per_join": "625K"
  },
  "used_columns": [
    "id",
    "a",
    "b"
  ],
  "attached_condition": "(`join_test`.`t2`.`b` = `join_test`.`t1`.`b`)"
}
```

图 3

3.3.4 根据执行计划成本分析

(1)t1 表查找 1 次，每次扫描 100 行；

(2)t3 表查找 20 次，每次扫描 200 行；

(3)t2 表查找 400 次，每次扫描 1000 行。

这样看，三表 JOIN 流程是：

(1) 全表扫描 t1，满足条件的有 20 行，先取第 1 行数据记为 R1；

(2) 从 R1 中取出 b 字段去 t3 表中查找；

(3) 取出 t3 中满足条件的行，跟 R1 组成一行，作为结果集的一部分，从结果集中取第 1 行数据记为 X1。①从 X1 中取出 b 字段去 t2 表中查找；②取出 t2 中满足条件的行，跟 X1 组成一行，作为结果集的一部分；③重复步骤①、②，直到结束。

(4) 重复第 2、3 步，直到结束。

以上流程如图 4 所示（这里展示的是索引嵌套循环算法三表 JOIN 的流程，块循环嵌套算法的不一样）。

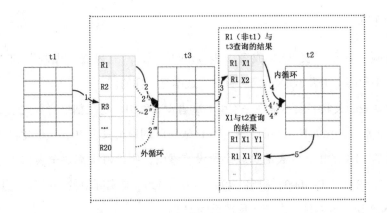

图 4

注意：由于例子中构造的数据比较特殊，第 3 步得出的中间结果集实际上只有 1 行，最终 t2 表的查找次数是 20×1=20，扫描总行数是 20×1000。所以单看 slow log 中显示的 24100 行，会误认为是先得出 t1 和 t3 表 JOIN 的结果，再去和 t2 进行 JOIN。

调整 t3 的数据，删除 20 行，再插入 20 行，使满足 b<21 的数据翻倍，这样第 3 步得出的中间结果集变成 2 行：

```
mysql> delete from t3 where id>180;
Query OK, 20 rows affected (0.00 sec)
mysql> insert into t3 select * from t3 where b<21;
Query OK, 20 rows affected (0.00 sec)
```

slow log 中扫描的总行数为 44100，t1、t3 的扫描行数不变，t2 的扫描行数变为 20×2×1000=40000：

```
# Query_time: 0.013848  Lock_time: 0.000100 Rows_sent: 40  Rows_examined: 44100
SET timestamp=1617354884;
select * from t1 join t2 on t1.b=t2.b  join t3 on t1.b=t3.b where t1.a<21;
```

为什么执行计划中分析得到的是 t2 表查找 400 次呢？

因为执行计划对 t1 JOIN t3 的扇出是个估算值，不准确；而 slow log 中的数据是真实执行后统计的，是个准确值。

为什么执行计划中 t2 表的执行次数是用 "t1 join t3 的扇出" 表示的？这不是说明 t1 先和 t3 JOIN，结果再和 t2 JOIN 吗？

其实拆解来看，"三表嵌套循环" 和 "前两表 JOIN 的结果和第三张表 JOIN" 两种算法，成本是一样的。而且如果要按三表嵌套循环的方式展示每张表的成本将非常复杂，可读性不强。所以执行计划中这么表示没有问题。

3.4 总结

总的来说，对于三表 JOIN 或者多表 JOIN 来说，"三表嵌套循环"和"两表先JOIN，其结果和第三张表 JOIN"两种算法，成本是一样的。

当被驱动表的关联字段不是唯一索引，或者没有索引，每次扫描行数大于 1 时，其扇出误差会非常大。比如在上面的示例中，t3 表实际的扇出只有 20，但优化器估算值是总扫描行数的 10%，由于 t3 表的关联字段没有索引，所以每次都要全表扫描 200 行，总的扫描行数 =20×200=4000，扇出 =4000×10%=400，是实际值的 20 倍。尤其对于后续表的 JOIN 来说，成本估算会产生更严重的偏差。

如果是 LEFT JOIN，每个被驱动表的 filtered 都会被优化器认定为 100%，误差更大！

通常建议 JOIN 不超过两张表，就是因为优化器估算成本误差大会导致选择不好的执行计划。如果非要这样用，一定要记住：关联字段必须有索引，最好是唯一性或者基数大的索引。

> 补充：MySQL 8.0 有 HASH JOIN 后这种情况会好很多。

4 如何通过 binlog 定位大事务？

作者：李彬

4.1 序

大事务想必大家都遇到过对大事务进行拆分，第一步当然就是要找到它。那么如何通过 binlog 来定位大事务呢？

首先，可通过 binlog 文件的大小来判断是否存在大事务，当一个 binlog 文件快被写完时，突然出现大事务，会突破 max_binlog_size 的大小继续写入。

官方文档中是这样描述的，"A transaction is written in one chunk to the binary log, so it is never split between several binary logs. Therefore, if you have big transactions, you might see binary log files larger than max_binlog_size"。

根据这个特点，只要进入 binlog 的存放目录，观察到文件大小异常的 binlog，那么你就可以去解析这个 binlog 获取大事务了。当然，需要注意的是，这只代表一部分情况，文件大小正常的 binlog 中也藏着大事务。

4.2 实践

既然要定位大事务的 SQL，针对已开启 GTID 的实例，只要定位到对应的 GTID 即可，下面我们开始对一个 binlog 进行解析（见表 4）。

表 4

环境	测试环境
binlog 格式	row
binlog 版本	v4
GTID	已开启
已验证的 MySQL 版本	MySQL 5.7.30、MySQL 8.0.28

首先，我们解析出一个 binlog 中按照事务大小排名前 N 的事务。

```
# 为了方便保存为脚本，这里定义几个基本的变量
BINLOG_FILE_NAME=$1                              # binlog 文件名
TRANS_NUM=$2                                     # 想要获取的事务数量
MYSQL_BIN_DIR='/data/mysql/3306/base/bin'        # basedir

# 获取前 TRANS_NUM 个大事务
${MYSQL_BIN_DIR}/mysqlbinlog ${BINLOG_FILE_NAME} | grep "GTID$(printf '\t')last_committed" -B 1 | grep -E '^# at' | awk '{print $3}' | awk 'NR==1 {tmp=$1} NR>1 {print ($1-tmp,tmp);tmp=$1}' | sort -n -r -k 1 | head -n ${TRANS_NUM} > binlog_init.tmp
```

经过对 binlog 的基本解析后，我们已经获取了对应事务的大小和可供定位 GTID 的 POS 信息，接下来对上述输出的临时文件进行逐行解析，针对每一个事务获取到相应的信息。

```
while read line
do
    # 事务大小这里取近似值，因为不是通过 (TRANS_END_POS-TRANS_START_POS) 计算出的
    TRANS_SIZE=$(echo ${line} | awk '{print $1}')
      logWriteWarning "TRANS_SIZE: $(echo | awk -v TRANS_SIZE=${TRANS_SIZE} '{ print (TRANS_SIZE/1024/1024) }')MB"
    FLAG_POS=$(echo ${line} | awk '{print $2}')
    # 获取 GTID
     ${MYSQL_BIN_DIR}/mysqlbinlog -vvv --base64-output=decode-rows ${BINLOG_FILE_NAME} | grep -m 1 -A3 -Ei "^# at ${FLAG_POS}" > binlog_parse.tmp
    GTID=$(cat binlog_parse.tmp | grep -i 'SESSION.GTID_NEXT' | awk -F "'" '{print $2}')
    # 通过 GTID 解析出事务的详细信息
     ${MYSQL_BIN_DIR}/mysqlbinlog --base64-output=decode-rows -vvv --include-gtids="${GTID}" ${BINLOG_FILE_NAME} > binlog_gtid.tmp
```

```
            START_TIME=$(grep -Ei '^BEGIN' -m 1 -A 3 binlog_gtid.tmp | grep -i 'serve
r id' | awk '{print $1,$2}' | sed 's/#//g')
            END_TIME=$(grep -Ei '^COMMIT' -m 1 -B 1 binlog_gtid.tmp | head -1 | awk '{
print $1,$2}' | sed 's/#//g')
            TRANS_START_POS=$(grep -Ei 'SESSION.GTID_NEXT' -m 1 -A 1 binlog_gtid.tmp
 | tail -1 | awk '{print $3}')
            TRANS_END_POS=$(grep -Ei '^COMMIT' -m 1 -B 1 binlog_gtid.tmp | head -1 | a
wk '{print $7}')
            # 输出
            logWrite "GTID: ${GTID}"
            logWrite "START_TIME: $(date -d "${START_TIME}" '+%F %T')"
            logWrite "END_TIME: $(date -d "${END_TIME}" '+%F %T')"
            logWrite "TRANS_START_POS: ${TRANS_START_POS}"
            logWrite "TRANS_END_POS: ${TRANS_END_POS}"
            # 统计对应的 DML 语句数量
            logWrite " 该事务的 DML 语句及相关表统计:"
            grep -Ei '^### insert' binlog_gtid.tmp | sort | uniq -c
            grep -Ei '^### delete' binlog_gtid.tmp | sort | uniq -c
            grep -Ei '^### update' binlog_gtid.tmp | sort | uniq -c

    done < binlog_init.tmp
```

至此，我们已经基本实现了通过解析一个 binlog 文件，从而获取对应的 GTID、事务开始和结束时间、事务开始和结束的 POS、对应的 DML 语句数量。为了不重复执行解析命令，我们可以将其封装为脚本，作为日常运维工具使用。

4.3 最终效果展示

```
    [root@localhost ~]$ sh parse_binlog.sh /opt/sandboxes/rsandbox_5_7_35/master/
data/mysql-bin.000003 2
    2023-12-12 15:15:40 [WARNING] 开始解析 BINLOG: /opt/sandboxes/rsandbox_5_7_35/
master/data/mysql-bin.000003
    2023-12-12 15:15:53 [WARNING] TRANS_SIZE: 0.00161743MB
    2023-12-12 15:16:06 [INFO] GTID: 00020236-1111-1111-1111-111111111111:362779
    2023-12-12 15:16:06 [INFO] START_TIME: 2023-12-12 15:14:35
    2023-12-12 15:16:06 [INFO] END_TIME: 2023-12-12 15:14:35
    2023-12-12 15:16:06 [INFO] TRANS_START_POS: 362096066
    2023-12-12 15:16:06 [INFO] TRANS_END_POS: 362097697
    2023-12-12 15:16:06 [INFO] 该事务的 DML 语句及相关表统计：
        1 ### INSERT INTO `sbtest`.`sbtest100`
        1 ### DELETE FROM `sbtest`.`sbtest100`
        2 ### UPDATE `sbtest`.`sbtest100`
    2023-12-12 15:16:06 [WARNING] TRANS_SIZE: 0.00161648MB
    2023-12-12 15:16:25 [INFO] GTID: 00020236-1111-1111-1111-111111111111:505503
```

```
2023-12-12 15:16:25 [INFO] START_TIME: 2023-12-12 15:15:36
2023-12-12 15:16:25 [INFO] END_TIME: 2023-12-12 15:15:36
2023-12-12 15:16:25 [INFO] TRANS_START_POS: 603539112
2023-12-12 15:16:25 [INFO] TRANS_END_POS: 603540742
2023-12-12 15:16:25 [INFO] 该事务的 DML 语句及相关表统计：
    1 ### INSERT INTO `sbtest`.`sbtest100`
    1 ### DELETE FROM `sbtest`.`sbtest100`
    1 ### UPDATE `sbtest`.`sbtest100`
    1 ### UPDATE `sbtest`.`sbtest87`
```

通过上述结果可以看到，这种解析方式是基于事务的大小进行排序的。有时我们还可能需要从时间维度进行排序，依据大致相同的思路写脚本也可以实现，这里提供一个开源的工具 my2sql。

my2sql 可指定 rows 和 time 进行过滤，在 mode 为 file 且 work-type 为 stats 时，连接任意一个 MySQL 实例（无需原库）均可对 binlog 中的事务进行解析。

```
# 统计指定 binlog 中各个表的 DML 操作数量(不加 row 和 time 限制)
[root@localhost ~]$ mkdir tmpdir
[root@localhost ~]$ ./my2sql -user root -password xxx -host 127.0.0.1 -port 3306 -mode file -local-binlog-file ./mysql-bin.005375 -work-type stats -start-file mysql-bin.005375 -output-dir ./tmpdir

# 按照事务的行数倒序排序
[root@localhost ~]$ less tmpdir/biglong_trx.txt | sort -nr -k 6 | less

# 按照事务的执行时间倒序排序
[root@localhost ~]$ less tmpdir/biglong_trx.txt | sort -nr -k 7 | less

# 输出示例 (binlog starttime stoptime startpos stoppos rows duration tables)
mysql-bin.005375 2023-12-12_16:04:06 2023-12-12_16:16:59 493014756 582840954 123336 53 [test.t1(inserts=61668, updates=0, deletes=0) test.t2(inserts=61668, updates=0, deletes=0)]
```

5 MySQL 8.0 字段信息统计机制

作者：杨奇龙

5.1 背景

前几天有同事咨询一个问题：某个业务基于 INFORMATION_SCHEMA 统计表的信息（比如最大值）向表里面插入数据，请问 INFORMATION_SCHEMA.TABLES 中的 AUTO_INCREMENT 会不会及时地更新呢？

先说结论：可以！

这里涉及的信息统计机制或者说频率问题，主要由参数 information_schema_stats_expiry 控制。

5.1.1 表信息更新的基本逻辑

默认情况下，MySQL 会高效地从系统表 mysql.index_stats 和 mysql.table_stats 中检索这些列的缓存值，而不是直接从存储引擎中获取统计信息。如果缓存的统计信息不可用或已过期，MySQL 将从存储引擎中检索最新的统计信息，并将其统计信息更新并缓存在 mysql.index_stats 和 mysql.table_stats 字典表中。后续查询将检索缓存的统计信息，直到缓存的统计数据过期。

值得注意的是，MySQL 重新启动或第一次打开 mysql.index_stats 和 mysql.table_stats 表时不会自动更新缓存的统计信息。

5.1.2 核心参数

核心参数 information_schema_stats_expiry 默认是 86400 秒，也就是说每隔一天自动收集一次相关统计信息到 information_schema 中的如下表字段中：

```
STATISTICS.CARDINALITY
TABLES.AUTO_INCREMENT
TABLES.AVG_ROW_LENGTH
TABLES.CHECKSUM
TABLES.CHECK_TIME
TABLES.CREATE_TIME
TABLES.DATA_FREE
TABLES.DATA_LENGTH
TABLES.INDEX_LENGTH
TABLES.MAX_DATA_LENGTH
```

```
TABLES.TABLE_ROWS
TABLES.UPDATE_TIME
```

参数 information_schema_stats_expiry 的值决定再次收集表的统计信息的时间间隔，默认 86400 秒。如果设置为 0，则表示实时更新统计信息，势必会影响一部分性能。

在以下情况中，查询统计信息列不会在 mysql.index_stats 和 mysql.table_stats 字典表中存储或更新统计信息：

(1) 缓存的统计信息尚未过期时。

(2) 当 information_schema_stas_expiry 设置为 0 时。

(3) 当 MySQL server 处于只读、超级只读、事务只读或 innodb_read_only 模式时。

(4) 查询还获取 Performance Schema 的数据时。

information_schema_stas_experity 支持全局和会话级别，每个会话都可以定义自己的过期值。从存储引擎中检索并由一个会话缓存的统计信息可用于其他会话。

5.2 测试

本文以 MySQL 8.0.30 为例进行分析。

5.2.1 测试准备

```
CREATE TABLE `sbtest1` (
`id` int NOT NULL AUTO_INCREMENT,
`k` int NOT NULL DEFAULT '0',
`c` char(120) NOT NULL DEFAULT '',
`pad` char(60) NOT NULL DEFAULT '',
PRIMARY KEY (`id`)
) ENGINE=InnoDB ;
```

5.2.2 测试

查看 information_schema.tables 中 sbtest1 的当前信息，最大值为 1200006，表结构定义中自增最大值也是 1200006。

```
master [localhost:22031] {msandbox} (test) > show variables like 'information_schema_stats_expiry' ;
+--------------------------------+--------+
| Variable_name                  | Value  |
+--------------------------------+--------+
| information_schema_stats_expiry| 86400  |
+--------------------------------+--------+
1 row in set (0.02 sec)

master [localhost:22031] {msandbox} (test) > select    table_name, AUTO_INCREMENT  from information_schema.tables where  table_name='sbtest1';
```

```
+------------+----------------+
| TABLE_NAME | AUTO_INCREMENT |
+------------+----------------+
| sbtest1    |        1200006 |
+------------+----------------+
1 row in set (0.01 sec)

master [localhost:22031] {msandbox} (test) > show create table  sbtest1 \G
*************************** 1. row ***************************
Table: sbtest1
Create Table: CREATE TABLE `sbtest1` (
  `id` int NOT NULL AUTO_INCREMENT,
  `k` int NOT NULL DEFAULT '0',
  `c` char(120) NOT NULL DEFAULT '',
  `pad` char(60) NOT NULL DEFAULT '',
  PRIMARY KEY (`id`)
) ENGINE=InnoDB AUTO_INCREMENT=1200006 DEFAULT CHARSET=utf8mb4 COLLATE=utf8mb4_0900_ai_ci
1 row in set (0.00 sec)
```

插入新的数据，自增值加 1。

```
master [localhost:22031] {msandbox} (test) > insert into sbtest1(k,c,pad) values(1,'c','cc');
Query OK, 1 row affected (0.00 sec)

master [localhost:22031] {msandbox} (test) > show create table  sbtest1 \G
*************************** 1. row ***************************
Table: sbtest1
Create Table: CREATE TABLE `sbtest1` (
  `id` int NOT NULL AUTO_INCREMENT,
  `k` int NOT NULL DEFAULT '0',
  `c` char(120) NOT NULL DEFAULT '',
  `pad` char(60) NOT NULL DEFAULT '',
  PRIMARY KEY (`id`)
) ENGINE=InnoDB AUTO_INCREMENT=1200007 DEFAULT CHARSET=utf8mb4 COLLATE=utf8mb4_0900_ai_ci
1 row in set (0.00 sec)
```

但是 information_schema.tables 中的值并未发生变化。

```
master [localhost:22031] {msandbox} (test) > select  table_name, AUTO_INCREMENT from information_schema.tables where table_name='sbtest1';
+------------+----------------+
| TABLE_NAME | AUTO_INCREMENT |
+------------+----------------+
| sbtest1    |        1200006 |
+------------+----------------+
1 row in set (0.00 sec)
```

设置为实时更新。

修改 information_schema_stats_expiry 为 0。

```
    master [localhost:22031] {msandbox} (test) > set  information_schema_stats_
expiry=0;
    Query OK, 0 rows affected (0.00 sec)

    master [localhost:22031] {msandbox} (test) > show variables like 'information_
schema_stats_expiry' ;
    +------------------------------+-------+
    | Variable_name                | Value |
    +------------------------------+-------+
    | information_schema_stats_expiry | 0  |
    +------------------------------+-------+
    1 row in set (0.00 sec)

    master [localhost:22031] {msandbox} (test) > select  table_name, AUTO_
INCREMENT from information_schema.tables where  table_name='sbtest1';
    +------------+----------------+
    | TABLE_NAME | AUTO_INCREMENT |
    +------------+----------------+
    | sbtest1    |        1200007 |
    +------------+----------------+
    1 row in set (0.00 sec)

    master [localhost:22031] {msandbox} (test) > show create table  sbtest1 \G
    *************************** 1. row ***************************
    Table: sbtest1
    Create Table: CREATE TABLE `sbtest1` (
    `id` int NOT NULL AUTO_INCREMENT,
    `k` int NOT NULL DEFAULT '0',
    `c` char(120) NOT NULL DEFAULT '',
    `pad` char(60) NOT NULL DEFAULT '',
    PRIMARY KEY (`id`)
    ) ENGINE=InnoDB AUTO_INCREMENT=1200007 DEFAULT CHARSET=utf8mb4
COLLATE=utf8mb4_0900_ai_ci
    1 row in set (0.00 sec)
```

插入数据自增加 1，查询 information_schema.tables 中自增列统计值也是实时更新的。

```
    master [localhost:22031] {msandbox} (test) > insert into sbtest1(k,c,pad)
values(1,'c','cc');
    Query OK, 1 row affected (0.00 sec)

    master [localhost:22031] {msandbox} (test) > select  table_name, AUTO_
INCREMENT from information_schema.tables where  table_name='sbtest1';
```

```
+------------+----------------+
| TABLE_NAME | AUTO_INCREMENT |
+------------+----------------+
| sbtest1    |        1200008 |
+------------+----------------+
1 row in set (0.00 sec)

master [localhost:22031] {msandbox} (test) > show create table  sbtest1 \G
*************************** 1. row ***************************
Table: sbtest1
Create Table: CREATE TABLE `sbtest1` (
`id` int NOT NULL AUTO_INCREMENT,
`k` int NOT NULL DEFAULT '0',
`c` char(120) NOT NULL DEFAULT '',
`pad` char(60) NOT NULL DEFAULT '',
PRIMARY KEY (`id`)
) ENGINE=InnoDB AUTO_INCREMENT=1200008 DEFAULT CHARSET=utf8mb4 COLLATE=utf8mb4_0900_ai_ci
1 row in set (0.00 sec)
```

手工修改自增列，也可以实时更新。

```
master [localhost:22031] {msandbox} (test) > alter table sbtest1  AUTO_INCREMENT=1200010;
Query OK, 0 rows affected (0.00 sec)
Records: 0  Duplicates: 0  Warnings: 0

master [localhost:22031] {msandbox} (test) > select  table_name, AUTO_INCREMENT  from information_schema.tables where  table_name='sbtest1';
+------------+----------------+
| TABLE_NAME | AUTO_INCREMENT |
+------------+----------------+
| sbtest1    |        1200010 |
+------------+----------------+
1 row in set (0.00 sec)
```

5.3 总结

MySQL 8.0 对于表字段的统计信息提供了更多的技术特性来支持，例如统计有效性时长、字段本身的直方图等，使用起来越来越便利。

回过头来看这个需求，其实如果是业务监控或者数据库监控信息（如监控表的主键最大值是否溢出），更为实际的建议还是查询具体的表的最大 id 值，查询频率可控。

6 InnoDB 全表扫描和全主键扫描一样吗？

作者：操盛春

本文基于 MySQL 8.0.32 源码，存储引擎为 InnoDB。

6.1 准备工作

创建测试表：

```
CREATE TABLE `t2` (
  `id` int unsigned NOT NULL AUTO_INCREMENT,
  `i1` int DEFAULT '0',
  PRIMARY KEY (`id`) USING BTREE
) ENGINE=InnoDB DEFAULT CHARSET=utf8mb3
```

插入测试数据：

```
INSERT INTO `t2` (`id`, `i1`) VALUES
(1, 20), (2, 21),
(3, 22), (4, 23),
(5, 23), (6, 33);
```

示例 SQL 1：

```
SELECT `i1` FROM `t2`
```

示例 SQL 2：

```
SELECT `id` FROM `t2`
```

6.2 执行计划对比

我们从执行计划入手，先来看看分别代表全表扫描、全主键扫描的示例 SQL 1 和 SQL 2 的执行计划。

示例 SQL 1（见图 5）：

table	partitions	type	possible_keys	key	key_len	ref
t2	<null>	ALL	<null>	<null>	<null>	

图 5

type = ALL，说明示例 SQL 1 读取数据的方式为全表扫描，这个方式我们应该都很熟悉了。

示例 SQL 2（见图 6）：

table	partitions	type	possible_keys	key	key_len	ref
t2	<null>	index	<null>	PRIMARY	4	

图 6

type = index、key = PRIMARY，表示示例 SQL 2 使用了覆盖索引，并且使用的是主键索引，这说明示例 SQL 2 读取数据的方式是全主键扫描。

InnoDB 表属于索引组织表，主键索引包含表中所有数据，也就是所谓的"索引即数据，数据即索引"。不管是全表扫描，还是全主键扫描，都需要从主键索引中读取数据。

既然都是从主键索引中读取数据，那怎么会得到两种不同的执行计划呢？

执行计划是代码给出的，我们去代码里找答案。

确定执行计划时，全表扫描和全主键扫描的分叉口位于 JOIN::adjust_access_methods() 方法中。

示例 SQL 1（见图 7）：

图 7

程序停留在 JOIN::adjust_access_methods() 方法的第 2949 行，说明第 2938 行的 if (tab->type() == JT_ALL) 条件成立，也就是说，示例 SQL 1 最初确定的读取数据方式是全表扫描。

接下来，我们通过调试控制台打印出 !tab->table()->covering_keys.is_clear_all() 表达式的值（见图 8）：

```
调试控制台
p thd->m_query_string
(LEX_CSTRING) $1 = (str = "SELECT `i1` FROM `t2`", length = 21)

p !tab->table()->covering_keys.is_clear_all()

(bool) $2 = false
```

图 8

如图 8 所示，该表达值的为 false，说明没有覆盖索引可用，不会进入对应的 if 分支，也就不会调用 tab->set_type(JT_INDEX_SCAN) 把全表扫描修改为覆盖索引扫描。

最终，示例 SQL 1 读取数据的方式就是全表扫描。

示例 SQL 2（见图 9）：

```
2932 void JOIN::adjust_access_methods() {
2933   ASSERT_BEST_REF_IN_JOIN_ORDER(this); this = {query_block:0x00000001329be748, thd:0x000000011086dc
2934   for (uint i = const_tables; i < tables; i++) { i = 0
2935     JOIN_TAB *const tab = best_ref[i]; tab = {table_ref:0x00000001329bfbe8, m_keyuse:0x000000000000
2936     Table_ref *const tl = tab->table_ref; tl = {next_local:0x0000000000000000, next_global:0x000000
2937
2938   1 if (tab->type() == JT_ALL) { tab = {table_ref:0x00000001329bfbe8, m_keyuse:0x0000000000000000,
2939 >     /*-
2947       */
2948       if (!tab->table()->no_keyread &&                  // 1 tab = {table_ref:0x00000001329bfbe8
2949         2 !tab->table()->covering_keys.is_clear_all() && // 2 tab = {table_ref:0x00000001329bfbe8
2950           !tl->uses_materialization())                 // 3 tl = {next_local:0x0000000000000000
2951       {
2952 >       /*-
2963         */
2964         if (tab->position()->sj_strategy != SJ_OPT_LOOSE_SCAN) tab = {table_ref:0x00000001329bfbe8,
2965           tab->set_index( tab = {table_ref:0x00000001329bfbe8, m_keyuse:0x0000000000000000, ...}
2966               i: find_shortest_key(table: tab->table(), usable_keys: &tab->table()->covering_keys));
2967         tab->set_type(t: JT_INDEX_SCAN);  // Read with index_first / index_next
2968         // From table scan to index scan, thus filter effect needs no recalc.
2969       }
```

图 9

程序停留在 JOIN::adjust_access_methods() 方法的第 2966 行：

(1) 说明 1 号框处的 if 条件成立，示例 SQL 2 最初确定的读取数据方式也是全表扫描。

(2) 说明 2 号框处的表达式值为 true，示例 SQL 2 可以使用覆盖索引扫描。

在 3 号框处，find_shortest_key() 函数从可用的覆盖索引中找到占用空间最小的索引。

然后，tab->set_type(JT_INDEX_SCAN) 把全表扫描修改为覆盖索引扫描。

t2 表只有主键索引,没有二级索引,可以使用的覆盖索引(tab->table()->covering_keys)也就只有主键索引了,我们可以打印 covering_keys 来验证(见图 10):

```
∨ 调试控制台
  p tab->table()->covering_keys
  (Key_map) $22 = (map = 1)
```

图 10

covering_keys 属性的类型是 Key_map,每个覆盖索引使用 1bit 作为标志位。map=1,说明只设置了一个标志位,这意味着可供示例 SQL 2 使用的覆盖索引只有一个(见图 11)。

```
∨ 调试控制台
  p thd->m_query_string
  (LEX_CSTRING) $18 = (str = "select id from t2", length = 17)

  p tab->table()->s->primary_key

  (uint) $19 = 0
  p tab->table()->covering_keys.is_set(0)

  (bool) $20 = true
```

图 11

主键索引的 ID(primary_key)为 0,covering_keys.is_set(0) 的输出结果为 true,说明 covering_keys 中设置的标志位对应的就是主键索引。

最终,示例 SQL 2 读取数据的方式由全表扫描修改为全主键扫描。

既然 InnoDB 主键索引中包含了全表数据,那代码中到底是怎么表示主键索引的(见图 12)?

```
∨ 调试控制台
  p tab->table()->key_info[0]->table->s->table_name
  (LEX_CSTRING) $24 = (str = "t2", length = 2)

  p tab->table()->key_info[0]->name

  (const char *) $25 = 0x000000011088b040 "PRIMARY"
  p tab->table()->key_info[0]->user_defined_key_parts

  (uint) $26 = 1
  p tab->table()->key_info[0]->key_part->field->field_name
  (const char *) $27 = 0x000000011088ac80 "id"
```

图 12

t2 表的主键索引（name 为 PRIMARY）有 1 个字段（user_defined_key_parts 为 1），字段名是 id，和表结构中 PRIMARY KEY 定义的主键字段一致。也就是说，虽然主键索引包含了表中全部数据，但是内存中的索引对象还是只包含表结构中定义的主键字段。

通过前面的介绍，我们可以对全表扫描和全主键扫描做个简单的总结：

- 如果 server 层只需要从 InnoDB 读取主键字段，主键索引可以充当覆盖索引的角色，读取数据的方式为全主键扫描。
- 如果 server 层需要从 InnoDB 读取主键之外的字段，主键索引就不能充当覆盖索引的角色了，读取数据的方式为全表扫描。

主键字段指的是表结构中 PRIMARY KEY 定义的字段。

6.3 执行流程对比

前面对全表扫描和全主键扫描两种执行计划做了简单的对比，接下来我们再从三个方面对两者的执行流程做对比，看看它们有什么异同。

首先，我们来看一下两者的主要堆栈。

全表扫描（见图 13）：

```
1    | > Query_expression::ExecuteIteratorQuery()
2    | + > TableScanIterator::Read()
3    | + - > handler::ha_rnd_next()
4    | + - x > ha_innobase::rnd_next()
5    | + - x = > ha_innobase::index_first()
6    | + - x = | > ha_innobase::index_read()
7    | + - x = | + > row_search_mvcc()
8    | + - x = | + - > // 定位到主键索引的第 1 条记录
9    | + - x = | + - > // 一条 SQL 只执行一次
10   | + - x = | + - > btr_pcur_t::open_at_side()
11   | + - x = | + - > // 页指针移动到下一条记录
12   | + - x = | + - > btr_pcur_t::move_to_next()
13   | + - x = | + - > // 获取下一条记录
14   | + - x = | + - > btr_pcur_t::get_rec()
15   | + - x = | + - > // 把 InnoDB 转换为 server 层的格式
16   | + - x = | + - > row_sel_store_mysql_rec()
17   | + - x = | + - x > row_sel_store_mysql_field()
18   | + - x = | + - x = > row_sel_field_store_in_mysql_format()
19   | + - x = | + - x = | > row_sel_field_store_in_mysql_format_func()
20   | + - > handler::ha_rnd_next()
21   | + - x > ha_innobase::rnd_next()
22   | + - x = > ha_innobase::general_fetch()
```

图 13

全主键扫描（见图 14）：

```
1     | > Query_expression::ExecuteIteratorQuery()
2     | + > IndexScanIterator<false>::Read()
3     | + - > handler::ha_index_first()
4     | + - x > // 为了和全表扫描堆栈位置对应，增加一行占位
5     | + - x > ha_innobase::index_first()
6     | + - x = > ha_innobase::index_read()
7     | + - x = | > row_search_mvcc()
8     | + - x = | + > // 定位到主键索引的第 1 条记录
9     | + - x = | + > // 一条 SQL 只执行一次
10    | + - x = | > btr_pcur_t::open_at_side()
11    | + - x = | + > // 移动到下一条记录
12    | + - x = | > btr_pcur_t::move_to_next()
13    | + - x = | + > // 获取下一条记录的地址
14    | + - x = | > btr_pcur_t::get_rec()
15    | + - x = | + > // 把 InnoDB 的记录格式转换为 server 层的记录格式
16    | + - x = | + > row_sel_store_mysql_rec()
17    | + - x = | + - > row_sel_store_mysql_field()
18    | + - x = | + - x > row_sel_field_store_in_mysql_format()
19    | + - x = | + - x = > row_sel_field_store_in_mysql_format_func()
20    | + - > handler::ha_index_next()
21    | + - x > ha_innobase::index_next()
22    | + - x = > ha_innobase::general_fetch()
```

图 14

第 2~4 行是读取第一条记录的入口，第 20~21 行是读取第二条及以后记录的入口。

入口之下，两者都会依次调用 ha_innbase::index_first() → ha_innobase::index_read() → row_search_mvcc() 读取第一条记录，依次调用 ha_innobase::general_fetch() → row_search_mvcc() 读取第二条及以后的记录。

row_search_mvcc() 读取记录最关键的步骤是：定位到要读取的第一条记录。因为读取第一条记录之后，只需要沿着记录之间的指针、数据页之间的指针就能依次读取到所有记录了。

row_search_mvcc() 调用 btr_pcur_t::open_at_side() 打开索引并定位到索引中的第一条记录。

程序执行到 row_search_mvcc() 的 pcur → open_at_side() 处，我们在调试控制台打印

出两者使用的索引。

全表扫描（见图 15）：

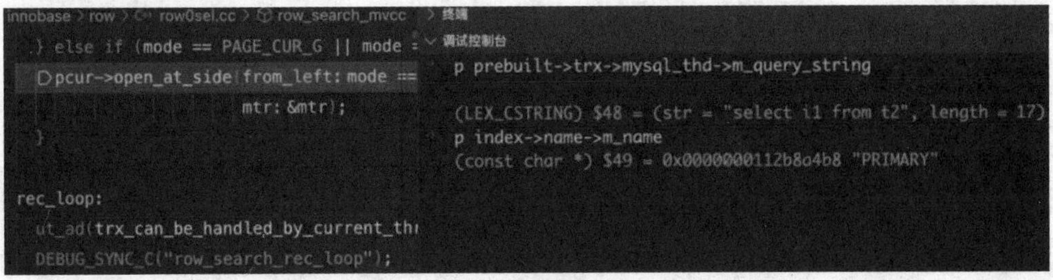

图 15

全主键扫描（见图 16）：

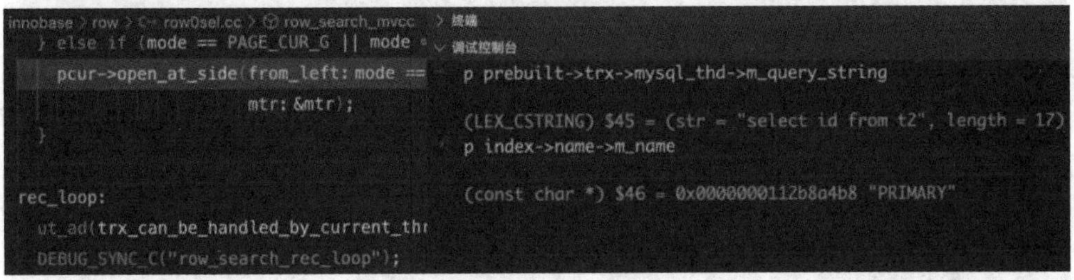

图 16

通过上面两张图中打印出的 SQL 和索引名，我们可以看到：全表扫描和全主键扫描都打开了主键索引，意味着都会从主键索引中读取数据。

pcur → open_at_side() 打开并定位到索引的第一条记录（infimum）之后，btr_pcur_t::move_to_next() 把数据页的记录指针移动到下一条记录（第一条用户记录），btr_pcur_t::get_rec() 获取第一条用户记录的地址。

然后，row_sel_store_mysql_rec() 把第一条用户记录从 InnoDB 格式转换为 server 层的格式。

我们通过第一条用户记录来看看全表扫描和全主键扫描读取了主键索引的哪些字段。

全表扫描（见图 17、图 18）：

图 17

图 18

row_sel_store_mysql_rec() 只读取 1 个字段（p prebuilt → n_template 输出 1），字段名为 i1（p rec_index → get_field(field_no) → name 输出 i1）。

全主键扫描（见图 19、图 20）：

```
innobase > row > C++ row0sel.cc > © row_sel_store_mysql_rec
  for (ulint i = 0; i < prebuilt->n_template; i++) { i = 0, prebuilt = {magic
    const auto templ: mysql_row_templ_t *const = &prebuilt->mysql_template[i]

    if (templ->is_virtual && rec_index->is_clustered()) { templ = {col_no:0,
    }

    ulint field_no = field_no = 0
    ulint sec_field_no = ULINT_UNDEFINED; sec_field_no = 18446744073709551615

    /* We should never deliver column prefixes to MySQL,
    except for evaluating innobase_index_cond() or
    row_search_end_range_check(). */
    ut_ad(rec_index->get_field(field_no)->prefix_len == 0); rec_index = {id:5

    if (clust_templ_for_sec) { clust_templ_for_sec = false
    }

    if (!row_sel_store_mysql_field(mysql_rec, prebuilt, rec, rec_index, mysql
    }
```

图 19

```
调试控制台
p prebuilt->trx->mysql_thd->m_query_string
(LEX_CSTRING) $64 = (str = "select id from t2", length = 17)

p prebuilt->n_template

(unsigned int) $65 = 1
p rec_index->get_field(field_no)->name

(id_name_t) $66 = (m_name = "id")
```

图 20

row_sel_store_mysql_rec() 也只读取 1 个字段（p prebuilt → n_template 输出 1），字段名为 id（p rec_index → get_field(field_no) → name 输出 id）。

通过前面三项对比，我们可以确定：全表扫描和全主键扫描都使用了主键索引，并且都是先定位到主键索引的第一条记录，然后沿着第一条记录一直读取到最后一条记录。

6.4 总结

从执行流程来看，全表扫描和全主键扫描都需要从主键索引中读取数据。

InnoDB 每读取一条记录，都需要把该记录所属的数据页从磁盘上的表空间文件读取到 Buffer Pool 中（如果数据页已在 Buffer Pool 中，不需要重复读取）。

全表扫描需要读取表中所有记录，全主键扫描需要读取主键索引的所有记录，由于 InnoDB 表的数据和主键索引合二为一了，两者都会把主键索引中所有叶子结点数据页全部读取到 Buffer Pool 中。

从这点可见，InnoDB 全主键扫描的执行效率并不会比全表扫描高。

全表扫描和全主键扫描读取时，都需要先定位到主键索引第 1 个叶子结点数据页中的第 1 条用户记录，然后沿着第 1 条用户记录依次读取，直到读完主键索引中的所有记录（或者说表中的所有记录）。

所以，我们可以认为全表扫描和全主键扫描本质上是一样的。

既然如此，执行计划中为什么要区分全表扫描和全主键扫描呢？这是因为 MySQL 支持多种存储引擎，对于使用堆表的存储引擎（例如 MyISAM），因为表中数据和索引是分开存储的，全表扫描和全主键扫描确实不同。server 层确定执行计划时，对于所有存储引擎一视同仁，InnoDB 自然也就区分全表扫描和全主键扫描了。

7 MySQL 权限变更何时生效？

作者：欧阳涵

本文讲述对三种级别权限的变更后，使其生效的方法。

Uproxy 是爱可生研发的云树®DMP 产品的一个高效的读写中间件，维护自身到后端 MySQL 数据库之间的连接池，用以保持到数据库后端的长连接。

7.1 背景

近期有客户反馈，通过 Uproxy 连接数据库，使用 REVOKE 回收全局库表 *.* 的某个权限后，却还能看到没有对应权限的库且能进行操作，FLUSH PRIVILEGES 也无效，难道这是 MySQL 的 bug？

7.2 MySQL 更改权限

其实不然。笔者在进行阐述前，先来说明一下 MySQL 更改权限的两种方式。

7.2.1 直接修改授权表

使用 INSERT、UPDATE 或 DELETE 等语句直接修改授权表（不推荐使用）。

```
update mysql.user set Select_priv='N' where user='ouyanghan' and host='%';
```

7.2.2 使用 GRANT/REVOKE 语句

使用 GRANT/REOVKE 来授予或回收权限（推荐使用）。

```
GRANT
    priv_type [(column_list)]
    [, priv_type [(column_list)]] ...
    ON [object_type] priv_level
    TO user [auth_option] [, user [auth_option]] ...
    [REQUIRE {NONE | tls_option [[AND] tls_option] ...}]
    [WITH {GRANT OPTION | resource_option} ...]
```

其中，第一种方式需要通过 FLUSH PRIVILEGES 来重新加载权限表。而第二种方式通过 MySQL 内部命令去更新权限，会自动重载权限表。但值得一提的是，刷新了权限表并不意味着你就拥有了对应的权限，具体的生效分为如下三种情况，官方文档早有说明，见图 21。

A grant table reload affects privileges for each existing client session as follows:
- Table and column privilege changes take effect with the client's next request.
- Database privilege changes take effect the next time the client executes a USE *db_name* statement.

> **Note**
> Client applications may cache the database name; thus, this effect may not be visible to them without actually changing to a different database.

- Global privileges and passwords are unaffected for a connected client. These changes take effect only in sessions for subsequent connections.

图 21

(1) 对表级别（db_name.table_name）和列级别的权限更改将在客户端下一次请求时生效，也就是立即生效。

(2) 对库级别权限（db_name.*）的更改在客户端执行 USE db_name 语句后生效。

(3) 对全局级别权限（*.*）的更改对于已连接的会话不受影响，仅在新连接的会话中生效。

对表、列和全局级别权限生效的方式，笔者在本地测试没有问题，上述内容也十分容易理解，这里就不占用大家的时间，但对库级权限的更改，官方文档说是要 USE db_ name 才能生效，但实际上却是立即生效的。

7.3 验证

创建 ouyanghan 用户，此时该用户只有 usage 权限，且只能看到 information_schema 库。

```
# root 用户登录，创建新用户
mysql> CREATE USER ouyanghan IDENTIFIED by 'oyh123';

# ouyanghan 用户登录，查看权限
mysql> SHOW GRANTS;
+---------------------------------------+
| Grants for ouyanghan@%                |
+---------------------------------------+
| GRANT USAGE ON *.* TO 'ouyanghan'@'%' |
+---------------------------------------+
1 row in set (0.00 sec)

mysql> SHOW DATABASES;
+--------------------+
| Database           |
+--------------------+
| information_schema |
+--------------------+
1 row in set (0.00 sec)
```

给 ouyanghan 用户授予库级的 SELECT 权限，发现对库级别的更改可以实时生效。

```
# root 用户授权
mysql> GRANT SELECT ON demp.* TO ouyanghan;
Query OK, 0 rows affected (0.00 sec)

# ouyanghan 用户登录查看权限（同一会话）
mysql> SHOW GRANTS;
+-----------------------------------------------+
| Grants for ouyanghan@%                        |
+-----------------------------------------------+
| GRANT USAGE ON *.* TO 'ouyanghan'@'%'         |
| GRANT SELECT ON `demp`.* TO 'ouyanghan'@'%'   |
+-----------------------------------------------+
2 rows in set (0.00 sec)

# 并且能查看到 demp 库
mysql> SHOW DATABASES;
+--------------------+
| Database           |
+--------------------+
| information_schema |
```

```
| demp              |
+-------------------+
2 rows in set (0.00 sec)
```

这是怎么回事，我也有找到官网错误的高光时刻了？其实不然。仔细一看，原来官网的说明里面还有一条注意事项：

Client applications may cache the database name; thus, this effect may not be visible to them without actually changing to a different database.（客户端应用程序可以缓存数据库名称；因此，如果不实际更改到另一个数据库，则可能无法看到此效果。）

7.4 开启缓存

那么我们把 MySQL 缓存开启一下，并赋予一定的缓存大小。

```
# 查看此时 ouyanghan 用户的权限
mysql> SHOW GRANTS FOR demo;
+----------------------------------------+
| Grants for demo@%                      |
+----------------------------------------+
| GRANT USAGE ON *.* TO 'demo'@'%'       |
| GRANT SELECT ON `demp`.* TO 'demo'@'%' |
| GRANT SELECT ON `db1`.* TO 'demo'@'%'  |
+----------------------------------------+
3 rows in set (0.00 sec)

# 开启缓存，并赋予大小
mysql> SET GLOBAL query_cache_type = 1;
Query OK, 0 rows affected, 1 warning (0.00 sec)

mysql> SET GLOBAL query_cache_size = 1000000;
Query OK, 0 rows affected, 2 warnings (0.00 sec)
ouyanghan 用户登录 MySQL，此时能查看 db1 库下表的具体信息
mysql> USE db1;
Database changed

mysql> SELECT * FROM t1;
+----+------+
| id | c    |
+----+------+
|  1 | a    |
+----+------+
1 row in set (0.00 sec)
```

root 用户回收权限。

```
mysql> REVOKE SELECT ON db1.* FROM ouyanghan;
Query OK, 0 rows affected (0.00 sec)
```

ouyanghan 用户查看权限。

```
# 发现权限已经被回收
mysql> SHOW GRANTS FOR ouyanghan;
+----------------------------------------------+
| Grants for ouyanghan@%                       |
+----------------------------------------------+
| GRANT USAGE ON *.* TO 'ouyanghan'@'%'        |
| GRANT SELECT ON `demp`.* TO 'ouyanghan'@'%'  |
+----------------------------------------------+
2 rows in set (0.00 sec)

# use db1 失败，报没有权限，但仍能查看里面的内容
mysql> USE db1;
ERROR 1044 (42000): Access denied for user 'ouyanghan'@'%' to database 'db1'

mysql> SELECT * FROM db1.t1;
+----+------+
| id | c    |
+----+------+
| 1  | a    |
+----+------+

# 切换不同的库后，此时才发现权限被真正回收了，不能查看对应的内容了
mysql> USE demp;
Reading table information for completion of table and column names
You can turn off this feature to get a quicker startup with -A

Database changed
mysql> SELECT * FROM db1.t1;
ERROR 1142 (42000): SELECT command denied to user 'ouyanghan'@'localhost' for table 't1'
```

可能有严谨的看官有疑问了："你对表、列级别的权限做更改的时候，也没见你开启 MySQL 查询缓存啊，说不定表级和列级的权限做更改的生效时间，也需要去 USE db_name 一下呢？"

嘿，你还别说，还真是，于是笔者火急火燎又去测试了一下，发现对表级和列级的权限做更改，就是立马生效的。不信你就去试试吧！

7.5 总结

不管是使用语句直接修改授权表，还是用 MySQL 内部命令去更改权限，都要遵守下面的生效规则：

(1) 对表级别（db_name.table_name）和列级别，权限更改将在客户端下一次请求时生效，也就是立即生效。

(2) 对库级别权限（db_name.*）的更改在客户端执行 USE db_name 语句后生效（需要开启 query_cache_type 参数。当然，通常为了 MySQL 性能，这个参数是不建议开启的，且在 MySQL 8.0 版本中已经被移除了）。

(3) 对全局级别权限（*.*）的更改对于已连接的会话不受影响，仅在新连接的会话中生效。

最后，在座各位已经知道如何解决笔者开始遇到的权限不生效的问题了吧？那就是刷新 Uproxy 连接池。

8 盘点 MySQL 创建内部临时表的所有场景

作者：刘嘉浩

临时表属于一种临时存放数据的表，这类表在会话结束时会被自动清理掉，但在 MySQL 中存在两种临时表，一种是外部临时表，另外一种是内部临时表。

外部临时表指的是用户使用 CREATE TEMPORARY TABLE 手动创建的临时表。而内部临时表用户是无法控制的，并不能像外部临时表一样使用 CREATE 语句创建，MySQL 的优化器会自动选择是否使用内部临时表。

那么由此引发一个问题，MySQL 到底在什么时候会使用内部临时表呢？我们将针对 UNION、GROUP BY 等场景进行分析。

8.1 UNION 场景

首先准备一个测试表。

```
CREATE TABLE `employees` (
`id` int NOT NULL AUTO_INCREMENT,
`first_name` varchar(100) COLLATE utf8mb4_bin DEFAULT NULL,
`last_name` varchar(100) COLLATE utf8mb4_bin DEFAULT NULL,
`sex` enum('M','F') COLLATE utf8mb4_bin DEFAULT NULL,
`age` int DEFAULT NULL,
`birth_date` date DEFAULT NULL,
`hire_date` date DEFAULT NULL,
PRIMARY KEY (`id`),
KEY `last_name` (`last_name`),
KEY `hire_date` (`hire_date`)
) ENGINE=InnoDB AUTO_INCREMENT=500002 DEFAULT CHARSET=utf8mb4 COLLATE=utf8mb4_bin;
```

准备插入数据的脚本。

```python
#! /usr/bin/python
#! coding=utf-8

import random
import pymysql
from faker import Faker
from datetime import datetime, timedelta

# 创建 Faker 实例
fake = Faker()

# MySQL 连接参数
db_params = {
'host': 'localhost',
'user': 'root',
'password': 'root',
'db': 'db1',
'port': 3311
}

# 连接数据库
connection = pymysql.connect(**db_params)

# 创建一个新的 Cursor 实例
cursor = connection.cursor()

# 生成并插入数据
for i in range(5000):
    id = (i+1)
    first_name = fake.first_name()
    last_name = fake.last_name()
    sex = random.choice(['M', 'F'])
    age = random.randint(20, 60)
    birth_date = fake.date_between(start_date='-60y', end_date='-20y')
    hire_date = fake.date_between(start_date='-30y', end_date='today')

    query = f"""INSERT INTO employees (id, first_name, last_name, sex, age, birth_date, hire_date)
    VALUES ('{id}', '{first_name}', '{last_name}', '{sex}', {age}, '{birth_date}', '{hire_date}');"""

    cursor.execute(query)

    # 每1000提交一次事务
    if (i+1) % 1000 == 0:
        connection.commit()
```

```
# 最后提交事务
connection.commit()

# 关闭连接
cursor.close()
connection.close()
```

在创建好测试数据后，执行一个带有 UNION 的语句。

```
root@localhost:mysqld.sock[db1]> explain (select 5000 as res from dual) union
(select id from employees order by id desc limit 2);
+----+-------------+------------+------------+-------+---------------+---------
--+---------+------+------+----------+-----------------------------------+
| id | select_type | table      | partitions | type  | possible_keys | key
| key_len | ref  | rows | filtered | Extra                             |
+----+-------------+------------+------------+-------+---------------+---------
--+---------+------+------+----------+-----------------------------------+
|  1 | PRIMARY     | NULL       | NULL       | NULL  | NULL          | NULL
| NULL    | NULL | NULL |     NULL | No tables used                    |
|  2 | UNION       | employees  | NULL       | index | NULL          | PRIMARY
| 4       | NULL |    2 |   100.00 | Backward index scan; Using index  |
| NULL | UNION RESULT | <union1,2> | NULL     | ALL   | NULL          | NULL
| NULL    | NULL | NULL |     NULL | Using temporary                   |
+----+-------------+------------+------------+-------+---------------+---------
--+---------+------+------+----------+-----------------------------------+
3 rows in set, 1 warning (0.00 sec)
```

可见第二行中 key 值是 PRIMARY，即第二个查询使用了主键 ID。第三行 extra 值是 Using temporary，表明在对上面两个查询的结果集做 UNION 的时候，使用了临时表。

UNION 操作是将两个结果集取并集，不包含重复项。要做到这一点，只需要先创建一个只有主键的内存内部临时表，并将第一个子查询的值插入这个表中，这样就避免了重复的问题。因为值 5000 早已存在临时表中，而第二个子查询的值 5000 就会因为二者冲突而无法插入，只能插入下一个值 4999。

UNION ALL 与 UNION 不同，并不会使用内存临时表，下面的例子是使用 UNION ALL 的执行计划。

```
root@localhost:mysqld.sock[db1]> explain (select 5000 as res from dual) union
all (select id from employees order by id desc limit 2);
+----+-------------+-----------+------------+------+---------------+---------
--+---------+------+------+----------+----------------------------------+
| id | select_type | table     | partitions | type | possible_keys | key
| key_len | ref  | rows | filtered | Extra                            |
+----+-------------+-----------+------------+------+---------------+---------
--+---------+------+------+----------+----------------------------------+
|  1 | PRIMARY     | NULL      | NULL       | NULL | NULL          | NULL
```

```
| NULL     | NULL    | NULL |    NULL | No tables used                       |
|     2 | UNION       | employees | NULL    | index | NULL              | PRIMARY
| 4        | NULL    |    2 |  100.00 | Backward index scan; Using index |
+----+-------------+-----------+------------+-------+-----------------+---------
+---------+------+----------+------------------------------------+
2 rows in set, 1 warning (0.01 sec)
```

因为 UNION ALL 并不需要去重，所以优化器不需要新建一个临时表做去重的动作，执行的时候只需要按顺序执行两个子查询并将子查询放在一个结果集里就好了。

可以看到，在实现 UNION 的语义上，临时表起到的是一个暂时存储数据并做去重的动作的作用。

8.2 GROUP BY

除了 UNION 之外，还有一个比较常用的子句 GROUP BY 也会使用内部临时表。下例展示的是使用 ID 列求余并进行分组统计，且按照余数大小排列。

```
root@localhost:mysqld.sock[db1]> explain select id%5 as complementation,count(*) from employees group by complementation order by 1;
+----+-------------+-----------+------------+-------+-------------------------
----+-----------+---------+------+------+----------+-------------------------
----------------+
| id | select_type | table     | partitions | type  | possible_keys                | key   | key_len | ref  | rows | filtered | Extra                              |
+----+-------------+-----------+------------+-------+-------------------------
----+-----------+---------+------+------+----------+-------------------------
----------------+
|  1 | SIMPLE      | employees | NULL       | index | PRIMARY,last_name,hire_date | hire_date | 4       | NULL | 5000 |   100.00 | Using index; Using temporary; Using filesort |
+----+-------------+-----------+------------+-------+-------------------------
----+-----------+---------+------+------+----------+-------------------------
----------------+
1 row in set, 1 warning (0.00 sec)
```

可以看到 extra 的值是 using index、using temporary、using filesort。这三个值分别是：使用索引、使用临时表、使用排序。

> 注意：在 MySQL 5.7 版本中 GROUP BY 会默认按照分组字段进行排序，在 MySQL 8.0 版本中取消了默认排序功能，所以此处使用了 ORDER BY 进行复现。

对于 GROUP BY 来说，上述的语句执行后，会先创建一个内存内部临时表，存储 complementation 与 count(*) 的值，主键为 complementation。然后按照索引 hire_date 对应的 ID 值依次计算 id%5 的值（记为 x），如果临时表中没有主键为 x 的值，那么将会在临时表中插入记录；如果存在则累加这一行的计数 count(*)。在遍历完成上述的操作后，再按照 ORDER BY 的规则对 complementation 进行排序。

在使用 GROUP BY 进行分组或使用 DISTINCT 进行去重时，MySQL 都给我们提供了使用 hint 去避免使用内存内部临时表的方法（见表 5）。

表 5

hint	解释
SQL_BIG_RESULT	显式指定该 SQL 语句使用磁盘内部临时表，适合大数据量的操作；适用于 InnoDB 引擎与 Memory 引擎
SQL_SMALL_ESULT	显式指定该 SQL 语句使用内存内部临时表，速度更快，适合小数据量的操作；适用于 Memory 引擎

下面是一个使用了 SQL_BIG_RESULT 的例子。

```
root@localhost:mysqld.sock[db1]> explain select SQL_BIG_RESULT id%5 as complementation,count(*) from employees group by complementation order by 1;
+----+-------------+-----------+------------+-------+---------------------+
| id | select_type | table     | partitions | type  | possible_keys       
| key         | key_len | ref  | rows | filtered | Extra                   |
+----+-------------+-----------+------------+-------+---------------------+
| 1  | SIMPLE      | employees | NULL       | index | PRIMARY,last_name,hire_date | hire_date | 4       | NULL | 5000 | 100.00   | Using index; Using filesort |
+----+-------------+-----------+------------+-------+---------------------+
1 row in set, 1 warning (0.00 sec)
```

从执行计划中我们可以看出，使用了 SQL_BIG_RESULT 这个 hint 进行查询后，在 extra 列中 Using Temporary 字样已经不见了，即避免了使用内存内部临时表。

8.3 其他场景

当然，除了上述两个例子外，MySQL 还会在下列情况下创建内部临时表：

(1) 对于 UNION 语句的评估，但有一些后续描述中的例外情况。

(2) 对于某些视图的评估，例如使用 TEMPTABLE 算法、UNION 或聚合的视图。

(3) 对派生表的评估。

(4) 对公共表达式的评估。

(5) 用于子查询或半连接材料化的表。

(6) 对包含 ORDER BY 子句和不同 GROUP BY 子句的语句的评估，特别是这些子句包含来自连接队列中第一个表以外的其他表的列。

(7) 对于 DISTINCT 与 ORDER BY 的组合，可能需要一个临时表。

(8) 对于使用 SQL_SMALL_RESULT 修饰符的查询，MySQL 使用内存中的临时表，除非查询还包含需要在磁盘上存储的元素。

(9) 为了评估从同一表中选取并插入的 INSERT…SELECT 语句，MySQL 创建一个内部临时表来保存 SELECT 的行，然后将这些行插入目标表中。

(10) 对于多表 UPDATE 语句的评估。

(11) 对于 GROUP_CONCAT() 或 COUNT(DISTINCT) 表达式的评估。

(12) 对于窗口函数的评估，根据需要使用临时表。

值得注意的是，某些查询条件 MySQL 不允许使用内存内部临时表，在这种情况下，服务器会使用磁盘内部临时表。

(1) 表中存在 BLOB 或 TEXT 列。MySQL 8.0 中用于内存内部临时表的默认存储引擎 TempTable 从 8.0.13 版本开始支持二进制大对象类型。

(2) 使用了 UNION 或 UNION ALL 且 SELECT 的列表中存在最大长度超过 512 的字符串列（对于二进制字符串，此长度指的是字节；对于非二进制字符串，为字符）。

(3) SHOW COLUMNS 和 DESCRIBE 语句使用 BLOB 作为某些列的类型，因此用于此结果的临时表是将会是磁盘内部临时表。

9 MySQL MGR 滚动升级可行吗？

作者：雷文霆

9.1 为什么要升级？

因为 MySQL 8.0.24 之前有很多 Bug，详见 MySQL 8.0 Release Notes，本文定稿时最新的版本是 2023 年 7 月 18 日发布的 8.0.34。

9.2 升级前 DBA 和业务侧需要做什么？

如果是 MySQL 8.0.x 到 MySQL 8.0.z 的小版本升级，可以直接升级。参数名称的变化和弃用情况，可以在 MySQL 服务启动后，从错误日志中看到，更新到配置文件中即可。

如果你没有做过升级或是需要进行从 MySQL 5.7 到 MySQL 8.0 的大版本升级，需要根据官方升级文档进行检查。

至少要做初步检查：

(1) 必须备份数据，只支持 GA。

(2) mysqlcheck -u root -p --all-databases --check-upgrade

(3) util.checkForServerUpgrade()

9.3 升级思路

前提条件：升级到 MySQL 8.0 的检查项都已经检查完毕。

根据官方的升级文档，升级分为如下 2 个步骤：

(1) 步骤 1：Data dictionary upgrade。

(2) 步骤 2：Server upgrade。

在 MySQL 8.0.16 之前，升级用 mysql_upgrade；从 MySQL 8.0.16 开始，mysqld 可以自动升级。

(1) 依次升级 MGR 的 Secondary 节点。

(2) 切换到新 Primary 节点。

(3) 最后升级旧 Primary 节点。

所以，理论上如果有 MySQL Router（它在应用程序和 MGR 集群之间，起到代理作用，自动发现后端可写主库），那么不影响业务。但是，实践出真知，升级时要通知业务侧，因为常会重启应用程序服务用于验证升级或是验证程序代码。所以申请一个变更窗口是最优解。

9.4 具体的升级步骤是什么？

环境信息：MySQL 8.0.27 升级到 MySQL 8.0.33。

从库升级时，建议设置 skip_slave_start=1 或 skip_replica_start=1

9.4.1 升级前的检查

```
# 以下三种方式均可
mysqlsh root@localhost:3306 -e "util.checkForServerUpgrade();"

MySQL    JS > util.checkForServerUpgrade("admin@127.0.0.1:3306")

util.checkForServerUpgrade("admin@127.0.0.1:3306", {"targetVersion":"8.0.33","outputFormat":"JSON","configPath":"/data/mysql/3306/my.cnf.3306"})[ 指定版本检查 ]

# 辅助检查
mysqlcheck -uadmin -p -h127.0.0.1 --all-databases --check-upgrade
```

9.4.2 备份数据和配置文件

逻辑或是物理备份。

```
cp -rp /data/mysql/data/3306   /data/mysql/data/3306_bak 「备份数据目录为可选」
cp my.cnf bak_8.0.xx_my.cnf
```

9.4.3 关闭服务和准备 basedir

查看表的状态是否正常（关库之前）。

```sql
select * from information_schema.tables where TABLE_COMMENT like "%repair%";
```

安全关闭服务。

```
set global innodb_fast_shutdown = 0;

# 数据刷到磁盘，关闭快速关机
shutdown;

# 关闭服务之后检查
ps -ef | grep mysql
```

如果是从库（无主从延迟）。

```
show slave status\G
stop slave;
```

则备份数据文件（如果磁盘充足）。

```
cp -rp /data/mysql/data/3366 /data/mysql/data/3366_bak_1013
```

准备 basedir，软连接或是直接替换。

```
show variables like "%basedir%"; 或 which mysql

# 备份原 base 目录
mv mysql_Basedir mysql_Basedir_bak
chown -R mysql:mysql mysql-8.0.33-linux-glibc2.12-x86_64

# ln -s / 目录 / 新软件包软连接的目录
ln -s mysql-8.0.xx新版本-linux-glibc2.12-x86_64 mysql_Basedir
chown -R mysql:mysql mysql_Basedir

# 示例
ln -s mysql-8.0.33-linux-glibc2.12-x86_64 base
lrwxrwxrwx 1 mysql mysql          35 Sep  7 16:14 base -> mysql-8.0.33-linux-glibc2.12-x86_64

# 环境变量可选
```

```
export PATH=$PATH:/usr/local/mysql/bin

# 或是 mv 替换，修改旧软件目录名称
mv mysql mysql_8.0.xx
mv mysql-8.0.xx-linux-glibc2.12-x86_64 mysql
```

9.4.4 启动数据库

启动就是升级，即数据字典升级和 server 升级。

记录 MySQL 服务的原启动方式。

查看和检查。

```
var cluster = dba.getCluster()
cluster.status()

SELECT * FROM performance_schema.replication_group_members;
```

启动。

```
# 两种启动方式
systemctl start

mysqld_safe --defaults-file=/*.cnf --user=mysql &
```

9.4.5 最后升级旧 Primary 节点之前，先切主

同版本时，执行切主命令（我们这次不能使用，因为不同版本不能切换）。

```
# shell 使用 8.0
# 不支持 5.7.31
cluster.setPrimaryInstance('clusteruser@10.186.65.181:6681')
# 简版
cluster.setPrimaryInstance('mgr01:3306')

# 手动指定主库 >8.0.13（推荐）
SELECT group_replication_set_as_primary('select @@server_uuid');
```

不能切主，怎么办？停止旧主组复制，使其自动切换。

```
# 低版本不能 SELECT 切换到高版本，通过退出组复制，自动切换
stop group_replication;
# 启动前配置文件添加防止应用更新
read_only=1;super_read_only=1;
```

9.4.6 升级组的通信协议（>MySQL 8.0.16）

group_replication_get_communication_protocol 函数返回组支持的最低 MySQL 版本（也就是说这个版本以下的 MySQL 无法加入集群）。

```
SELECT group_replication_get_communication_protocol();
```

```
SELECT group_replication_set_communication_protocol("8.0.33");
```

提示：虽然我们 set 成功了，但是依然显示如下低版本号，因为它查到的是最低支持的版本。官方建议升级之后升级对应通信协议，所以别忘记要 set 一下。

```
SELECT group_replication_get_communication_protocol();
+------------------------------------------------+
| group_replication_get_communication_protocol() |
+------------------------------------------------+
| 8.0.27                                         |
+------------------------------------------------+
```

9.4.7 验证

已升级成功。

```
select version();
+-----------+
| version() |
+-----------+
| 8.0.33    |
+-----------+
```

从库需要启动复制。

```
stop group_replication;
```

9.5 关于回退

如果启动失败，重新以原来的软件版本启动。

(1) 清理 redo log 文件 rm -f ib_logfile{0,1,2,3}。

(2) 清理 link，启动旧版本。

参考步骤：

```
unlink mysql
mv mysql_Basedir mysql_Basedir_bak
ln -s mysql-8.0.xx旧版本-linux-glibc2.12-x86_64 mysql_Basedir
# 启动服务
```

9.6 关于节点切换

变相自定义切换（虽说节点切换是自动的）。

(1) 节点权重 group_replication_member_weight 值越大越优先。

(2) erver_uuid 越小越容易被选为主节点。

9.7 关于升级中遇到的问题

启动数据库时，查看 error 日志，替换或是去除 Warning。

```
2023-09-04T08:54:54.092942-00:00 0 [Warning] [MY-011068] [Server] The syntax '--admin-ssl=off' is deprecated and will be removed in a future release. Please use --admin-tls-version='' instead.
2023-09-04T08:54:54.092975-00:00 0 [Warning] [MY-011068] [Server] The syntax 'skip_slave_start' is deprecated and will be removed in a future release. Please use skip_replica_start instead.
```

正确的升级提示信息为：

```
2023-09-04T08:55:10.136353-00:00 7 [System] [MY-013381] [Server] Server upgrade from '80027' to '80033' started.
2023-09-04T08:55:12.621536-00:00 7 [System] [MY-013381] [Server] Server upgrade from '80027' to '80033' completed.
```

以下为低版本不能切换到高版本：

```
023-09-07T07:41:09.211541-00:00 59 [ERROR] [MY-013223] [Repl] The function 'group_replication_set_as_primary' failed. Error processing configuration start message: The appointed primary member has a version that is greater than the one of some of the members in the group.
2023-09-07T07:50:34.708283-00:00 0` [ERROR] [MY-013212] [Repl] Plugin group_replication reported: 'Error while executing a group configuration operation: Error processing configuration start message: The appointed primary member has a version that is greater than the one of some of the members in the group.'

# 低版本离开集群之后无法重新加回来 stop group_replication;start group_replication; 好在集群会自动切换
2023-09-07T07:51:47.747695-00:00 0 [ERROR] [MY-011521] [Repl] Plugin group_replication reported: 'Member version is incompatible with the group.'
```

原因：当新成员加入复制组时，它会检查组的现有成员宣布的通信协议版本。如果新成员支持该版本，则它将加入组并使用组已宣布的通信协议，即使该成员支持其他通信功能也是如此。如果新成员不支持通信协议版本，则会将其从组中逐出。

一个 MySQL 服务器 5.7.24 实例无法成功加入使用通信协议版本 8.0.16 的组。（高版本服务器可以加入低版本的组）

一种不推荐的方式：如果需要更改组的通信协议版本，以便早期版本中的成员可以加入，请使用 group_replication_set_communication_protocol("5.7.25") 函数指定允许加入组的最低 MySQL Server 版本。

9.8 总结

(1) 此方法属于 in-place upgrade，核心步骤是用新软件包替换旧软件包（basedir），适合小版本升级。

(2) 其他的升级方式还有逻辑备份（MysqlDump 和 MyDumper）之后导入新版本。

(3) 如果是大版本升级，数据量又大，可以建立 Master 低版本到 Slave 高版本的复制，最后主从切换实现升级。

充分测试之后，请选择自己擅长的方式吧！

10 你知道 MySQL 函数 sysdate() 与 now() 的区别吗？

作者：余振兴，陈伟

10.1 背景

在客户现场优化一批监控 SQL 时，发现部分 SQL 使用 sysdate() 作为统计数据的查询范围值，执行效率十分低下。查看执行计划发现用 sysdate() 时 SQL 不能使用索引，而改为 now() 函数后则可以正常使用索引，以下是对该现象的分析。

sysdate() 的和 now() 的区别是个老问题了。

10.2 函数 sysdate() 与 ()now 的区别

下面我们来详细了解一下函数 sysdate() 与 now() 的区别，我们可以去官方文档查找两者之间的详细说明。

官方说明如下：

(1)now() 函数返回的是一个常量时间，该时间为语句开始执行的时间。当存储函数或触发器中调用 now() 函数时，now() 会返回存储函数或触发器语句开始执行的时间。

(2)sysdate() 函数则返回的是该语句执行的确切时间。

下面我们通过官方提供的案例直观展现两者区别。

```
mysql> SELECT NOW(), SLEEP(2), NOW();
+---------------------+----------+---------------------+
| NOW()               | SLEEP(2) | NOW()               |
+---------------------+----------+---------------------+
```

```
| 2023-12-14 15:13:09 |       0 | 2023-12-14 15:13:09 |
+---------------------+---------+---------------------+
1 row in set (2.00 sec)

mysql> SELECT SYSDATE(), SLEEP(2), SYSDATE();
+---------------------+----------+---------------------+
| SYSDATE()           | SLEEP(2) | SYSDATE()           |
+---------------------+----------+---------------------+
| 2023-12-14 15:13:19 |        0 | 2023-12-14 15:13:21 |
+---------------------+----------+---------------------+
1 row in set (2.00 sec)
```

通过上面的两条 SQL 语句我们可以发现，当 SQL 语句两次调用 now() 函数时，前后两次 now() 函数返回的是相同的时间，而当 SQL 语句两次调用 sysdate() 函数时，前后两次 sysdate() 函数返回的时间在更新。

到这里我们根据官方文档的说明加上自己的推测大概可以知道，函数 sysdate() 之所以不能使用索引是因为 sysdate() 的不确定性导致索引不能用于评估引用它的表达式。

10.3 测试示例

以下通过示例模拟客户的类似场景。

我们先创建一张测试表，对 create_time 字段创建索引并插入数据，观测函数 sysdate() 和 now() 使用索引的情况。

```
mysql> create table t1(
    -> id int primary key auto_increment,
    -> create_time datetime default current_timestamp,
    -> uname varchar(20),
    -> key idx_create_time(create_time)
    -> );
Query OK, 0 rows affected (0.02 sec)

mysql> insert into t1(id) values(null),(null),(null);
Query OK, 3 rows affected (0.01 sec)
Records: 3  Duplicates: 0  Warnings: 0

mysql> insert into t1(id) values(null),(null),(null);
Query OK, 3 rows affected (0.00 sec)
Records: 3  Duplicates: 0  Warnings: 0

mysql> select * from t1;
+----+---------------------+-------+
| id | create_time         | uname |
+----+---------------------+-------+
|  1 | 2023-12-14 15:34:30 | NULL  |
```

```
|  2 | 2023-12-14 15:34:30 | NULL  |
|  3 | 2023-12-14 15:34:30 | NULL  |
|  4 | 2023-12-14 15:34:37 | NULL  |
|  5 | 2023-12-14 15:34:37 | NULL  |
|  6 | 2023-12-14 15:34:37 | NULL  |
+----+---------------------+-------+
6 rows in set (0.00 sec)
```

先来看看函数 sysdate() 使用索引的情况。可以发现 possible_keys 和 key 均为 NULL，确实使用不了索引。

```
mysql> explain select * from t1 where create_time<sysdate()\G
*************************** 1. row ***************************
           id: 1
  select_type: SIMPLE
        table: t1
   partitions: NULL
         type: ALL
possible_keys: NULL
          key: NULL
      key_len: NULL
          ref: NULL
         rows: 6
     filtered: 33.33
        Extra: Using where
1 row in set, 1 warning (0.00 sec)
```

再来看看函数 now() 使用索引的情况，可以看到 key 使用到了 idx_create_time 这个索引。

```
mysql> explain select * from t1 where create_time<now()\G
*************************** 1. row ***************************
           id: 1
  select_type: SIMPLE
        table: t1
   partitions: NULL
         type: range
possible_keys: idx_create_time
          key: idx_create_time
      key_len: 6
          ref: NULL
         rows: 6
     filtered: 100.00
        Extra: Using index condition
1 row in set, 1 warning (0.00 sec)
```

10.4 示例详解

下面我们进一步通过 trace 去分析优化器对于函数 now() 和 sysdate() 具体是如何优化的。

10.4.1 函数 sysdate() 部分关键 trace 输出

```
"rows_estimation": [
## 估算使用各个索引进行范围扫描的成本
  {
    "table": "`t1`",
    "range_analysis": {
      "table_scan": {
        "rows": 6,
        "cost": 2.95
      },
      "potential_range_indexes": [
        {
          "index": "PRIMARY",
          "usable": false,
          "cause": "not_applicable"
        },
        {
          "index": "idx_create_time",
          "usable": true,
          "key_parts": [
            "create_time",
            "id"
..........................................
      "setup_range_conditions": [
      ],
      "group_index_range": {
        "chosen": false,
        "cause": "not_group_by_or_distinct"
      },
      "skip_scan_range": {
        "chosen": false,
        "cause": "disjuntive_predicate_present"
      }
..........................................
"considered_execution_plans": [
## 对比各可行计划的代价，选择相对最优的执行计划
  {
    "plan_prefix": [
    ],
    "table": "`t1`",
    "best_access_path": {
```

```
          "considered_access_paths": [
            {
              "rows_to_scan": 6,
              "access_type": "scan",
              "resulting_rows": 6,
              "cost": 0.85,
              "chosen": true
            }
          ]
        },
        "condition_filtering_pct": 100,
        "rows_for_plan": 6,
        "cost_for_plan": 0.85,
        "chosen": true
.........................................
```

10.4.2 函数 now() 部分关键 trace 输出

```
  "rows_estimation": [
## 估算使用各个索引进行范围扫描的成本
.........................................
        "analyzing_range_alternatives": {
          "range_scan_alternatives": [
            {
              "index": "idx_create_time",
              "ranges": [
                "NULL < create_time < '2023-12-14 15:48:39'"
              ],
              "index_dives_for_eq_ranges": true,
              "rowid_ordered": false,
              "using_mrr": false,
              "index_only": false,
              "in_memory": 1,
              "rows": 6,
              "cost": 2.36,
              "chosen": true
            }
          ],
.........................................
        },
        "chosen_range_access_summary": {
          "range_access_plan": {
            "type": "range_scan",
            "index": "idx_create_time",
            "rows": 6,
            "ranges": [
              "NULL < create_time < '2023-12-14 15:48:39'"
            ]
          },
```

```
          "rows_for_plan": 6,
          "cost_for_plan": 2.36,
          "chosen": true
........................................
"considered_execution_plans": [
## 对比各可行计划的代价，选择相对最优的执行计划
  {
    "plan_prefix": [
    ],
    "table": "`t1`",
    "best_access_path": {
      "considered_access_paths": [
        {
          "rows_to_scan": 6,
          "access_type": "range",
          "range_details": {
            "used_index": "idx_create_time"
          },
          "resulting_rows": 6,
          "cost": 2.96,
          "chosen": true
        }
      ]
    },
    "condition_filtering_pct": 100,
    "rows_for_plan": 6,
    "cost_for_plan": 2.96,
    "chosen": true
........................................
```

通过上述 trace 输出，我们可以发现对于函数 now()，优化器在 rows_estimation 时（即估算使用各个索引进行范围扫描的成本时）可以将 now() 的值转换为一个常量，最终在 considered_execution_plans 这一步去对比各可行计划的代价，选择相对最优的执行计划。而通过函数 sysdate() 时则无法做到该优化，因为 sysdate() 是动态获取时间的。

10.5 总结

通过实际验证执行计划和 trace 记录并结合官方文档的说明，我们可以得出以下理解。

(1) 函数 now() 在语句一开始执行时就获取时间（常量时间），优化器进行 SQL 解析时，已经能确认 now() 的具体返回值并可以将其当作一个已确定的常量去做优化。

(2) 函数 sysdate() 则在执行时动态获取时间（为该语句执行的确切时间），所以在优化器对 SQL 解析时不能确定其返回值是多少，因而不能做 SQL 优化和评估，也就导致优化器只能选择对该条件做全表扫描。

11 细说 MySQL 用户安全加固策略

作者：余振兴，官永强

11.1 背景

基于安全的背景下，客户对 MySQL 的用户安全提出了一系列需求，希望能对 MySQL 进行安全加固，具体的需求如下所述。

11.1.1 用户密码类

(1) 密码需要至少 25 个字符。

①密码必须包含至少 2 个大写字母。

②密码必须包含至少 2 个小写字母。

③密码必须包含至少 2 个数字。

④密码必须包含至少 2 个特殊字符。

⑤密码中不能包含用户名。

⑥密码不能是简单的重复字符（例如，AAA、wuwuwuwu、dsadsadsa、111）。

(2) 密码需要有过期时间，需要 365 天修改一次，否则过期并锁定用户。

(3) 密码不得使用历史 5 次内曾用过的老密码。

(4) 密码在 24 小时内最多只能修改一次。

(5) 密码不能包含指定的字符，如公司名称、业务名称等。

11.1.2 用户连接类

登录时如果连续 10 次失败，需要等待 10 分钟，且每次失败都持续增加等待时间。

11.2 需求分析

基于背景描述我们可以把需求分为三大块：

(1) 密码复杂度策略。

(2) 连接控制的策略。

(3) 密码变更的策略。

MySQL 有以下功能插件/组件、配置可实现以上需求：

(1) 密码校验插件/组件。

(2) 连接控制插件。

(3) 用户密码属性配置。

11.3 环境信息

MySQL 版本：8.0.33、5.7.41。

11.4 安装配置

11.4.1 密码校验组件配置

MySQL 5.7 版本使用的是密码校验插件，虽然安装方式和变量在语法上有些许差异，但功能基本相同。以下操作以 MySQL 8.0 版本为例说明，具体细节可参考官方文档。

```
## MySQL8.0 版本安装密码校验组件
INSTALL COMPONENT 'file://component_validate_password';

## 查看插件默认配置
show variables like 'validate_password%';
+--------------------------------------+--------+
| Variable_name                        | Value  |
+--------------------------------------+--------+
| validate_password.check_user_name    | ON     |   ## 密码不能包含用户名
| validate_password.dictionary_file    |        |   ## 指定密码匹配字典文件，在文件中
的字符串不能包含在设置的密码中，policy 为 STRONG 时有效
| validate_password.length             | 8      |   ## 密码最小长度，默认至少 8 位
| validate_password.mixed_case_count   | 1      |   ## 密码至少包含 1 个大小写字母
| validate_password.number_count       | 1      |   ## 密码至少包含 1 个数字
| validate_password.policy             | MEDIUM |   ## 密码默认复杂度策略
| validate_password.special_char_count | 1      |   ## 密码至少包含 1 个特殊字符
+--------------------------------------+--------+
7 rows in set (0.0042 sec)

## 修改配置以便符合背景需求
set global validate_password.length=25;
set global validate_password.mixed_case_count=2;
set global validate_password.number_count=2;
set global validate_password.special_char_count=2;

## 查看调整后的配置（动态生效）
show variables like 'validate_password%';
+--------------------------------------+--------+
| Variable_name                        | Value  |
+--------------------------------------+--------+
```

```
| validate_password.check_user_name       | ON     |
| validate_password.dictionary_file       |        |
| validate_password.length                | 25     |
| validate_password.mixed_case_count      | 2      |
| validate_password.number_count          | 2      |
| validate_password.policy                | MEDIUM |
| validate_password.special_char_count    | 2      |
+-----------------------------------------+--------+
7 rows in set (0.0056 sec)

## 持久化配置到 my.cnf 配置文件（永久生效）
## 在 [mysqld] 标签下增加配置
vim /data/mysql/3306/my.cnf.3306
[mysqld]
## 密码校验组件参数配置
validate_password.check_user_name       = ON
validate_password.policy                = MEDIUM
validate_password.length                = 25
validate_password.mixed_case_count      = 2
validate_password.number_count          = 2
validate_password.special_char_count    = 2
```

11.4.2 连接控制插件配置

连接控制插件在 MySQL 5.7 和 MySQL 8.0 中基本无变化，均以插件形式提供。以下操作仅以 MySQL 8.0 版本为例，具体细节可参考官方文档。

```
## 安装连接控制插件
INSTALL PLUGIN CONNECTION_CONTROL SONAME 'connection_control.so';
INSTALL PLUGIN CONNECTION_CONTROL_FAILED_LOGIN_ATTEMPTS SONAME 'connection_control.so';

## 查看插件默认配置
show variables like 'connection_control%';
+-------------------------------------------------+------------+
| Variable_name                                   | Value      |
+-------------------------------------------------+------------+
| connection_control_failed_connections_threshold | 3          | ## 登录失败后尝
试的次数，默认为3，表示当连接失败3次后启用连接控制，0表示不开启
| connection_control_max_connection_delay         | 2147483647 | ## 响应延迟的最
大时间，默认约25天
| connection_control_min_connection_delay         | 1000       | ## 登录失败后响
应延迟的最小时间，默认为1000毫秒，即1秒，每失败一次逐步累加，直到最大值
+-------------------------------------------------+------------+

## 修改配置以便符合背景需求，响应延迟的最大时间设置为1天
set global connection_control_max_connection_delay=24*60*60*1000;

## 查看调整后的配置（动态生效）
```

```
show variables like 'connection_control%';
+-------------------------------------------------------+----------+
| Variable_name                                         | Value    |
+-------------------------------------------------------+----------+
| connection_control_failed_connections_threshold       | 3        |
| connection_control_max_connection_delay               | 86400000 | ## 调整为最大一天
| connection_control_min_connection_delay               | 1000     |
+-------------------------------------------------------+----------+

## 持久化配置到 my.cnf 配置文件（永久生效）
## 在 [mysqld] 标签下增加配置
vim /data/mysql/3306/my.cnf.3306
[mysqld]
## 配置连接控制插件
connection-control                                  = FORCE
connection-control-failed-login-attempts            = FORCE
connection_control_min_connection_delay             = 1000
connection_control_max_connection_delay             = 86400000
connection_control_failed_connections_threshold     = 3
```

11.4.3 密码变更策略配置

MySQL 密码变更策略配置记录在 mysql.user 表中，MySQL5.7 和 8.0 版本支持的配置略有差异，具体细节可参考官方文档关于 CREATE USER 和 ALTER USER 语法中 password_option 部分属性的说明。

11.4.4 相关配置参数含义说明

11.4.4.1 MySQL 5.7 版本下仅支持的参数

default_password_lifetime：密码有效期（默认为 0 或 NULL），表示密码永久有效。

注意：线上环境配置密码过期策略虽然可提升安全性，但如果没及时更新密码会导致业务中断问题，需要综合评估后配置！

11.4.4.2 MySQL 8.0 版本支持的参数

(1)default_password_lifetime：密码有效期（默认为 0 或 NULL），表示密码永久有效。

注意：线上环境配置密码过期策略虽然可提升安全性，但如果没及时更新密码会导致业务中断问题，需要综合评估后配置！

(2)password_history：历史密码可重用的循环，表示记录历史上前多少次密码不允许被重复使用，历史密码信息记录在 mysql.password_history 表中。

(3)password_reuse_interval：指定历史密码要经过多长时间才能被重用，单位为天。

关于 default_password_lifetime、password_history 以及 password_reuse_interval 在 my.cnf 配置后，创建用户默认属性不生效的问题，我提了 MySQL Bug，等待官方反馈是否符合预期。

结论：不是 Bug，而是对参数的理解有误。

以密码过期配置为示例说明，password_reuse_interval、password_history 均遵循相同逻辑。

```
ALTER USER eee PASSWORD EXPIRE;
```

eee 用户密码立即过期，mysql.user 表中 password_expired 字段标记为 Y。

```
ALTER USER eee PASSWORD EXPIRE DEFAULT;
```

eee 用户密码采用全局参数 default_password_lifetime 指定的值作为过期策略，mysql.user 表中的 password_lifetime 字段为 NULL。

```
ALTER USER eee PASSWORD EXPIRE NEVER;
```

eee 用户密码过期策略设置为永不过期，mysql.user 表中的 password_lifetime 字段值为 0。

```
ALTER USER eee PASSWORD EXPIRE INTERVAL 3 DAY;
```

eee 用户密码过期策略设置为 3 天后过期，mysql.user 表中的 password_lifetime 字段值为 3。

具体如图 22 所示。

全局参数配置	用户属性	生效值	mysql.user表字段值	备注
default_password_lifetime=365	PASSWORD EXPIRE DEFAULT	密码365天后过期	password_lifetime=NULL	用户属性DEFAULT表示继承全局参数配置，表对应字段值为NULL
default_password_lifetime=365	PASSWORD EXPIRE NEVER	密码永不过期	password_lifetime=0	
default_password_lifetime=365	PASSWORD EXPIRE INTERVAL 3 DAY	密码3天后过期	password_lifetime=3	
password_history=5	PASSWORD HISTORY DEFAULT	前5次旧密码不可使用	password_reuse_history=NULL	用户属性DEFAULT表示继承全局参数配置，表对应字段值为NULL
password_history=5	PASSWORD HISTORY 0	不限制旧密码使用	password_reuse_history=0	
password_history=5	PASSWORD HISTORY 6	前6次旧密码不可使用	password_reuse_history=6	
password_reuse_interval=5	PASSWORD REUSE INTERVAL DEFAULT	5天内的旧密码不可使用	password_reuse_interval=NULL	用户属性DEFAULT表示继承全局参数配置，表对应字段值为NULL
password_reuse_interval=5	PASSWORD REUSE INTERVAL 0 DAY	不限制旧密码使用	password_reuse_interval=0	
password_reuse_interval=5	PASSWORD REUSE INTERVAL 30 DAY	30天内旧密码不可使用	password_reuse_interval=30	
password_require_current=ON	PASSWORD REQUIRE CURRENT DEFAULT	修改密码需要给定当前密码	password_require_current=NULL	用户属性DEFAULT表示继承全局参数配置，表对应字段值为NULL
password_require_current=ON	PASSWORD REQUIRE CURRENT	修改密码需要给定当前密码	password_require_current=Y	
password_require_current=ON	PASSWORD REQUIRE CURRENT OPTIONAL	修改密码不需要给定当前密码	password_require_current=N	
password_require_current=OFF	PASSWORD REQUIRE CURRENT DEFAULT	修改密码不需要给定当前密码	password_require_current=NULL	用户属性DEFAULT表示继承全局参数配置，表对应字段值为NULL
password_require_current=OFF	PASSWORD REQUIRE CURRENT	修改密码需要给定当前密码	password_require_current=Y	
password_require_current=OFF	PASSWORD REQUIRE CURRENT OPTIONAL	修改密码不需要给定当前密码	password_require_current=N	

图 22

关于 password_history 和 password_reuse_interval 参数同时使用时，实际只有 password_reuse_interval 参数有效的问题，我提了 MySQL Bug，等待官方反馈是否符合预期。

```
################## 以 5.7 版本为例 ##################
## 1.修改当前默认密码策略为需求所需配置(动态生效)
set global default_password_lifetime=365;
```

2. 查看当前密码有效期配置。这里一定注意，表中的 password_lifetime 为 NULL 不是表示无策略，而是表示使用 default_password_lifetime 参数指定的全局策略

```
select user,host,password_lifetime from mysql.user where user not in ('mysql.session','mysql.sys');
+----------+-----------+-------------------+
| user     | host      | password_lifetime |
+----------+-----------+-------------------+
| root     | localhost |              NULL |
| zhenxing | %         |              NULL |
| sysbench | %         |              NULL |
| aaa      | %         |              NULL |
| bbb      | %         |              NULL |
+----------+-----------+-------------------+
```

3. 持久化当前默认密码策略为需求所需配置（永久生效）
在 [mysqld] 标签下增加配置
```
vim /data/mysql/3306/my.cnf.3306
[mysqld]
## 密码策略配置
default_password_lifetime = 365
```

################## 以 8.0 版本为例 ##################
1. 修改当前默认密码策略为需求所需配置(动态生效)
```
set global default_password_lifetime=365;
set global password_history=5;
set global password_reuse_interval=1;
```

2. 查看当前已存在用户的密码策略配置（默认均为 NULL，表示使用 default_password_lifetime、password_history、password_reuse_interval 参数指定的全局策略）

```
select user,host,password_lifetime,Password_reuse_history,Password_reuse_time from mysql.user where user not in ('mysql.infoschema','mysql.session','mysql.sys');
+----------+-----------+-------------------+------------------------+---------------------+
| user     | host      | password_lifetime | Password_reuse_history | Password_reuse_time |
+----------+-----------+-------------------+------------------------+---------------------+
| aaa      | %         |              NULL |                   NULL |                NULL |
| sysbench | %         |              NULL |                   NULL |                NULL |
| zhenxing | %         |              NULL |                   NULL |                NULL |
| backup   | 127.0.0.1 |              NULL |                   NULL |                NULL |
| backup   | localhost |              NULL |                   NULL |                NULL |
| root     | localhost |              NULL |                   NULL |                NULL |
+----------+-----------+-------------------+------------------------+---------------------+
```

3. 持久化当前默认密码策略为需求所需配置（永久生效）
在 [mysqld] 标签下增加配置
```
vim /data/mysql/3306/my.cnf.3306
[mysqld]
## 密码策略配置
default_password_lifetime = 365
```

```
password_history              = 5
password_reuse_interval       = 1
```

11.5 功能验证

11.5.1 密码校验组件

MySQL 5.7 版本使用的是密码校验插件，安装方式和变量语法有差异，功能基本相同，以下操作仅以 MySQL 5.7 版本为例。

```
################## 以 5.7 版本为例 ##################
## 1.查看当前密码校验插件配置信息
mysql> show variables like 'validate%';
+--------------------------------------+----------------------+
| Variable_name                        | Value                |
+--------------------------------------+----------------------+
| validate_password_check_user_name    | ON                   |
| validate_password_dictionary_file    | /usr/share/dict/words|
| validate_password_length             | 25                   |
| validate_password_mixed_case_count   | 2                    |
| validate_password_number_count       | 2                    |
| validate_password_policy             | STRONG               |
| validate_password_special_char_count | 2                    |
+--------------------------------------+----------------------+
7 rows in set (0.00 sec)

## 2.进行验证
## 2.1 验证密码不符合validate_password_mixed_case_count≥2 时，是否可以成功新建用户
mysql> create user test33@'%' identified WITH 'mysql_native_password' by'1234567890@#$tyuiopasdfg';
ERROR 1819 (HY000): Your password does not satisfy the current policy requirements

## 2.2 验证密码不符合validate_password_number_count≥2 时，是否可以成功新建用户
mysql> create user test33@'%' identified WITH 'mysql_native_password' by'qazwsxEDCRFVtgb%$#ujmnbgf';
ERROR 1819 (HY000): Your password does not satisfy the current policy requirements

## 2.3 验证密码不符合validate_password_special_char_count≥2 时，是否可以成功新建用户
mysql> create user test33@'%' identified WITH 'mysql_native_password' by'1qazWSXEDCdsa321321321dsadwq';
ERROR 1819 (HY000): Your password does not satisfy the current policy requirements

## 2.4 验证密码不符合validate_password_length≥25 时，是否可以成功新建用户
mysql> create user test33@'%' identified WITH 'mysql_native_password' by'123!@#qazWWSX';
```

```
ERROR 1819 (HY000): Your password does not satisfy the current policy
requirements
```

2.5 关闭大小写字母数量验证功能,验证使用的密码记录在密码字典中时(2167sags$er24sfwjdtegcfaskvc),是否可以成功新建用户

```
[root@10-186-60-13 dict]# cat words
12!@qwqw
12qw!@qw
2167sags$er24sfwjdtegcfaskvc
mysql> set global validate_password_mixed_case_count=0;
Query OK, 0 rows affected (0.00 sec)
mysql> show variables like 'validate%';
+--------------------------------------+----------------------+
| Variable_name                        | Value                |
+--------------------------------------+----------------------+
| validate_password_check_user_name    | ON                   |
| validate_password_dictionary_file    | /usr/share/dict/words|
| validate_password_length             | 25                   |
| validate_password_mixed_case_count   | 0                    |
| validate_password_number_count       | 2                    |
| validate_password_policy             | STRONG               |
| validate_password_special_char_count | 2                    |
+--------------------------------------+----------------------+
7 rows in set (0.00 sec)
mysql> create user test33@'%' identified WITH 'mysql_native_password' by'2167sags$er24sfwjdtegcfaskvc';
ERROR 1819 (HY000): Your password does not satisfy the current policy
requirements
mysql> set global validate_password_mixed_case_count=2;
Query OK, 0 rows affected (0.00 sec)
mysql> show variables like 'validate%';
+--------------------------------------+----------------------+
| Variable_name                        | Value                |
+--------------------------------------+----------------------+
| validate_password_check_user_name    | ON                   |
| validate_password_dictionary_file    | /usr/share/dict/words|
| validate_password_length             | 25                   |
| validate_password_mixed_case_count   | 2                    |
| validate_password_number_count       | 2                    |
| validate_password_policy             | STRONG               |
| validate_password_special_char_count | 2                    |
+--------------------------------------+----------------------+
7 rows in set (0.00 sec)
```

11.5.2 连接控制插件

MySQL 5.7 版本使用的是连接控制插件,功能基本相同,以下操作仅以 MySQL 5.7 版本为例。

```
################## 以 5.7 版本为例 ##################
## 1. 查看当前连接控制插件配置信息
mysql> show variables like 'connection_control%';
+-------------------------------------------------+-------+
| Variable_name                                   | Value |
+-------------------------------------------------+-------+
| connection_control_failed_connections_threshold | 3     |
| connection_control_max_connection_delay         | 86400 |
| connection_control_min_connection_delay         | 1000  |
+-------------------------------------------------+-------+
3 rows in set (0.00 sec)

## 2. 新建测试用户后，尝试三次连接失败，观察等待时间，观察到等待反馈登录失败的时间会越来越长
mysql> create user test33@'%' identified WITH 'mysql_native_password' by'1qaz@WSX#EDC4rfv5tgb6yhnZVAF';
Query OK, 0 rows affected (0.00 sec)
mysql> ^DBye
    [root@10-186-60-13 ~]# /opt/mysql/base/5.7.25/bin/mysql -utest33 -p -S /opt/mysql/data/3306/mysqld.sock
    Enter password:
    ERROR 1045 (28000): Access denied for user 'test33'@'localhost' (using password: YES)
    [root@10-186-60-13 ~]# /opt/mysql/base/5.7.25/bin/mysql -utest33 -p -S /opt/mysql/data/3306/mysqld.sock
    Enter password:
    ERROR 1045 (28000): Access denied for user 'test33'@'localhost' (using password: YES)
    [root@10-186-60-13 ~]# /opt/mysql/base/5.7.25/bin/mysql -utest33 -p -S /opt/mysql/data/3306/mysqld.sock
    Enter password:
    ERROR 1045 (28000): Access denied for user 'test33'@'localhost' (using password: YES)
    [root@10-186-60-13 ~]# /opt/mysql/base/5.7.25/bin/mysql -utest33 -p -S /opt/mysql/data/3306/mysqld.sock
    Enter password:
    ERROR 1045 (28000): Access denied for user 'test33'@'localhost' (using password: YES)
    [root@10-186-60-13 ~]# /opt/mysql/base/5.7.25/bin/mysql -utest33 -p -S /opt/mysql/data/3306/mysqld.sock
    Enter password:
    ERROR 1045 (28000): Access denied for user 'test33'@'localhost' (using password: YES)
    [root@10-186-60-13 ~]# /opt/mysql/base/5.7.25/bin/mysql -utest33 -p -S /opt/mysql/data/3306/mysqld.sock
    Enter password:
```

11.5.3 密码变更策略

MySQL 密码变更策略配置记录在 mysql.user 表中，MySQL 5.7 和 8.0 版本支持的配置略有差异，以下将展示两个版本的测试过程和测试结果。

```
################## 以 5.7 版本为例 ##################
## 1. 检查当前密码变更策略相关信息
mysql> select @@default_password_lifetime\G
*************************** 1. row ***************************
@@default_password_lifetime: 365
1 row in set (0.00 sec)

## 2. 新建 test33 用户，设置其密码过期时间为一天，修改机器时间后重启，再使用该用户进行登录
操作，观察到提示密码过期
mysql> CREATE USER 'test33'@'%'
    -> IDENTIFIED WITH 'mysql_native_password'
    -> BY
    -> '1qaz@WSX#EDC4rfv5tgb6yhnZVAF'
    -> REQUIRE NONE
    -> PASSWORD EXPIRE INTERVAL 1 DAY;
Query OK, 0 rows affected (0.00 sec)

mysql> select user,host,password_lifetime from mysql.user where
host!='localhost' and user not in ('mysql.infoschema','mysql.session','mysql.
sys');
+-------------+-----------+-------------------+
| user        | host      | password_lifetime |
+-------------+-----------+-------------------+
| root        | 127.0.0.1 |              NULL |
| universe_op | %         |              NULL |
| test4       | %         |              NULL |
| tt          | %         |              NULL |
| test33      | %         |                 1 |
+-------------+-----------+-------------------+
5 rows in set (0.00 sec)

[root@10-186-60-13 dict]# date
Tue Aug 22 16:53:22 CST 2023
[root@10-186-60-13 dict]# timedatectl set-time '2023-8-23 16:53:20'
[root@10-186-60-13 dict]# date
Wed Aug 23 16:53:20 CST 2023

[root@10-186-60-13 ~]# /opt/mysql/base/5.7.25/bin/mysql -utest33 -p1qaz@WSX#EDC4rfv5tgb6yhnZVAF -S /opt/mysql/data/3306/mysqld.sock
mysql> select now();
ERROR 1820 (HY000): You must reset your password using ALTER USER statement
before executing this statement.
```

```
################## 以 8.0 版本为例 ##################
## 1.检查当前密码变更策略相关信息
mysql> select @@default_password_lifetime,@@password_history,@@password_reuse_
interval,@@password_require_current\G
*************************** 1. row ***************************
@@default_password_lifetime: 365
@@password_history: 5
@@password_reuse_interval: 1
@@password_require_current: 1
1 row in set (0.00 sec)

## 2.进行验证
## 2.1 密码过期验证参考 5.7 版，此处不再赘述

## 2.2 新建 test33 用户，并进行密码修改，测试当使用近 5 次密码时，是否能修改密码成功
mysql> create user test33@'%' identified WITH 'mysql_native_password' by'1qaz@WSX#EDC4rfv5tgb6yhnZVAF';
mysql> alter user test33@'%' identified WITH 'mysql_native_password' by'1qaz@WSX#EDC4rfv5tgb6yhnZVAF';
ERROR 3638 (HY000): Cannot use these credentials for 'test33@%' because they contradict the password history policy
mysql> alter user test33@'%' identified WITH 'mysql_native_password' by'1qaz@WSX#EDC4rfv5tgb6yhnZVAG';
Query OK, 0 rows affected (0.00 sec)
mysql> alter user test33@'%' identified WITH 'mysql_native_password' by'1qaz@WSX#EDC4rfv5tgb6yhnZVAF';
ERROR 3638 (HY000): Cannot use these credentials for 'test33@%' because they contradict the password history policy
```

11.6 关于 validate_password.dictionary_file 的配置说明

validate_password.dictionary_file 参数指定的密码字典文件遵循以下逻辑：

(1) 该文件最大为 1MB，一行作为一个字符串。

(2) 该文件仅在 validate_password.policy 参数设置为 2 或者 STRONG 时生效。

(3) 每行至少 4 位长，最多 100 位长，长度不在此范围均不生效。

(4) 该文件中的英文字母必须均为小写，但匹配密码时会忽略大小写。

(5) 对文件中每行字符采用模糊匹配，也就是密码中不允许出现这些字符串。例如，文件中一行为 zhenxing，则适配规则如下：

```
-- 密码字段文件内容
cat /data/mysql/3306/tmp/password_list.txt
zhenxing

create user demo identified by 'aaBB11__zhenxing';      -- 不支持，包含完整的 zhenxing
```

```
字符串
    create user demo identified by 'aaBB1zhenxing1__';  -- 不支持,包含完整的 zhenxing
字符串
    create user demo identified by 'zhenxingaaBB11__';  -- 不支持,包含完整的 zhenxing
字符串
    create user demo identified by 'aaBB1zhen00xing1__'; -- 支持
```

(6) 对该文件新增或删除数据都需要重新配置才可动态生效,如:先将参数调整为默认值再重新设置为指定文件值。

① set global validate_password.dictionary_file=default;。

② set global validate_password.dictionary_file='/data/mysql/3306/tmp/password_list.txt';。

③ 可以观测 show global status like 'validate_password.dictionary_file%'; 的输出查看文件最新生效时间。

(7) 该文件主要功能实际类似背景需求中的场景:密码不能包含指定的字符,如公司名称、业务名称等,可以将公司名称、业务名称等在该文件中配置。

11.7 总结

(1) 在使用以上功能前需确定不同 MySQL 版本的支持度。

(2) MySQL 5.7 版本上的部分插件在 MySQL 8.0 中调整为组件,使用时需注意语法和参数名称的变化。

(3) MySQL 8.0 版本对密码进行了更精细化的配置,如增加了 password_history、password_reuse_interval 等参数。

(4) 在配置 default_password_lifetime 时需要注意对业务的影响,防止密码过期导致业务中断的风险。

(5) 连接控制插件的使用需要注意避免大量错误异常导致账号连接等待时间拉长,具体是否启用也需结合业务场景和安全性综合判断。

11.8 需求中未实现的功能

(1) 密码在 24 小时内最多只能修改一次。

(2) 密码不能是简单的重复字符(例如,AAA、wuwuwuwu、dsadsadsa、111)。

12 MySQL 创建表后神秘消失？揭秘零宽字符陷阱

作者：秦福朗

12.1 引言

在 MySQL 的使用过程中，有时候一个小小的字符也能带来大麻烦。在未发现真相之前，以为这是见了鬼了，而发现真相时，却没想到是一个字符带来的问题，零宽字符像个幽灵隐藏在 IT 行业的各个角落。本文分享一个关于 MySQL 中"消失的表"的复现案例。

12.2 问题描述

通过某种方式（如命令行或数据库开发工具等）在数据库 test 中创建一个名为 lang 的表。表结构如下：

```
CREATE TABLE `lang` (
`id` int(11) NOT NULL AUTO_INCREMENT COMMENT '主键自增',
`create_time` datetime NOT NULL DEFAULT CURRENT_TIMESTAMP COMMENT '创建时间',
PRIMARY KEY (`id`)
) ENGINE=InnoDB DEFAULT CHARSET=utf8;
```

通过 SHOW TABLES; 命令，我们可以确认这个表的存在。

```
mysql> show tables;
+----------------+
| Tables_in_test |
+----------------+
| a1             |
| lang           |
| t1             |
| z1             |
+----------------+
4 rows in set (0.00 sec)
```

但是，当你尝试在 MySQL 客户端执行 SELECT * FROM lang; 查询或者业务程序连接该表时，却收到了错误信息：

```
mysql> select * from lang;
ERROR 1146 (42S02): Table 'test.lang' doesn't exist
```

表就这么神奇地消失了。

12.3 原因分析

像见了鬼一样,即使反复多次手动输入查询语法,也无法查询到这个表。将建表语句复制到 Sublime Text 文本工具中(见图 23):

```
1  CREATE TABLE `lang<0x200b>` (
2    `id` int(11) NOT NULL AUTO_INCREMENT COMMENT '主键自增',
3    `create_time` datetime NOT NULL DEFAULT CURRENT_TIMESTAMP COMMENT '创建时间',
4    PRIMARY KEY (`id`)
5  ) ENGINE=InnoDB DEFAULT CHARSET=utf8;
```

图 23

此时,我发现了问题:表名后面跟了一个"<0×200b>"的字符。这就是零宽空格,是零宽字符的一种。

12.4 什么是零宽字符?

零宽字符是一种特殊的 Unicode 字符,它不占用任何可见空间,因此在大多数情况下是不可见的。然而,它们可以存在于文本中,并且可能对计算机程序产生影响,包括数据库管理系统。在 Unicode 中,U+200B 代表零宽空格,常用于可能需要换行的地方。除此之外,还有其他零宽字符,这里不再赘述。

那么,这像幽灵一样的字符为何会存在?所谓存在即合理,零宽字符常常被用于数据防爬虫、信息加密传递、防止敏感词扫描等场景。但在数据库系统里使用它们,有时候就会出现让人头疼的现象,本文提到的就是其中之一。这些字符虽然不占用任何空间,但可能会破坏 SQL 命令的正确结构,导致后续使用出错。

12.5 如何解决?

(1) 在创建表之前,将建表语句复制到多个文本编辑工具中,检查是否有异常符号提示(一般文本工具可能无法显示零宽字符)。经过尝试,Sublime Text、Visual Studio Code 等工具或插件有提醒零宽字符的功能,还有一些在线网页工具可以查看 Unicode 字符。

(2) 在创建表之后,使用 SHOW CREATE TABLE; 命令查看表结构,然后将输出结果复制到上述文本编辑工具中,检查是否有异常符号。

(3) 经过多次测试发现,在 MySQL 客户端上执行 SHOW TABLES; 命令时,含有零宽空格的表名后面的边框线"|"与其他行是不对齐的。这可以快速发现问题表,但并不

显示具体字符。当然这种方式一般不适用于第三方开发工具、业务程序等。

12.6 总结

零宽字符是一个隐形的陷阱，可能在 MySQL 的使用过程中引发一些看似无解的问题。通过了解其本质，仔细检查 SQL 命令，避免从不可靠的来源复制和粘贴，使用适当的工具，并遵循最佳实践，可以确保我们的数据库顺利运行，不会出现类似问题。

13 MySQL VARCHAR 最佳长度评估实践

作者：官永强，李富强

你的 VARCHAR 长度合适吗？

13.1 背景描述

有客户反馈，他们对一个 VARCHAR 类型的字段进行长度扩容。第一次很快就可以修改好，但是第二次却需要执行很久。明明表中的数据量是差不多的，为什么从 VARCHAR(20) 调整为 VARCHAR(50) 就比较快，但是从 VARCHAR(50) 调整为 VARCHAR(100) 就需要执行很久呢？于是，我们对该情况进行场景复现并进行问题分析。

13.2 环境信息

本次验证涉及的产品及版本信息如表 6 所示。

表 6

产品	版本
MySQL	5.7.25-log MySQL Community Server (GPL)
Sysbench	sysbench 1.0.17

13.3 场景复现

13.3.1 数据准备

```
mysql> show create table test.sbtest1;
+--------+-----------------------------------------+
| Table  | Create Table                            |
+--------+-----------------------------------------+
```

```
| sbtest1 | CREATE TABLE `sbtest1` (
  `id` int(11) NOT NULL AUTO_INCREMENT,
  `k` int(11) NOT NULL DEFAULT '0',
  `c` varchar(20) COLLATE utf8mb4_bin NOT NULL DEFAULT '',
  `pad` varchar(20) COLLATE utf8mb4_bin NOT NULL DEFAULT '',
  PRIMARY KEY (`id`),
  KEY `k_1` (`k`)
) ENGINE=InnoDB AUTO_INCREMENT=1000001 DEFAULT CHARSET=utf8mb4 COLLATE=utf8mb4_bin |
+---------+-----------------------------------------+
1 row in set (0.00 sec)
mysql> select count(*) from test.sbtest1;
+----------+
| count(*) |
+----------+
|  1000000 |
+----------+
1 row in set (0.10 sec)
```

13.3.2 问题验证

为模拟客户的场景,我们对字段 c 进行修改,将 VARCHAR(20) 修改为 VARCHAR(50) 后再修改为 VARCHAR(100),并观察其执行所需时间。以下是相关的操作命令以及执行结果:

```
mysql> ALTER TABLE test.sbtest1 MODIFY c VARCHAR(50);
Query OK, 0 rows affected (0.01 sec)
Records: 0  Duplicates: 0  Warnings: 0

mysql> show create table test.sbtest1;
+---------+-------------------------------+
| Table   | Create Table                  |
+---------+-------------------------------+
| sbtest1 | CREATE TABLE `sbtest1` (
  `id` int(11) NOT NULL AUTO_INCREMENT,
  `k` int(11) NOT NULL DEFAULT '0',
  `c` varchar(50) COLLATE utf8mb4_bin DEFAULT NULL,
  `pad` varchar(20) COLLATE utf8mb4_bin NOT NULL DEFAULT '',
  PRIMARY KEY (`id`),
  KEY `k_1` (`k`)
) ENGINE=InnoDB AUTO_INCREMENT=1000001 DEFAULT CHARSET=utf8mb4 COLLATE=utf8mb4_bin |
+---------+-------------------------------------------------------------+
1 row in set (0.00 sec)

mysql> ALTER TABLE test.sbtest1 MODIFY c VARCHAR(100);
Query OK, 1000000 rows affected (4.80 sec)
Records: 1000000  Duplicates: 0  Warnings: 0
```

```
mysql> show create table test.sbtest1;
+---------+--------------------------+
| Table   | Create Table             |
+---------+--------------------------+
| sbtest1 | CREATE TABLE `sbtest1` (
  `id` int(11) NOT NULL AUTO_INCREMENT,
  `k` int(11) NOT NULL DEFAULT '0',
  `c` varchar(100) COLLATE utf8mb4_bin DEFAULT NULL,
  `pad` varchar(20) COLLATE utf8mb4_bin NOT NULL DEFAULT '',
  PRIMARY KEY (`id`),
  KEY `k_1` (`k`)
) ENGINE=InnoDB AUTO_INCREMENT=1000001 DEFAULT CHARSET=utf8mb4 COLLATE=utf8mb4_bin |
+---------+----------------------------------------------------------------------+
1 row in set (0.00 sec)
```

通过验证发现，该问题会稳定复现，故继续尝试去修改，最终发现在修改 VARCHAR(63) 为 VARCHAR(64) 时需要执行很久，但在此后继续进行长度扩容则可以很快完成。

```
mysql> ALTER TABLE test.sbtest1 MODIFY c VARCHAR(63);
Query OK, 0 rows affected (0.01 sec)
Records: 0  Duplicates: 0  Warnings: 0

mysql> ALTER TABLE test.sbtest1 MODIFY c VARCHAR(64);
Query OK, 1000000 rows affected (4.87 sec)
Records: 1000000  Duplicates: 0  Warnings: 0

mysql> show create table test.sbtest1;
+---------+---------------+
| Table   | Create Table  |
+---------+---------------+
| sbtest1 | CREATE TABLE `sbtest1` (
  `id` int(11) NOT NULL AUTO_INCREMENT,
  `k` int(11) NOT NULL DEFAULT '0',
  `c` varchar(64) COLLATE utf8mb4_bin DEFAULT NULL,
  `pad` varchar(20) COLLATE utf8mb4_bin NOT NULL DEFAULT '',
  PRIMARY KEY (`id`),
  KEY `k_1` (`k`)
) ENGINE=InnoDB AUTO_INCREMENT=1000001 DEFAULT CHARSET=utf8mb4 COLLATE=utf8mb4_bin |
+---------+----------------------------------------------------------------------+
1 row in set (0.00 sec)

mysql> ALTER TABLE test.sbtest1 MODIFY c VARCHAR(65);
Query OK, 0 rows affected (0.01 sec)
```

```
Records: 0  Duplicates: 0  Warnings: 0

mysql> ALTER TABLE test.sbtest1 MODIFY c VARCHAR(66);
Query OK, 0 rows affected (0.01 sec)
Records: 0  Duplicates: 0  Warnings: 0
```

13.3.3 问题分析

对于 VARCHAR(63) 修改为 VARCHAR(64) 需要执行很久的这个情况进行分析，通过查阅官方文档发现，VARCHAR 字符类型在字节长度为 1 时可存储的字符为 0~255。当前字符集类型为 UTF8MB4，由于 UTF8MB4 为四字节编码字符集，即一个字节长度可存储 63.75（255/4）个字符，所以当我们将 VARCHAR(63) 修改为 VARCHAR(64) 时，需要增加一个字节去进行数据的存储，就要通过建立临时表的方式去完成本次长度扩容，故需要花费大量时间。

13.4 拓展验证

13.4.1 数据准备

```
mysql> show create table test_utf8.sbtest1;
+---------+----------------------------------------+
| Table   | Create Table                           |
+---------+----------------------------------------+
| sbtest1 | CREATE TABLE `sbtest1` (
`id` int(11) NOT NULL AUTO_INCREMENT,
`k` int(11) NOT NULL DEFAULT '0',
`c` varchar(20) NOT NULL DEFAULT '',
`pad` varchar(20) NOT NULL DEFAULT '',
PRIMARY KEY (`id`),
KEY `k_1` (`k`)
) ENGINE=InnoDB AUTO_INCREMENT=1000001 DEFAULT CHARSET=utf8 |
+---------+-------------------+
1 row in set (0.00 sec)

mysql> select count(*) from test_utf8.sbtest1;
+----------+
| count(*) |
+----------+
|  1000000 |
+----------+
1 row in set (0.10 sec)
```

13.4.2 UTF8 场景验证

由于 UTF8 为三字节编码字符集，即一个字节可存储 85（255/3=85）个字符。

本次修改顺序：VARCHAR(20) → VARCHAR(50) → VARCHAR(85)，观察其执行

所需时间。以下是相关的操作命令以及执行结果：

```
mysql> ALTER TABLE test_utf8.sbtest1 MODIFY c VARCHAR(50)
,algorithm=inplace,lock=none;
Query OK, 0 rows affected (0.01 sec)
Records: 0  Duplicates: 0  Warnings: 0

mysql> ALTER TABLE test_utf8.sbtest1 MODIFY c VARCHAR(85)
,algorithm=inplace,lock=none;
Query OK, 0 rows affected (0.00 sec)
Records: 0  Duplicates: 0  Warnings: 0

mysql> show create table test_utf8.sbtest1;
+---------+------------------------------+
| Table   | Create Table                 |
+---------+------------------------------+
| sbtest1 | CREATE TABLE `sbtest1` (
 `id` int(11) NOT NULL AUTO_INCREMENT,
 `k` int(11) NOT NULL DEFAULT '0',
 `c` varchar(85) DEFAULT NULL,
 `pad` varchar(20) NOT NULL DEFAULT '',
 PRIMARY KEY (`id`),
 KEY `k_1` (`k`)
) ENGINE=InnoDB AUTO_INCREMENT=1000001 DEFAULT CHARSET=utf8 |
+---------+-------------------------------------------------+
1 row in set (0.00 sec)
```

修改顺序为 VARCHAR(85) → VARCHAR(86) → VARCHAR(100) 时，我们观察到执行的 SQL 语句直接返回报错。于是我们删除 algorithm=inplace、lock=none 这两个参数，即允许本次 SQL 创建临时表以及给目标表上锁，然后重新执行 SQL。以下是相关的操作命令以及执行结果：

```
mysql> ALTER TABLE test_utf8.sbtest1 MODIFY c VARCHAR(86)
,algorithm=inplace,lock=none;
ERROR 1846 (0A000): ALGORITHM=INPLACE is not supported. Reason: Cannot change column type INPLACE. Try ALGORITHM=COPY.

mysql> ALTER TABLE test_utf8.sbtest1 MODIFY c VARCHAR(86);
Query OK, 1000000 rows affected (4.94 sec)
Records: 1000000  Duplicates: 0  Warnings: 0

mysql> show create table test_utf8.sbtest1;
+---------+------------------------------+
| Table   | Create Table                 |
+---------+------------------------------+
| sbtest1 | CREATE TABLE `sbtest1` (
 `id` int(11) NOT NULL AUTO_INCREMENT,
```

```
`k` int(11) NOT NULL DEFAULT '0',
`c` varchar(86) DEFAULT NULL,
`pad` varchar(20) NOT NULL DEFAULT '',
PRIMARY KEY (`id`),
KEY `k_1` (`k`)
) ENGINE=InnoDB AUTO_INCREMENT=1000001 DEFAULT CHARSET=utf8 |
+---------+-----------------------------------------------------+
1 row in set (0.00 sec)

mysql> ALTER TABLE test_utf8.sbtest1 MODIFY c VARCHAR(100)
,algorithm=inplace,lock=none;
Query OK, 0 rows affected (0.00 sec)
Records: 0  Duplicates: 0  Warnings: 0
```

13.4.3 UTF8MB4 场景验证

由于 UTF8MB4 为四字节编码字符集,即一个字节长度可存储 63(255/4=63.75)个字符。

本次修改顺序:VARCHAR(20) → VARCHAR(50) → VARCHAR(63),观察其执行所需时间。以下是相关的操作命令以及执行结果:

```
mysql> ALTER TABLE test.sbtest1 MODIFY c VARCHAR(50)
,algorithm=inplace,lock=none;
Query OK, 0 rows affected (0.00 sec)
Records: 0  Duplicates: 0  Warnings: 0

mysql> ALTER TABLE test.sbtest1 MODIFY c VARCHAR(63)
,algorithm=inplace,lock=none;
Query OK, 0 rows affected (0.00 sec)
Records: 0  Duplicates: 0  Warnings: 0

mysql> show create table test.sbtest1;
+---------+-------------------------+
| Table   | Create Table            |
+---------+-------------------------+
| sbtest1 | CREATE TABLE `sbtest1` (
`id` int(11) NOT NULL AUTO_INCREMENT,
`k` int(11) NOT NULL DEFAULT '0',
`c` varchar(63) COLLATE utf8mb4_bin DEFAULT NULL,
`pad` varchar(20) COLLATE utf8mb4_bin NOT NULL DEFAULT '',
PRIMARY KEY (`id`),
KEY `k_1` (`k`)
) ENGINE=InnoDB AUTO_INCREMENT=1000001 DEFAULT CHARSET=utf8mb4
COLLATE=utf8mb4_bin |
+---------+-------------------------------------------------------------
------+
1 row in set (0.00 sec)
```

本次修改顺序：VARCHAR(63) → VARCHAR(64) → VARCHAR(100)。我们观察到执行的 SQL 语句直接返回报错。于是我们删除 algorithm=inplace、lock=none 这两个参数，即允许本次 SQL 创建临时表以及给目标表上锁，然后重新执行 SQL。以下是相关的操作命令以及执行结果：

```
mysql> ALTER TABLE test.sbtest1 MODIFY c VARCHAR(64)
,algorithm=inplace,lock=none;
ERROR 1846 (0A000): ALGORITHM=INPLACE is not supported. Reason: Cannot change column type INPLACE. Try ALGORITHM=COPY.

mysql> ALTER TABLE test.sbtest1 MODIFY c VARCHAR(64) ;
Query OK, 1000000 rows affected (4.93 sec)
Records: 1000000  Duplicates: 0  Warnings: 0

mysql> show create table test.sbtest1;
+----------+---------------------------+
| Table    | Create Table              |
+----------+---------------------------+
| sbtest1 | CREATE TABLE `sbtest1` (
 `id` int(11) NOT NULL AUTO_INCREMENT,
 `k` int(11) NOT NULL DEFAULT '0',
 `c` varchar(64) COLLATE utf8mb4_bin DEFAULT NULL,
 `pad` varchar(20) COLLATE utf8mb4_bin NOT NULL DEFAULT '',
 PRIMARY KEY (`id`),
 KEY `k_1` (`k`)
) ENGINE=InnoDB AUTO_INCREMENT=1000001 DEFAULT CHARSET=utf8mb4 COLLATE=utf8mb4_bin |
+----------+-------------------------------------------------------------------------+
1 row in set (0.00 sec)

mysql> ALTER TABLE test.sbtest1 MODIFY c VARCHAR(100)
,algorithm=inplace,lock=none;
Query OK, 0 rows affected (0.00 sec)
Records: 0  Duplicates: 0  Warnings: 0
```

13.4.4 对比分析

UTF8 与 UTF8MB4 对比如表 7 所示。

表 7

字符长度修改	UTF8	UTF8MB4
20→50	online ddl (inplace)	online ddl (inplace)
50→100	online ddl (copy)	online ddl (copy)
X→Y	当 Y*3 < 256 时，inplace 当 X*3 ≥ 256，inplace	当 Y*4 < 256 时，inplace 当 X*4 ≥ 256，inplace
备注	一个字符最大占用 3 个字节	一个字符最大占用 4 个字节

13.5 结论

当一个字段的最大字节长度≥256 字符时，需要 2 个字节来表示字段长度。

使用 UTF8MB4 举例：

(1) 对于字段的最大字节长度在 256 字符内变化（即 x*4<256 且 Y*4<256），online ddl 为 inplace 模式，效率高。

(2) 对于字段的最大字节长度在 256 字符外变化（即 x*4 ≥ 256 且 Y*4 ≥ 256），online ddl 为 inplace 模式，效率高。

(3) 否则，online ddl 为 copy 模式，效率低.
UTF8(MB3) 同理。

13.6 建议

为避免由于后期字段长度扩容，online ddl 采用效率低的 copy 模式，建议：

(1) 对于 UTF8(MB3) 字符类型：

① 字符个数小于 50 个，建议设置为 VARCHAR(50) 或更小的字符长度。

② 字符个数接近 84（256/3=83.33）个,建议设置为 VARCHAR(84) 或更大的字符长度。

(2) 对于 UTF8MB4 字符类型：

① 字符个数小于 50 个，建议设置为 VARCHAR(50)，或更小的字符长度。

② 字符个数接近 64（256/4=64）个，建议设置为 VARCHAR(64) 或更大的字符长度。

本次验证结果仅供参考，若需要在生产环境中进行操作，请结合实际情况合理定义 VARCHAR 的长度，避免造成经济损失。

14 什么情况下 MySQL 连查询都能被阻塞？

作者：贾特特

MySQL 的锁不少，在哪种情况下会连查询都能被阻塞？这是一个有意思的问题。

在工作中，很多开发者和 DBA 可能接触较多的锁也就是行锁。对于行锁，阻塞写能理解，阻塞读实在是想不到。能阻塞读的肯定是颗粒度更大的锁了，比如表级别的。

> 本文操作环境为 MySQL 8.0。

14.1 MySQL 表级锁有两种实现

(1) 服务器（SERVER）层：本层的锁主要是元数据锁（metadata lock，MDL）。

(2) 存储引擎（ENGINE）层：本层不同的存储引擎可能会实现不同的锁定策略。例如 MyISAM 引擎实现了表级锁，InnoDB 存储引擎实现了行级锁和表级锁，其中表级锁是通过意向锁体现的。

元数据锁（MDL）由 SERVER 层管理，用于锁定数据库对象的元数据信息，如：表结构、索引等。元数据锁可以阻止对表结构的改变，以确保数据定义的一致性。

14.1.1 元数据锁的类型

元数据锁的类型如图 24 所示。

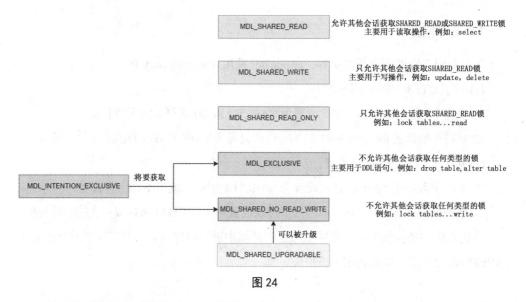

图 24

每种锁类型后面会详细介绍。简单来说，对于元数据锁而言，当对一个表进行增删改查操作的时候，会加元数据读锁。当对表数据结构进行变更的时候会加元数据写锁。它读写互斥，写写互斥，只有读读不冲突。

意向锁是在存储引擎层实现的，用于协调不同事务对表级锁和行级锁的请求。当一个事务在某个层次（表级或行级）上获取锁时，会首先获取对应层次的意向锁，以提示其他事务该事务在该层次上有锁的意向。这样可以在更高层次上减少锁冲突，提高并发性能。

14.1.2 InnoDB 存储引擎的意向锁种类

(1) 意向共享锁（Intention Shared Lock，IS）：事务打算给数据行加共享锁（S 锁）。

(2) 意向排他锁（Intention Exclusive Lock，IX）：事务打算给数据行加排他锁（X 锁）。

这样看来，表对象不可读写也可能就是元数据锁互斥所导致的。

14.2 Waiting for table metadata lock

本节中未完成的读写事务,在实际中可能是未完成的大事务,也可能是未显式结束的事务。

14.2.1 元数据锁互斥(未完成的读事务)

会话 1 执行时有未完成的读事务,此时获取了元数据共享读锁(见图 25)。

MDL_SHARED_READ 锁允许会话读取表的数据,并允许其他会话获取 SHARED_READ 或 SHARED_WRITE 锁,但不允许获取 SHARED_NO_READ_WRITE 或 EXCLUSIVE 锁。

图 25

会话 2 执行时,尝试运时 ALTER 表结构变更语句,此时 ALTER 语句要获取元数据排它锁(见图 26)。

MDL_EXCLUSIVE 锁允许会话读取和修改表的数据和结构,但不允许其他会话获取任何类型的锁。

图 26

元数据锁互斥等待,之后其他会话对于所涉及的表不可读写。

14.2.2 元数据锁互斥(未完成的写事务)

会话 1 执行时有未完成的写事务,此时获取了元数据写锁(见图 27)。

MDL_SHARED_WRITE 锁允许会话读取和修改表的数据,并允许其他会话获取

SHARED_READ 锁，但不允许获取 SHARED_WRITE、SHARED_NO_READ_WRITE 或 EXCLUSIVE 锁。

```
mysql> begin;
Query OK, 0 rows affected (0.00 sec)

mysql> update db_version set name='c' where id=3;
Query OK, 1 row affected (0.00 sec)
Rows matched: 1  Changed: 1  Warnings: 0

mysql>
```

图 27

会话 2 执行时，尝试运时 ALTER 表结构变更语句，此时 ALTER 语句要获取元数据排它锁（见图 28）。

MDL_EXCLUSIVE 锁允许会话读取和修改表的数据和结构，但不允许其他会话获取任何类型的锁。

```
mysql> show processlist;
+-------+-----------------+------------+------+---------+----------+---------------------------------+-------------------------------------+
| Id    | User            | Host       | db   | Command | Time     | State                           | Info                                |
+-------+-----------------+------------+------+---------+----------+---------------------------------+-------------------------------------+
|     5 | event_scheduler | localhost  | NULL | Daemon  | 10382598 | Waiting on empty queue          | NULL                                |
| 14519 | root            | localhost  | tmp  | Sleep   |       15 |                                 | NULL                                |
| 14520 | root            | localhost  | tmp  | Query   |        6 | Waiting for table metadata lock | alter table db_version drop column name |
| 14521 | root            | localhost  | tmp  | Query   |        3 | Waiting for table metadata lock | select * from db_version            |
| 14523 | root            | [redacted] | xzts | Sleep   |    20786 |                                 | NULL                                |
| 14526 | root            | [redacted] | xzts | Sleep   |    20149 |                                 | NULL                                |
| 14533 | root            | [redacted] | tmp  | Sleep   |    17015 |                                 | NULL                                |
| 14560 | root            | localhost  | tmp  | Query   |        0 | init                            | show processlist                    |
+-------+-----------------+------------+------+---------+----------+---------------------------------+-------------------------------------+
```

图 28

元数据锁互斥等待，之后其他会话对于所涉及的表不可读写。

14.2.3 LOCK TABLES…READ/WRITE

LOCK TABLES 可以显式获取表锁，需要注意的是会话只能自己获取和释放表锁。UNLOCK TABLES 可以显式释放当前会话的表锁。

14.2.3.1 LOCK TABLES…READ

会话 1 执行 lock tables db_version read；（见图 29）。

MDL_SHARED_READ_ONLY 锁允许会话读取表的数据，并允许其他会话获取 SHARED_READ 锁，但不允许获取 SHARED_WRITE、SHARED_NO_READ_WRITE 或 EXCLUSIVE 锁。

```
mysql> unlock tables;
Query OK, 0 rows affected (0.00 sec)

mysql> lock tables db_version read;
Query OK, 0 rows affected (0.01 sec)

mysql>
```

图 29

此时 db_version 加了元数据共享只读锁（见图 30）。

会话 2 执行 ALTER 表结构变更语句，此时 ALTER 语句要获取元数据排它锁，元数据锁互斥等待。

图 30

之后所涉及的表对象将不可读写。

14.2.3.2 LOCK TABLES…WRITE

会话 1 执行 lock tables db_version write；（见图 31）。

MDL_SHARED_NO_READ_WRITE 锁允许当前会话读取和修改表的数据，但不允许其他会话获取任何类型的锁。

图 31

此时 db_version 加上了独占写锁，只能在会话 1 中读写，它会阻止其他会话获取任何类型的锁（见图 32）。

图 32

因此其他会话既不能读也不能写，当然查询也会被阻塞了（见图 33）。

图 33

需要注意的是，此时会话 1 对其他表也不可读写。

14.3 FLUSH TABLES 以及 WAITING FOR TABLE FLUSH

FLUSH TABLES 关闭所有打开的表，强制关闭所有正在使用的表，并刷新准备好的语句缓存。当存在活动的 LOCK TABLES 时，不允许执行 FLUSH TABLES 操作。

（1）当 ALTER 表结构时，执行 FLUSH TABLES 阻塞，从而导致表对象不可读写（见图 34）。

图 34

（2）当 LOCK TABLES 后，执行 FLUSH TABLES 会被阻塞，从而导致表对象不可读写（见图 35）。

① 会话 1 执行 lock tables db_version read;

② 会话 2 执行 flush tables。

图 35

此时，会话 2 会被阻塞，其他会话对所涉及的表将不可读写。SHOW PROCESSLIST 中会提示 Waiting for table flush。

需要说明的是，会话 1 执行完 lock tables...read lock 后，其他会话执行 DML 增删改语句，虽然会因获取不到元数据锁而阻塞，但不会阻塞其他会话执行 SELECT 查询（见图 36）。

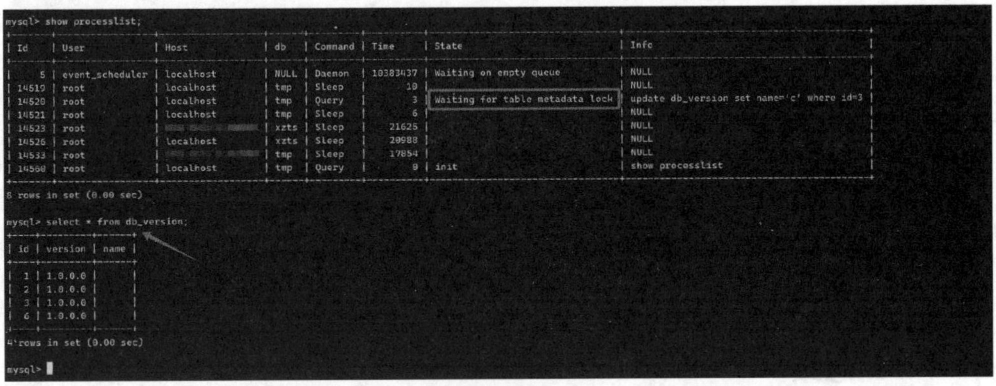

图 36

换言之，执行 lock tables…read 后，当遇到元数据锁排它锁互斥阻塞（ALTER 语句）或者 FLUSH TABLES 发生阻塞后，才会发生所涉及的表对象不可读写。

14.4 处理延伸

如何处理并找到源头 SQL 呢？

对于因元数据锁互斥而导致的表不可读写，一般可以通过 sys 库下的内置视图来查看。可能会涉及的表如下：

(1)sys.schema_table_lock_waits：可直接通过 sys 下的内置视图，看到元数据锁互斥的相关信息。

(2)information_schema.innodb_trx：找到长时间未提交的事务。

对于因 FLUSH TABLE 等待而导致的表不可读写的场景，通过上述视图 / 表不一定能获取数据，大致会有以下两种情况：

(1)Waiting for table flush：这种情况主要出现在因 FLUSH TABLES 而等待后，执行 DML 语句时。可以按如下方式寻找源头。

```
SELECT
    b.PROCESSLIST_ID,
    b.THREAD_ID,
    a.OBJECT_NAME,
    a.LOCK_TYPE,
    a.LOCK_STATUS,
    b.PROCESSLIST_STATE
```

```
FROM
`performance_schema`.metadata_locks a
LEFT JOIN `performance_schema`.threads b ON a.OWNER_THREAD_ID = b.THREAD_ID
WHERE
a.OBJECT_SCHEMA = 'tmp';
```

也可以通过线程 ID 找到该会话最近的 10 条语句进一步判断确认。

```
select THREAD_ID,event_id,sql_text from
`performance_schema`.events_statements_history
where THREAD_ID =  14503
order by event_id;
```

(2)Waiting for table metadata lock：这种情况主要出现于因 FLUSH TABLES 而等待后，执行 DDL 语句（如 ALTER TABLE）时。可以参考元数据锁互斥而导致的表不可读写处理。

14.5 总结

以下情况会导致表对象不可读写：

(1) 因 Waiting for table metadata lock 而导致的表对象不可读写。

(2) 因 Waiting for table flush 而导致的表对象不可读写。

15 MySQL 生产环境 GROUP BY 优化实践

作者：张洛丹

15.1 案例介绍

首先，我们看一个生产环境上 GROUP BY 语句的优化案例。

SQL 优化前，执行时间为 3s。

```
SELECT taskUniqueId,
max(reportTime) AS reportTime
FROM task_log_info
WHERE reportTime > '2024-04-07'
GROUP BY  taskUniqueId
SQL 优化后：执行时间 30ms！
SELECT a.taskUniqueId,
reportTime
FROM task_log_info a
JOIN
```

```
(SELECT taskUniqueId,
max(id) AS id
FROM task_log_info
GROUP BY  taskUniqueId ) tmp
ON a.id=tmp.id
AND reportTime>='2024-04-07'
```

注意：id 和 reporttime 字段值具有相关性的情况才可以这样修改。

两条 SQL 语句的 GROUP BY 使用了同一个索引，但是效率却相差很多，这到底是为什么呢？

15.2 环境准备

对于 GROUP BY 在使用索引上的优化，可分为两种情况讨论：

(1) 表上无索引。执行时，会生成临时表进行分组。可以通过索引来优化，避免使用临时表。

(2) 表上有索引。GROUP BY 语句有几种扫描算法：

① 松散索引扫描（Loose Index Scan）；

② 紧凑索引扫描（Tight Index Scan）；

③ 两种算法结合。

准备测试数据如下：

```
CREATE TABLE t2 (
id INT AUTO_INCREMENT,
c1 CHAR(64) NOT NULL,
c2 CHAR(64) NOT NULL,
c3 CHAR(64) NOT NULL,
c4 CHAR(64) NOT NULL,
PRIMARY KEY(id),
KEY c1_c2_c3_idx (c1, c2,c3)
) ENGINE=INNODB;

INSERT INTO t2 VALUES (null,'a','b','a','a'), (null,'a','b','a','a'),
(null,'a','c','a','a'), (null,'a','c','a','a'),
(null,'a','d','b','b'), (null,'a','b','b','b'),
(null,'d','b','c','c'), (null,'e','b','c','c'),
(null,'f','c','d','d'), (null,'k','c','d','d'),
(null,'y','d','y','y'), (null,'f','b','f','y'),
(null,'a','b','a','a'), (null,'a','b','a','a'),
(null,'a','c','a','a'), (null,'a','c','a','a'),
(null,'a','d','b','b'), (null,'a','b','b','b'),
(null,'d','b','c','c'), (null,'e','b','c','c'),
(null,'f','c','d','d'), (null,'k','c','d','d'),
```

```
(null,'y','d','y','y'), (null,'f','b','f','y');

-- 收集统计信息，否则可能影响测试
ANALYZE TABLE t2;
```

15.3 无索引的情况

不使用索引的 GROUP BY：

```
mysql> explain select c4,count(*) from t2 group by c4 order by null;
+----+-------------+-------+------------+------+---------------+------+---------+------+------+----------+-----------------+
| id | select_type | table | partitions | type | possible_keys | key  | key_len | ref  | rows | filtered | Extra           |
+----+-------------+-------+------------+------+---------------+------+---------+------+------+----------+-----------------+
|  1 | SIMPLE      | t2    | NULL       | ALL  | NULL          | NULL | NULL    | NULL |   24 |   100.00 | Using temporary |
+----+-------------+-------+------------+------+---------------+------+---------+------+------+----------+-----------------+
1 row in set, 1 warning (0.00 sec)

mysql> explain select c4,count(*) from t2 group by c4;
+----+-------------+-------+------------+------+---------------+------+---------+------+------+----------+---------------------------------+
| id | select_type | table | partitions | type | possible_keys | key  | key_len | ref  | rows | filtered | Extra                           |
+----+-------------+-------+------------+------+---------------+------+---------+------+------+----------+---------------------------------+
|  1 | SIMPLE      | t2    | NULL       | ALL  | NULL          | NULL | NULL    | NULL |   24 |   100.00 | Using temporary; Using filesort |
+----+-------------+-------+------------+------+---------------+------+---------+------+------+----------+---------------------------------+
1 row in set, 1 warning (0.00 sec)
```

Extra: Using temporary 可以看到这里使用到了临时表。

使用索引的 GROUP BY：

```
mysql> explain select c1,count(*) from t2 group by c1;
+----+-------------+-------+------------+-------+---------------+-------------+---------+------+------+----------+-------------+
| id | select_type | table | partitions | type  | possible_keys | key         | key_len | ref  | rows | filtered | Extra       |
+----+-------------+-------+------------+-------+---------------+-------------+---------+------+------+----------+-------------+
|  1 | SIMPLE      | t2    | NULL       | index | c1_c2_c3_idx  | c1_c2_c3_idx|     768 | NULL |   24 |   100.00 | Using index |
+----+-------------+-------+------------+-------+---------------+-------------+---------+------+------+----------+-------------+
```

```
-+----------+------+------+----------+--------------+
    1 row in set, 1 warning (0.01 sec)
```

Extra: Using index & type: index 表示全索引扫描。这种情况下，如果表数据量很大，还是会比较耗时的。

15.4 有索引的情况

有索引并正常使用时，索引的访问有两种算法：

(1) 松散索引扫描（Loose Index Scan）

① 不需要扫描所有的索引，根据分组前缀（GROUY BY 的字段）跳跃扫描部分索引。

② Extra: Using index for group-by。

(2) 紧凑索引扫描（Tight Index Scan）

① 需要扫描范围或全部的索引。

② Extra: Using index。

另外还有一种将两种算法结合使用的方式，我们在后文说明。

下面是两条 SQL 分别使用 Loose Index Scan 和 Tight Index Scan：

```
mysql> explain SELECT c1,MIN(c2) FROM t2 GROUP BY c1;
+----+-------------+-------+------------+-------+---------------+-------------
-+---------+------+------+----------+--------------------------+
| id | select_type | table | partitions | type  | possible_keys | key
| key_len | ref  | rows | filtered | Extra                    |
+----+-------------+-------+------------+-------+---------------+-------------
-+---------+------+------+----------+--------------------------+
|  1 | SIMPLE      | t2    | NULL       | range | c1_c2_c3_idx  | c1_c2_c3_idx
| 256     | NULL |    7 |   100.00 | Using index for group-by |
+----+-------------+-------+------------+-------+---------------+-------------
-+---------+------+------+----------+--------------------------+
    1 row in set, 1 warning (0.00 sec)

mysql> explain SELECT c1,COUNT(*) FROM t2 GROUP BY c1;
+----+-------------+-------+------------+-------+---------------+-------------
-+---------+------+------+----------+-------------+
| id | select_type | table | partitions | type  | possible_keys | key
| key_len | ref  | rows | filtered | Extra       |
+----+-------------+-------+------------+-------+---------------+-------------
-+---------+------+------+----------+-------------+
|  1 | SIMPLE      | t2    | NULL       | index | c1_c2_c3_idx  | c1_c2_c3_idx
| 768     | NULL |   24 |   100.00 | Using index |
+----+-------------+-------+------------+-------+---------------+-------------
-+---------+------+------+----------+-------------+
```

```
1 row in set, 1 warning (0.00 sec)


mysql> SELECT c1,MIN(c2),COUNT(*) FROM t2 GROUP BY c1;
+----+---------+----------+
| c1 | MIN(c2) | COUNT(*) |
+----+---------+----------+
| a  | b       |       12 |
| d  | b       |        2 |
| e  | b       |        2 |
| f  | b       |        4 |
| k  | c       |        2 |
| y  | d       |        2 |
+----+---------+----------+
6 rows in set (0.00 sec)
```

第一条 SQL 扫描示意图如图 37 所示。第二条 SQL 扫描示意图如图 38 所示。

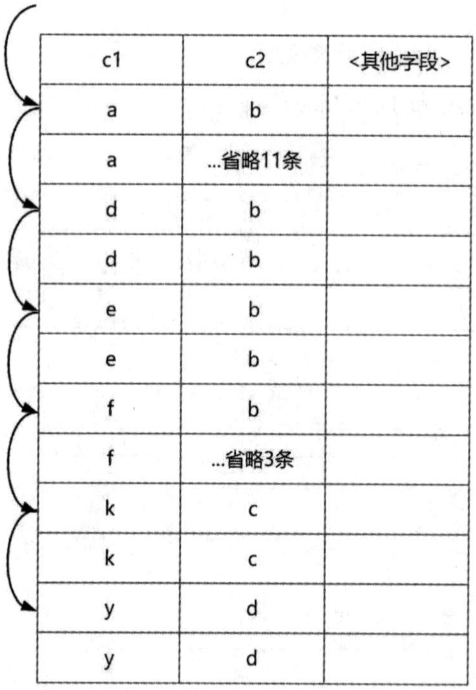

c1	c2	<其他字段>
a	b	
a	…省略11条	
d	b	
d	b	
e	b	
e	b	
f	b	
f	…省略3条	
k	c	
k	c	
y	d	
y	d	

图 37

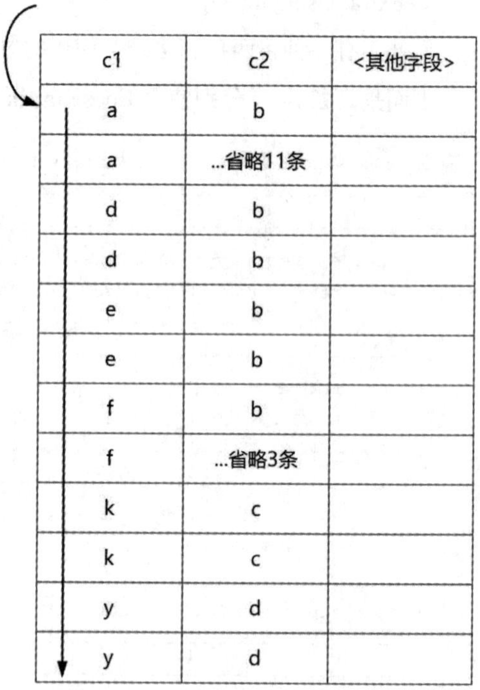

c1	c2	<其他字段>
a	b	
a	…省略11条	
d	b	
d	b	
e	b	
e	b	
f	b	
f	…省略3条	
k	c	
k	c	
y	d	
y	d	

图 38

下面我们详细说明一下两种扫描方式。

15.4.1 Loose Index Scan

跳跃扫描部分索引，而不需要扫描全部。

举例：

```
mysql> explain SELECT c1,MIN(c2) FROM t2 GROUP BY c1;
+----+-------------+-------+------------+-------+---------------+-------------
-+---------+------+------+----------+--------------------------+
| id | select_type | table | partitions | type  | possible_keys | key
| key_len | ref  | rows | filtered | Extra                    |
+----+-------------+-------+------------+-------+---------------+-------------
-+---------+------+------+----------+--------------------------+
|  1 | SIMPLE      | t2    | NULL       | range | c1_c2_c3_idx  | c1_c2_c3_idx
| 256     | NULL |    7 |   100.00 | Using index for group-by |
+----+-------------+-------+------------+-------+---------------+-------------
-+---------+------+------+----------+--------------------------+
1 row in set, 1 warning (0.00 sec)

ysql> SELECT c1,MIN(c2),COUNT(*) FROM t2 GROUP BY c1;
+----+---------+----------+
| c1 | MIN(c2) | COUNT(*) |
+----+---------+----------+
| a  | b       |       12 |
| d  | b       |        2 |
| e  | b       |        2 |
| f  | b       |        4 |
| k  | c       |        2 |
| y  | d       |        2 |
+----+---------+----------+
6 rows in set (0.00 sec)
```

Extra: Using index for group-by 表示使用松散索引扫描。

使用场景如下：

(1) 当需要获取每个分组的某条记录，而非对全部记录做聚合运算时。比如：

① 求最小值或最大值：MIN()、MAX()。

② 统计类：COUNT(distinct)、SUM(distinct)、AVG(distinct)。

注意：如果 SQL 语句中既有 1~2 个 MIN\MAX，也有 1~3 个 COUNT(distinct)\SUM(distinct)\AVG(distinct) 时，无法用到 Loose Index Scan；两组函数分别出现的时候才可能会用到。

(2) distinct 可以转换为 GROUP BY 进行处理。

(3) 使用 Loose Index Scan 的其他必要条件：

① 查询基于一个表。

② GROUP BY 的字段满足索引的最左匹配原则。

③ 聚合函数使用的列必须包含在索引上；且使用多个聚合函数时，必须使用相同的字段；且 GROUP BY 字段 + 聚合函数字段也满足最左匹配原则。

④ 索引中字段必须是全字段索引，而不能是前缀索引，例如 INDEX(c1(10))。

以上条件结合索引的结构就很好理解了。

另外，在选择是否使用 Loose Index Scan 时，也会受到 SQL、统计信息、成本等因素的影响。

举例：

```
-- 场景 1
-- MIN()、MAX()
SELECT c1,MIN(c2),MAX(c2) FROM t2 GROUP BY c1;
SELECT c1,c2,MAX(c3),MIN(c3)  FROM t2 WHERE c2 > 'k' GROUP BY c1, c2;
SELECT c1,c2,MAX(c3),MIN(c3)  FROM t2 WHERE c3 > 'k' GROUP BY c1, c2;
SELECT c1,c2,c3,MAX(id) FROM t2  GROUP BY c1,c2,c3;
SELECT c1, c2 FROM t2 WHERE c3 = 'd' GROUP BY c1, c2;

-- 以下几种情况在当前数据量和数据分布下没有用到，和成本计算有关，结合后文成本对比的章节改变数
据量和数据分布就能测试出来
SELECT c1,c2,MAX(c3),MIN(c3)  FROM t2 WHERE c1>='k' and c2 > 'f' GROUP BY c1, c2;
SELECT DISTINCT c1, c2 FROM t2 where c1>'k';
SELECT c1,c2,count(distinct c3) FROM t2 where c1>='k' and c2>'k' GROUP BY c1,c2;

-- count(distinct)、sum(distinct)、avg(distinct)
SELECT c1,count(distinct c2,c3) FROM t2 GROUP BY c1;
SELECT c1,c2,sum(distinct c3) FROM t2 GROUP BY c1,c2;
SELECT c1,c2,sum(distinct c3) FROM t2 where c2>'k' GROUP BY c1,c2;

-- 场景 2
SELECT DISTINCT c1, c2 FROM t2;

-- 无法使用 Loose Index Scan
SELECT c1,count(distinct c2,c3),sum(distinct c2) FROM t2 GROUP BY c1;
```

15.4.2 Tight Index Scan

对于无法使用 Loose Index Scan 的一些 GROUP BY，在满足索引最左匹配原则情况下可能会用到 Tight Index Scan。

该种方式实际上是范围索引扫描或全部索引扫描，在数据量大的情况下性能仍然可能会比较差，但是相比无索引还是可以避免使用临时表和全表扫描的，在某些情况下有一定的优化作用。

15.4.3 两种算法结合

对于统计类 AGG(DISTINCT) 即 SUM|COUNT|AVG(distinct)，可能会出现使用松散索引扫描（Loose Index Scan）成本大于紧凑索引扫描（Tight Index Scan）的情况。

两种方式在引擎层主要包含的成本如下。

(1)Loose Index Scan 的成本：

①读取分组的第一条记录,得到分组前缀。

②根据分组前缀读取分组的第一条或最后一条记录返回给 SERVER 层。

(2)Tight Index Scan 的成本:

①从 ENGINE 层读取数据,返回给 SERVER 层。

② SERVER 层判断是否存在符合 WHERE 条件的记录,并根据聚合函数进行处理。

可以看到,对于 ENGINE 层的访问,Loose Index Scan 的成本有可能会高于 Tight Index Scan,而且在 MySQL 中,ENGINE 层读取数据页的成本常数是 1,SERVER 层判断一条记录的成本常数是 0.2。

至于 MIN/MAX 为什么不会出现 Loose Index Scan 成本 >Tight Index Scan 成本的情况,我理解只有到组内值都是唯一的情况下才会出现,那这样也没有必要去分组求最值了。

在某些情况下,Loose Index Scan 的成本会高于 Tight Index Scan,比如:当分组较多,但组内的记录数并不多或唯一值较高的情况;对于每一个分组,都需要扫描两次,能跳过的记录数很少的情况。即 Loose Index Scan 在分组字段的选择性相对不太高,组内的数据量相对较多的情况更适用。

举例:该 SQL 在当前的测试数据中,Loose Index Scan 的成本还是要低于 Tight Index Scan。

```
select count(distinct c1,c2) from t2;

mysql> explain select count(distinct c1,c2) from t2;
+----+-------------+-------+------------+-------+---------------+--------------
-+---------+------+------+----------+---------------------------+
| id | select_type | table | partitions | type  | possible_keys | key
| key_len | ref  | rows | filtered | Extra                     |
+----+-------------+-------+------------+-------+---------------+--------------
-+---------+------+------+----------+---------------------------+
|  1 | SIMPLE      | t2    | NULL       | range | c1_c2_c3_idx  | c1_c2_c3_idx
| 512     | NULL |   10 |   100.00 | Using index for group-by  |
+----+-------------+-------+------------+-------+---------------+--------------
-+---------+------+------+----------+---------------------------+
1 row in set, 1 warning (0.01 sec)
```

新建一个相同表结构的表,插入下面的测试数据。

```
INSERT INTO t2 VALUES (null,'a','b','a','a'), (null,'a','b','a','a'),
(null,'k','c','a','a'), (null,'k','g','a','a'),
(null,'a','d','b','b'), (null,'a','b','b','b'),
(null,'d','b','c','c'), (null,'e','b','c','c'),
(null,'f','c','d','d'), (null,'k','c','d','d'),
(null,'y','d','y','y'), (null,'f','b','f','y'),
```

```
(null,'j','b','a','a'), (null,'m','b','a','a'),
(null,'z','c','a','a'), (null,'t','c','a','a'),
(null,'x','d','b','b'), (null,'x','b','b','b'),
(null,'d','b','c','c'), (null,'e','b','c','c'),
(null,'f','c','d','d'), (null,'k','c','d','d'),
(null,'y','d','y','y'), (null,'f','b','f','y');

-- 其数据分布
mysql> select c1,c2,count(*) from t2 group by c1,c2;
+----+----+----------+
| c1 | c2 | count(*) |
+----+----+----------+
| a  | b  |        3 |
| a  | d  |        1 |
| d  | b  |        2 |
| e  | b  |        2 |
| f  | b  |        2 |
| f  | c  |        2 |
| j  | b  |        1 |
| k  | c  |        3 |
| k  | g  |        1 |
| m  | b  |        1 |
| t  | c  |        1 |
| x  | b  |        1 |
| x  | d  |        1 |
| y  | d  |        2 |
| z  | c  |        1 |
+----+----+----------+
15 rows in set (0.00 sec)

mysql> explain select count(distinct c1,c2) from t2;
+----+-------------+-------+------------+-------+---------------+-------------+---------+------+------+----------+---------------------------------------+
| id | select_type | table | partitions | type  | possible_keys | key         | key_len | ref  | rows | filtered | Extra                                 |
+----+-------------+-------+------------+-------+---------------+-------------+---------+------+------+----------+---------------------------------------+
|  1 | SIMPLE      | t2    | NULL       | range | c1_c2_c3_idx  | c1_c2_c3_idx| 512     | NULL |   16 |   100.00 | Using index for group-by (scanning)   |
+----+-------------+-------+------------+-------+---------------+-------------+---------+------+------+----------+---------------------------------------+
1 row in set, 1 warning (0.00 sec)
Extra: Using index for group-by (scanning)
```

该方式可以理解为 Loose Index Scan 的扩展或两种方式的结合（索引顺序扫描的同时进行去重）。

示意图如图 39 所示。

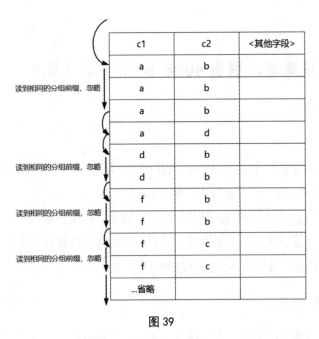

图 39

最后，再回到文章开头的案例，其执行计划如图 40 及 41 所示。

id	select_type	table	partitions	type	possible_keys	key	key_len	ref	rows	filtered	Extra
1	SIMPLE	task_log_info	NULL	index	index_taskUniqueId,index_reportTime	index_taskUniqueId	403	NULL	1001708	50	Using where

图 40　优化前

id	select_type	table	partitions	type	possible_keys	key	key_len	ref	rows	filtered	Extra
1	PRIMARY	<derived2>	NULL	ALL	NULL	NULL	NULL	NULL	1080	100	Using where
1	PRIMARY	a	NULL	eq_ref	PRIMARY,index_reportTime	PRIMARY	8	tmp.id	1	50	Using where
2	DERIVED	task_log_info	NULL	range	index_taskUniqueId	index_taskUniqueId	403	NULL	1080	100	Using index for group-by

图 41　优化后

其核心就是将 Tight Index Scan 转化为 Loose Index Scan。

15.5　总结

对于 GROUP BY 可以使用索引进行优化，Loose Index Scan 相对于 Tight Index Scan 在一些情况下可以大大减少扫描的行数，使用 Loose Index Scan 时，采用 Extra: Using index for group-by。

在 Loose Index Scan 的成本大于 Tight Index Scan 的一些情况下，可以尝试用两者的结合的方式，Extra: Using index for group-by (scanning)。

Loose Index Scan 更适用于分组内重复值相对较多，分组个数相对较少的情况。

16 MySQL 从库可以设置 sync_binlog 非 1 吗?

作者：胡呈清

众所周知，为防止断电丢失 Binlog 以及在故障恢复过程中丢失数据，MySQL 主库必须设置 sync_binlog=1。那么作为从库可以例外吗？

我们的第一反应当然是不行，既然主库会丢数据，从库自然一样。但其实不然，从库丢了数据是可以重新从主库上复制的，只要这个复制的位置和从库本身数据的位置一致就可以了，它们能一致吗？本文将对这个问题进行讨论。

16.1 背景知识

为了更好地说明这个问题，下面回顾一下相关的知识点：

(1) 在 InnoDB 的二阶段提交中，Prepare 阶段写 Redo Log，Commit 阶段写 Binlog。在故障恢复时保证：

① 所有已提交事务的 Binlog 一定存在。

② 所有未提交事务一定不记录 Binlog。

(2) 从库设置 relay_log_info_repository = table 时，slave_relay_log_info（即从库回放位置）的更新与 Relay Log 回放的 SQL 在同一个事务中提交。

(3) GTID 持久化在 Binlog 中，从库在某些条件下启动复制时会从 Executed_Gtid_Set 开始到主库复制数据。

根据以上 3 点，从库如果设置 sync_binlog 不为 1，在做故障恢复时就会发生以下情况：

(1) 事务状态为 TRX_COMMITTED_IN_MEMORY、TRX_NOT_STARTED。如果 Binlog 未落盘，事务会重做，数据将比 Binlog 多，slave_relay_log_info 表记录的复制位置也将领先于 Executed_Gtid_Set。

(2) 事务状态为 TRX_PREPARED。由于 Binlog 未刷盘，Recovery 时会回滚事务，数据与 Binlog 是一致的，slave_relay_log_info 表记录的复制位置等于 Executed_Gtid_Set。

如果从库断电恢复后，启动复制时的位置由 slave_relay_log_info 决定，则从库还是能正常复制数据，并且能与主库保持一致，只是 GTID 会出现跳号。

反之，如果启动复制时的位置由 Executed_Gtid_Set 决定，则从库复制会因为重复回放事务而报错，需要进行修复。下面设计一个实验来进行验证。

16.2 实验过程

16.2.1 设置从库参数并制造"故障"

从库参数设置如下，主库用工具并发写入数据（这里用的是 mysqlslap），然后从库强制关机（reboot-f）。

```
sync_binlog = 1000
innodb_flush_log_at_trx_commit = 1
relay_log_info_repository = table   ##slave_relay_log_info 表为 innodb 表
relay_log_recovery = on
gtid_mode = on
```

16.2.2 重启从库

从库服务器开机后重启 MySQL，查看信息。show master status 输出的 Executed_Gtid_Set 如下：

```
fb9b7d78-6eb5-11ec-985a-0242ac101704:1-167216

mysql> select * from slave_relay_log_info\G
*************************** 1. row ***************************
       Number_of_lines: 7
        Relay_log_name: ./localhost-relay-bin.000004
         Relay_log_pos: 4
        Master_log_name: mysql-bin.000001
         Master_log_pos: 48159613
             Sql_delay: 0
     Number_of_workers: 0
                    Id: 1
          Channel_name:
1 row in set (0.00 sec)
```

根据输出内容可知，从库的数据确实回放到了 mysql-bin.000001:48159613，对应的 GTID 为 fb9b7d78-6eb5-11ec-985a-0242ac101704:167222，

只是从库的 Binlog 有丢失，GTID 为 fb9b7d78-6eb5-11ec-985a-0242ac101704:1-167216。

```
...
SET @@SESSION.GTID_NEXT= 'fb9b7d78-6eb5-11ec-985a-0242ac101704:167222'/*!*/;
...
### INSERT INTO `mysqlslap`.`t`
### SET
###   @1=167216 /* INT meta=0 nullable=0 is_null=0 */
# at 48159586
#220407 14:10:34 server id 123456  end_log_pos 48159613     Xid = 169239
```

```
COMMIT/*!*/;
# at 48159613
...
```

从库已经有 167222 事务对应的数据。

```
mysql> select * from t where id=167216;
+--------+
| id     |
+--------+
| 167216 |
+--------+
1 row in set (0.00 sec)
```

16.2.3 从库启动复制

Error Log 显示的起始位置和 slave_relay_log_info 内容一样，从主库的 mysql-bin.000001:48159613 开始，对应 GTID 为 167222+1。

```
Slave I/O thread: Start asynchronous replication to master
'repl@10.186.61.32:3308' in log 'mysql-bin.000001' at position 48159613
```

但接下来 SQL 线程报错位置却是 mysql-bin.000001:48158146，比开始位置还靠前，这个位置对应的 GTID 为 167217（即 167216+1）：

```
2022-04-07T06:33:18.611181-00:00 4 [ERROR] Slave SQL for channel '': Could
not execute Write_rows event on table mysqlslap.t; Duplicate entry '167212' for
key 'PRIMARY', Error_code: 1062; handler error HA_ERR_FOUND_DUPP_KEY; the event's
master log mysql-bin.000001, end_log_pos 48158146, Error_code: 1062
```

而且解析从库 Relay Log（因为设置了 relay_log_recovery = on，启动复制时会丢弃旧的未 Relay Log 重新到主库取 Binlog），第一个事务也是 SET @@SESSION.GTID_NEXT= 'fb9b7d78-6eb5-11ec-985a-0242ac101704:167217'/*!*/;，而不是 167223。这说明了启动复制的位置并不是 slave_relay_log_info 记录的位置，而是从库的 GTID。

16.2.4 重复以上测试

在启动从库复制前执行 change master to master_auto_position=0;，这回未报错，从 167223 这个 GTID 开始复制数据，但从库 GTID 会出现跳号。

```
mysql> show master status;
+------------------+----------+--------------+------------------+--------------------------------------------------+
| File             | Position | Binlog_Do_DB | Binlog_Ignore_DB | Executed_Gtid_Set                                |
+------------------+----------+--------------+------------------+--------------------------------------------------+
```

```
| mysql-bin.000006 |   9976340 |                  |  fb9b7d78-
6eb5-11ec-985a-0242ac101704:1-167216:167223-200670 |
+------------------+-----------+------------------+---------------
------------------------------------------------+
1 row in set (0.01 sec)
```

16.3 结论

从库 sync_binlog 设置不为 1，发生断电会丢失 Binlog，因为 GTID 持久化在 Binlog 中，因此也会丢失 GTID。但是数据和 slave_relay_log_info 表中保存的 SQL 线程回放位置一致。

此时：

(1) 如果 master_auto_position=0，则从库重启复制时可以从正确的位置开始复制数据，从而与主库数据一致。不过，从库会产生 GTID 跳号。

(2) 如果 master_auto_position=1，则从库重启复制时会从 GTID 处开始复制数据，由于 GTID 有丢失，所以会重复回放事务，产生报错。

17 从实现原理来看为什么 Clone 插件比 Xtrabackup 更好用？

作者：戴骏贤

从 MySQL 8.0.17 版本开始，官方实现了 Clone 的功能，允许用户通过简单的 SQL 命令把远端或本地的数据库实例复制到其他实例后，快速拉起一个新的实例。

该功能由一系列的 WL 组成：

(1)Clone local replica(WL#9209)：实现了数据本地 Clone。

(2)Clone remote replica(WL#9210)：在本地 Clone 的基础上，实现了远程 Clone。将数据保存到一个远程的目录中，解决跨节点部署 MySQL 的问题。

(3)Clone Remote provisioning(WL#11636)：将数据直接复制到需要重新初始化的 MySQL 实例中。此外这个 WL 还增加了预检查的功能。

(4)Clone Replication Coordinates(WL#9211)：完成了获取和保存 Clone 点位的功能，方便 Clone 实例正常地加入集群中。

(5)Support cloning encrypted database (WL#9682)：最后一个 worklog 解决了数据加密情况下的数据复制问题。

本文初步地介绍 Clone 插件的原理以及和 Xtrabackup 的异同，以及整体实现的框架。

17.1 Xtrabackup 备份的不足

在使用 Xtrabackup 备份的过程中，可能遇到的最大的问题在于复制 Redo Log 的速度跟不上线上生产 Redo Log 的速度（如图 42）。

因为 Redo Log 是会循环利用的，在 Checkpoint 之后，旧的 Redo Log 可能会被新的 Redo Log 覆盖，而此时如果 Xtrabackup 没有完成旧的 Redo Log 的复制，就没法保证备份过程中的数据一致性（如图 43）。

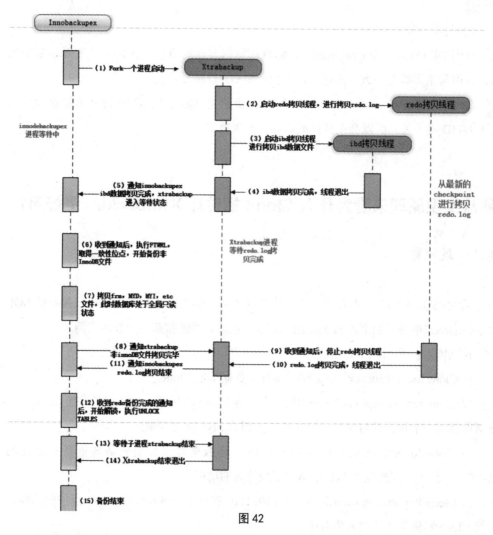

图 42

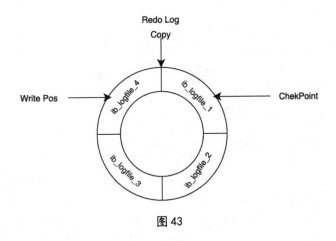

图 43

17.2 Clone 实现的基本原理

那么在 Clone 插件中如何去解决这个问题？从 WL#9209 中可以看到官方整体的设计思路。Clone 的过程分为 5 步：

(1)INIT：克隆初始化。

(2)FILE COPY：文件复制。

(3)PAGE COPY：数据页复制。

(4)REDO COPY：重做日志复制。

(5)DONE：克隆结束。

这中间最重要的便是：

(1)FILE COPY：跟 Xtrabackup 一样，会物理地复制所有的 InnoDB 表空间文件，同时启动一个 Page Tracking 进程，从 CLONE START LSN 开始监控所有 InnoDB PAGE 的改动。

(2)PAGE COPY：PAGE COPY 是在 Xtrabackup 中没有的一个阶段，主要完成 2 项工作。

① 在完成数据库库文件复制之后，会开启 Redo Archiving，同时停止 Page Tracking 进程（开始前会做一次 checkpoint）。Redo Archiving 会从指定的 LSN 位置开始复制 Redo Log。

② 将 Page Tracking 记录的脏页发送到指定位置。为了保持高效，会基于 spaceid 和 page id 进行排序，尽可能确保磁盘读写的顺序性。

(3)Redo Copy：这个阶段会加锁获取 Binlog 文件及当前偏移位置和 gtid_executed 信息，并停止 Redo Archiving 进程。之后将所有归档的 Redo Log 日志文件发往目标端（见图 44）。

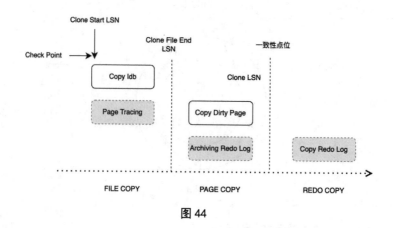

图 44

17.3 代码结构和调用逻辑

整体实现上分为三个部分。

17.3.1 SQL/Server 层

(1)sql/sql_lex.h。

(2)sql/sql_yacc.yy：增加了对 Clone 语法的支持。

(3)sql_admin.cc：增加了客户端处理 SQL(clone instance) 和服务端处理 COM_XXX 命令。

(4)clone_handler.cc：增加调用插件的具体实现响应 SQL 层处理。

17.3.2 Plugin 插件层

plugin 插件层如图 45 所示。

(1)clone_plugin.cc：plugin interface

(2)clone_local.cc：具体的 Clone 操作。

(3)clone_os.cc：系统层面的一些具体操作函数，包括 OS [sendfile/read/write]。

(4)clone_hton.cc：与存储引擎层的接口。

(5)clone_client.cc 和 clone_server.cc：Clone 的客户端和服务端。

(6)clone_status.cc：Clone 时的整体任务的进度和状态，会有一个 Clone_Task_Manager 去记录状态信息。

(7)clone_plugin.cc：Clone 插件的入口以及初始化和系统变量等内容。

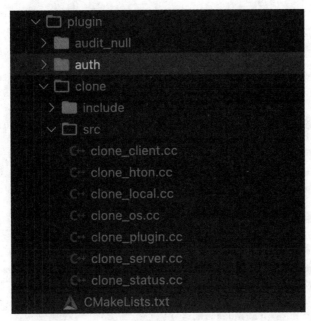

图 45

17.3.3 InnoDB 引擎层

(1)Clone：位于 storage/innobase/clone 下。

① clone0clone.cc：clone task and runtime operation。

② clone0snapshot.cc：snapshot management。

③ clone0copy.cc：copy specific methods。

④ clone0apply.cc：apply specific methods。

⑤ clone0desc.cc：serialized data descriptor。

(2)Archiver：位于 storage/innobase/arch 下，与 Page tracing 相关。

① arch0arch.cc。

② arch0page.cc。

③ arch0log.cc。

本地 Clone 的函数调用栈（如图 46）：

```
    Clone_Handle::process_chunk(Clone_Task*, unsigned int, unsigned int, Ha_clone_
cbk*) (/mysql-8.0.33/storage/innobase/clone/clone0copy.cc:1440)
    Clone_Handle::copy(unsigned int, Ha_clone_cbk*) (/mysql-8.0.33/storage/
innobase/clone/clone0copy.cc:1379)
    innodb_clone_copy(handlerton*, THD*, unsigned char const*, unsigned int,
unsigned int, Ha_clone_cbk*) (/mysql-8.0.33/storage/innobase/clone/clone0api.
cc:561)
    hton_clone_copy(THD*, std::__1::vector<myclone::Locator, std::__1::allocator
```

```
<myclone::Locator>>&, std::__1::vector<unsigned int, std::__1::allocator<unsigned 
int>>&, Ha_clone_cbk*) (/mysql-8.0.33/plugin/clone/src/clone_hton.cc:152)
    myclone::Local::clone_exec() (/mysql-8.0.33/plugin/clone/src/clone_local.
cc:172)
    myclone::Local::clone() (/mysql-8.0.33/plugin/clone/src/clone_local.cc:73)
    plugin_clone_local(THD*, char const*) (/mysql-8.0.33/plugin/clone/src/clone_
plugin.cc:456)
    Clone_handler::clone_local(THD*, char const*) (/mysql-8.0.33/sql/clone_
handler.cc:135)
    Sql_cmd_clone::execute(THD*) (/mysql-8.0.33/sql/sql_admin.cc:2017)
    mysql_execute_command(THD*, bool) (/mysql-8.0.33/sql/sql_parse.cc:4714)
    dispatch_sql_command(THD*, Parser_state*) (/mysql-8.0.33/sql/sql_parse.
cc:5363)
    dispatch_command(THD*, COM_DATA const*, enum_server_command) (/mysql-8.0.33/
sql/sql_parse.cc:2050)
    do_command(THD*) (/mysql-8.0.33/sql/sql_parse.cc:1439)
    handle_connection(void*) (/mysql-8.0.33/sql/conn_handler/connection_handler_
per_thread.cc:302)
    pfs_spawn_thread(void*) (/mysql-8.0.33/storage/perfschema/pfs.cc:3042)
    _pthread_start (@_pthread_start:40)
```

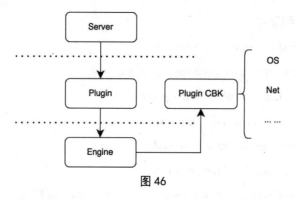

图 46

17.4 Page Archiving 系统

Page Archiving 是之前 Xtrabackup 中没有的部分,因此在这里特别介绍一下其整体实现的过程。

为了减少在 Clone 过程中 Redo Log 的复制量,Clone 插件使用了对脏页进行跟踪和收集的方法,在复制表空间的过程中追踪脏页,并在 File Copy 结束的阶段将脏页打包发送到目标端。

Page Tracking 脏页监控的方式可以有两种实现方案:

(1) 在 mtr 提交的时候收集。

(2) 在 purge 进程刷脏的时候收集。

为了不阻塞 MySQL 事务的提交，当前 Clone 插件选择的是方案 2。

Purge 进程刷脏的入口是 buf_flush_page 函数。

```
buf0flu.cc
if (flush) {
/* We are committed to flushing by the time we get here */

mutex_enter(&buf_pool->flush_state_mutex);
...

arch_page_sys->track_page(bpage, buf_pool->track_page_lsn, frame_lsn,
false);
}
```

在将脏页刷回到磁盘的时候，会将需要追踪的脏页加入 arch_page_sys 中。如果在加入脏页的过程中 block 满了，就要开辟新的空间，这会阻塞刷脏的进程。

```
/** Check and add page ID to archived data.
Check for duplicate page.
@param[in]      bpage           page to track
@param[in]      track_lsn       LSN when tracking started
@param[in]      frame_lsn       current LSN of the page
@param[in]      force           if true, add page ID without check */
void Arch_Page_Sys::track_page(buf_page_t *bpage, lsn_t track_lsn,
lsn_t frame_lsn, bool force) {
Arch_Block *cur_blk;
uint count = 0;
... ...

/* We need to track this page. */
arch_oper_mutex_enter();

while (true) {
if (m_state != ARCH_STATE_ACTIVE) {
break;
}
... ...

cur_blk = m_data.get_block(&m_write_pos, ARCH_DATA_BLOCK);

if (cur_blk->get_state() == ARCH_BLOCK_ACTIVE) {
if (cur_blk->add_page(bpage, &m_write_pos)) {
/* page added successfully. */
break;
}

/* Current block is full. Move to next block. */
```

```
cur_blk->end_write();

m_write_pos.set_next();

/* Writing to a new file so move to the next reset block. */
if (m_write_pos.m_block_num % ARCH_PAGE_FILE_DATA_CAPACITY == 0) {
Arch_Block *reset_block =
m_data.get_block(&m_reset_pos, ARCH_RESET_BLOCK);
reset_block->end_write();

m_reset_pos.set_next();
}

os_event_set(page_archiver_thread_event);

++count;
continue;

} else if (cur_blk->get_state() == ARCH_BLOCK_INIT ||
cur_blk->get_state() == ARCH_BLOCK_FLUSHED) {
ut_ad(m_write_pos.m_offset == ARCH_PAGE_BLK_HEADER_LENGTH);

cur_blk->begin_write(m_write_pos);

if (!cur_blk->add_page(bpage, &m_write_pos)) {
/* Should always succeed. */
ut_d(ut_error);
}

/* page added successfully. */
break;

} else {
bool success;
... ...
/* Might release operation mutex temporarily. Need to
loop again verifying the state. */
success = wait_flush_archiver(cbk);
count = success ? 0 : 2;

continue;
}
}
arch_oper_mutex_exit();
}
```

脏页收集的整体入口为 Page_Arch_Client_Ctx::start 和 Arch_Page_Sys::start。

这里需要注意的是，在开启 Page Archiving 之前需要强制一次 checkpoint，因此如果系统处于比较高的负载(比如 IO Wait 很高)情况下可能会卡顿。

```
int Page_Arch_Client_Ctx::start(bool recovery, uint64_t *start_id) {
... ...
/* Start archiving. */
err = arch_page_sys->start(&m_group, &m_last_reset_lsn, &m_start_pos,
m_is_durable, reset, recovery);
... ...
}

int Arch_Page_Sys::start(Arch_Group **group, lsn_t *start_lsn,
Arch_Page_Pos *start_pos, bool is_durable,
bool restart, bool recovery) {
... ...

log_sys_lsn = (recovery ? m_last_lsn : log_get_lsn(*log_sys));
/* Enable/Reset buffer pool page tracking. */
set_tracking_buf_pool(log_sys_lsn); // page_id
... ...
auto err = start_page_archiver_background();  sp_id, page_id
... ...
if (!recovery) {
/* Request checkpoint */
log_request_checkpoint(*log_sys, true);  checkpoint
}
}
```

脏页的归档由 page_archiver_thread 线程进行:

```
/** Archiver background thread */
void page_archiver_thread() {
bool page_wait = false;

... ...

while (true) {
/* Archive in memory data blocks to disk. */
auto page_abort = arch_page_sys->archive(&page_wait);

if (page_abort) {
ib::info(ER_IB_MSG_14) << "Exiting Page Archiver";
break;
}

if (page_wait) {
/* Nothing to archive. Wait until next trigger. */
os_event_wait(page_archiver_thread_event);
os_event_reset(page_archiver_thread_event);
```

```
    }
  }
}

bool Arch_Page_Sys::archive(bool *wait) {
... ...

db_err = flush_blocks(wait);

if (db_err != DB_SUCCESS) {
is_abort = true;
}

... ...
return (is_abort);
}

dberr_t Arch_Page_Sys::flush_blocks(bool *wait) {
... ...
err = flush_inactive_blocks(cur_pos, end_pos);
... ...
}

dberr_t Arch_Page_Sys::flush_inactive_blocks(Arch_Page_Pos &cur_pos,
Arch_Page_Pos end_pos) {

/* Write all blocks that are ready for flushing. */
while (cur_pos.m_block_num < end_pos.m_block_num) {
cur_blk = m_data.get_block(&cur_pos, ARCH_DATA_BLOCK);

err = cur_blk->flush(m_current_group, ARCH_FLUSH_NORMAL);

if (err != DB_SUCCESS) {
break;
}

... ...
}

return (err);
}
```

最后，调用 Arch_Block 归档脏页。这里脏页归档的时候也需要使用 doublewrite buffer。

```
/** Flush this block to the file group.
@param[in]      file_group      current archive group
@param[in]      type            flush type
```

```cpp
@return error code. */
dberr_t Arch_Block::flush(Arch_Group *file_group, Arch_Blk_Flush_Type type) {
... ...
switch (m_type) {
case ARCH_RESET_BLOCK:
err = file_group->write_file_header(m_data, m_size);
break;

case ARCH_DATA_BLOCK: {
bool is_partial_flush = (type == ARCH_FLUSH_PARTIAL);

/* Callback responsible for setting up file's header starting at offset 0.
This header is left empty within this flush operation. */
auto get_empty_file_header_cbk = [](uint64_t, byte *) {
return DB_SUCCESS;
};

/* We allow partial flush to happen even if there were no pages added
since the last partial flush as the block's header might contain some
useful info required during recovery. */
err = file_group->write_to_file(nullptr, m_data, m_size, is_partial_flush,
true, get_empty_file_header_cbk);
break;
}

default:
ut_d(ut_error);
}

return (err);
}

dberr_t Arch_Group::write_to_file(Arch_File_Ctx *from_file, byte *from_buffer,
uint length, bool partial_write,
bool do_persist,
Get_file_header_callback get_header) {
... ...
if (do_persist) {
Arch_Page_Dblwr_Offset dblwr_offset =
(partial_write ? ARCH_PAGE_DBLWR_PARTIAL_FLUSH_PAGE
: ARCH_PAGE_DBLWR_FULL_FLUSH_PAGE);

/** Write to the doublewrite buffer before writing archived data to a file.
The source is either a file context or buffer. Caller must ensure that data
is in single file in source file context. **/
Arch_Group::write_to_doublewrite_file(from_file, from_buffer, write_size,
dblwr_offset);
}
```

```
... ...
return (DB_SUCCESS);
}
```

17.5 总结

(1)Clone 功能相对于使用 Xtrabackup 拉起一个从库，更加方便。

(2)Clone 功能相对于 Xtrabackup，复制的 Redo Log 日志量更少，也更不容易遇到失败的问题（arch_log_sys 会控制日志写入以避免未归档的日志被覆盖）。

(3) 从源码的分析来看，启动 Clone 的时候会强制做一次 checkpoint，在 Redo Log Archiving 的时候会控制日志写入量。因此从原理上看，如果处于高负载的主库做 Clone 操作，可能会对系统有影响。

18 基于 MySQL 的机房容灾建设和切换演练

作者：徐良

18.1 容灾需求背景简述

在社会数字化转型过程中，IT 系统的容灾能力将影响社会运行稳定。2023 年工业和信息化部等六部门联合印发《算力基础设施高质量发展行动计划》，提出到 2025 年，重点行业核心数据、重要数据灾备覆盖率达到 100%。

为保障容灾架构的有效性，定期实施灾难恢复演练是不可或缺的。企业数字化转型过程中积累了大量数据库实例，而传统的 MySQL 数据库容灾方案搭建易、演练难。通过定制 MySQL 8.0.39 版本以支持多通道主主双向复制数据防回路机制，可以实现大规模数据库实例下较简单和低成本的容灾高可用架构，结合内部自动化平台和技术，实现容灾切换运维流程的优化，从而显著降低日常运维的工作量。

18.2 多通道主主复制功能说明

在跨机房容灾场景中，同时开启多源复制和主主双向复制时，可能出现数据回路问题。为此，新增了 replicate_server_mode 选项，用于控制只应用多源复制管道内邻近主节点产生的 binlog，避免应用其他的非邻近节点产生的 binlog，从而防止出现数据回路问题。在多通道主主双向复制时，只有主机房内的主节点是可写的，其他节点是只读的，这样能减少机房容灾演练和切换时的主从配置变更，保障数据一致性，降低切换运维的工作量，减少误操作风险。

18.2.1 新增系统选项

新增系统选项见表 8。

表 8

System Variable Name	replicate_server_mode
Command-Line Format	replicate_server_mode[={0|1}]
System Variable	replicate_server_mode
Variable Scope	Global
Dynamic Variable	No
Type	bool
Permitted Values	[0, 1]
Default	0

replicate_server_mode 设置为 1，表示从库的 SQL_thread 只应用多源复制管道内邻近主节点上产生的 binlog，不会应用其他的非邻近节点产生的 binlog；设置为 0，表示应用 binlog 时不做此限制，即默认状态下数据可以环形复制。

该选项是个全局选项，且不可在线动态修改，因此需要在实例启动前就设置好。如果尝试在线修改的话，会提示报错，例如：

```
mysql> select @@session.replicate_server_mode;
ERROR 1238 (HY000): Variable 'replicate_server_mode' is a GLOBAL variable

mysql> select @@global.replicate_server_mode;+--------------------------------
------+| @@global.replicate_server_mode |+--------------------------------+|
0 |+--------------------------------+1 row in set (0.00 sec)

mysql> set global replicate_server_mode = 0;
ERROR 1238 (HY000): Variable 'replicate_server_mode' is a read only variable
```

18.2.2 安装依赖

在龙蜥 8 操作系统下编译安装 mysql-8.0.39。

安装依赖包：

```
yum install gcc gcc-c++.x86_64 binutils.x86_64 libevent libevent-devel jemalloc jemalloc-devel -y
```

```
yum install openssl-devel.x86_64 -y
yum install -y diffutils git lbzip2 libaio-devel libbison-devel libcurl4-openssl-devel libevent-devel libexpat1-devel libffi-devel libgflags-devel libgtest-devel libjemalloc-devel libldap2-devel liblz4-devel libncurses-devel libnuma-devel libreadline-devel libsnappy-devel libssh-devel libtirpc-devel libtool libxml2-devel libzstd-devel net-tools numactl pkg-config psmisc vim wget patchelf
yum install ncurses-devel readline readline-devel libtirpc-devel rpcgen -y
yum install libgudev1-devel.x86_64 libgudev1.x86_64 -y
```

下载 boost_1_77_0：

```
cd /usr/local/src/
wget https://boostorg.jfrog.io/artifactory/main/release/1.77.0/source/boost_1_77_0.tar.gz
tar -zxf boost_1_77_0.tar.gz
```

18.2.3 定制 MySQL 8

```
cd /usr/local/src/
wget https://downloads.mysql.com/archives/get/p/23/file/mysql-8.0.39.tar.gz
tar -zxvf mysql-8.0.39.tar.gz
cd mysql-8.0.39
```

在 sys_vars.cc 文件中增加 replicate_server_mode 配置项（见图 47）。

```
vi sql/sys_vars.cc
6337 static Sys_var_deprecated_alias Sys_slave_skip_errors("slave_skip_errors",
6338                                                      Sys_replica_skip_errors);
6339
6340 //add by xuliang 20240728 begin
6341 static Sys_var_uint Sys_replicate_server_mode(
6342     "replicate_server_id_mode",
6343     "In replication, if set to 1, do only replicate events having the server id of master,skip
6344     "Default value is 0 (to replicate all the events from master).",
6345     READ_ONLY GLOBAL_VAR(replicate_server_id_mode), CMD_LINE(OPT_ARG),
6346     VALID_RANGE(0, 2),
6347     DEFAULT(0),
6348     BLOCK_SIZE(1), NO_MUTEX_GUARD, NOT_IN_BINLOG,
6349     ON_CHECK(nullptr), ON_UPDATE(nullptr));
6350 //add by xuliang 20240728 end
6351
6352 static Sys_var_ulonglong Sys_relay_log_space_limit(
6353     "relay_log_space_limit", "Maximum space to use for all relay logs",
6354     READ_ONLY GLOBAL_VAR(relay_log_space_limit), CMD_LINE(REQUIRED_ARG),
6355     VALID_RANGE(0, ULLONG_MAX), DEFAULT(0), BLOCK_SIZE(1));
```

图 47

在 log_event.cc 文件中增加多通道 binlog 复制过滤邻近主日志逻辑（见图 48）。

```
vi ./sql/log_event.cc
1082    if ((server_id == ::server_id && !rli->replicate_same_server_id) ||
1083        (rli->slave_skip_counter == 1 && rli->is_in_group()))
1084      return EVENT_SKIP_IGNORE;
1085
1086    //add by xuliang 20240728 begin
1087    else if (server_id != rli->mi->master_id && rli->replicate_server_id_mode>0)
1088      return EVENT_SKIP_IGNORE;
1089    //add by xuliang 20240728 end
1090  }
1091    else if (rli->slave_skip_counter > 0)
1092      return EVENT_SKIP_COUNT;
1093    else
1094      return EVENT_SKIP_NOT;
1095  }

6352  static Sys_var_ulonglong Sys_relay_log_space_limit(
6353      "relay_log_space_limit", "Maximum space to use for all relay logs",
6354      READ_ONLY GLOBAL_VAR(relay_log_space_limit), CMD_LINE(REQUIRED_ARG),
6355      VALID_RANGE(0, ULLONG_MAX), DEFAULT(0), BLOCK_SIZE(1));
```

图 48

18.2.4 编译安装

```
    cmake3 -DCMAKE_INSTALL_PREFIX=/usr/local/mysql -DBOOST_INCLUDE_DIR=../
boost_1_77_0/ -DMYSQL_UNIX_ADDR=/data/mysql/datanode1/mysql.sock -DDEFAULT_
CHARSET=utf8mb4 -DDEFAULT_COLLATION=utf8mb4_general_ci -DWITH_EXTRA_
CHARSETS:STRING=all -DWITH_MYISAM_STORAGE_ENGINE=1 -DWITH_INNOBASE_STORAGE_ENGINE=1
-DWITH_MEMORY_STORAGE_ENGINE=1 -DWITH_PARTITION_STORAGE_ENGINE=1 -DENABLED_LOCAL_
INFILE=1 -DMYSQL_DATADIR=/data/mysql/datanode1 -DMYSQL_USER=mysql -DMYSQL_TCP_
PORT=3306 -DCMAKE_EXE_LINKER_FLAGS="-ljemalloc" -DWITH_BOOST=.. -DFORCE_INSOURCE_
BUILD=1 -DWITH_ZLIB=bundled -DWITH_AUTHENTICATION_LDAP=OFF -DCMAKE_EXE_LINKER_
FLAGS="-ljemalloc" -DWITH_LIBEVENT="bundled" -DCMAKE_BUILD_TYPE=RelWithDebInfo
-DBUILD_CONFIG=mysql_release -DWITH_TOKUDB=OFF -DWITH_ROCKSDB=OFF -DWITH_
COREDUMPER=OFF make -j 4  && make install
```

验证数据库版本：

```
    [xuliang@mysql01]# /usr/local/mysql/bin/mysqld --version
    /usr/local/mysql/bin/mysqld  Ver 8.0.39 for Linux on x86_64
    [xuliang@mysql01]# /usr/local/mysql/bin/mysqld --verbose --help|grep
replicate-server-mode
      --replicate-server-mode
    replicate-server-mode                                          FALSE
```

18.3 多机房容灾部署

18.3.1 部署架构图

部署架构如图 49 所示。

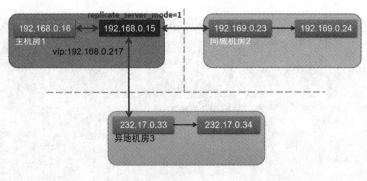

图 49

以传统的模式开启多源复制和主主复制会出现数据同步回路的问题。新增 replicate_server_mode 配置项后,当主机房边界节点配置该值为 1 时,该主库的 3 个主从复制通道(红色)仅同步邻近主库的 binlog 日志,丢弃级联上层主库或者其他数据库实例产生的 binlog,就能够解决数据回路问题。

18.3.2 数据库信息

数据库信息如表 9 所示。

表 9

IP 地址	hostname	角色	auto_increment_offset	read_only	replicate_server_mode
192.168.0.15	mysql01	主机房主库	1	0	1
192.168.0.16	mysql02	主机房备库	2	1	0
192.169.0.23	mysql03	同城主库	3	1	0
192.169.0.24	mysql04	同城从库	1	1	0
232.17.0.33	mysql05	异地主库	4	1	0
232.17.0.34	mysql06	异地从库	1	1	0

18.3.2.1 主机房 IP 地址

数据库主库 A:192.168.0.15

数据库备库 B:192.168.0.16

数据库 VIP:192.168.0.217

replicate_server_mode 默认为 0,主机房边界节点数据库 A(192.168.0.15)的 my.cnf 配置 replicate_server_mode=1,主从同步链路仅同步邻近主库产生的 binlog,其他节点的 replicate_server_mode 项保持默认值。

18.3.2.2 同城机房 IP 地址

同城容灾数据主库 C:192.169.0.23

同城容灾数据库从库 D：192.169.0.24

18.3.2.3 异地机房 IP 地址

异地容灾数据主库 E：232.17.0.33

异地容灾数据库从库 F：232.17.0.34

为了降低数据库机房切换时双写数据错乱的风险，auto_increment_increment 统一配置为 4。对于 auto_increment_offset 项，主机房主库配置为 1，备库配置为 2，同程机房主库配置为 3，异地机房主库配置为 4，其他从库默认不承担写流量，因而配置为 1。

18.3.3 主机房内建立主主关系

机房内 2 台数据库建立主从关系，同时修改 my.cnf 配置，错开各个实例的自增长 ID 值间隔，降低容灾切换瞬间可能出现的多写冲突风险。replicate_server_mode 默认为 0，除了主机房边界节点 192.168.0.15 将 replicate_server_mode 配置为 1，其他节点无需配置该选项。

主机房 1 数据库 A（192.168.0.15）：

```
auto_increment_increment = 4
auto_increment_offset = 1
replicate_server_mode=1
```

主机房 1 数据库 B（192.168.0.16）：

```
auto_increment_increment = 4
auto_increment_offset = 2
```

主机房备库与主库建立主从关系：

```
change master to master_host='192.168.0.15',master_user='repl',master_password='xxxxxx',MASTER_LOG_FILE='mysql-bin.000003',MASTER_LOG_POS=4 ;
start slave;
set global read_only=1;
```

主机房主库与备库建立主主关系：

```
change master to master_host='192.168.0.16',master_user='repl',master_password='xxxxxx',MASTER_LOG_FILE='mysql-bin.000003',MASTER_LOG_POS=4 for channel 'dr_center1_s0_16';
start slave for channel 'dr_center1_s0_16';
set global read_only=0;
```

机房内高可用 keepalived 配置切换时，将备库修改为可写的。

```
 notify_master /etc/keepalived/scripts/set_mysql_writable.sh
# 脚本文件 set_mysql_writable.sh 内容
#!/bin/bash
```

```
MYSQL_USER="your_user"
MYSQL_PASSWORD="your_password"
MYSQL_SOCKET="/var/run/mysqld/mysqld.sock"
echo "Setting MySQL read_only to 0..."
mysql -u $MYSQL_USER -p$MYSQL_PASSWORD -e "SET GLOBAL read_only = 0;"
--socket=$MYSQL_SOCKET

echo "MySQL is now writable."
```

18.3.4 机房间建立主从关系

同城机房 2 容灾数据库 C（192.169.0.23）：

```
auto_increment_increment = 4
auto_increment_offset = 3

change master to master_host='192.168.0.15',master_user='repl',master_password='xxxxxx',MASTER_LOG_FILE='mysql-bin.000003',MASTER_LOG_POS=4 ;
start slave;
```

同城机房 2 容灾数据库从库 D（192.169.0.24）：

```
auto_increment_increment =4
auto_increment_offset = 1
change master to master_host='192.169.0.24',master_user='repl',master_password='xxxxxx',MASTER_LOG_FILE='mysql-bin.000003',MASTER_LOG_POS=4 ;
start slave;
```

异地机房 3 容灾数据库 E（232.17.0.33）：

```
auto_increment_increment = 4
auto_increment_offset = 4

master_host='192.168.0.15',master_user='repl',master_password='xxxxxx',MASTER_LOG_FILE='mysql-bin.000003',MASTER_LOG_POS=4 ;
start slave;
```

异地机房 3 容灾数据库从库 F（232.17.0.34）：

```
auto_increment_increment = 4
auto_increment_offset = 1

master_host='232.17.0.33',master_user='repl',master_password='xxxxxx',MASTER_LOG_FILE='mysql-bin.000003',MASTER_LOG_POS=4 ;
start slave;
```

18.3.5 建立多通道主主复制关系

主机房边界节点数据主库 A（192.168.0.15）和其他节点建立多源复制，同时设置主机房主库和备库可写、其他机房只读。

验证主机房边界节点是否开启多通道主主复制：

```
mysql> show variables like 'replicate_server_mode';
+-----------------------+-------+
| Variable_name         | Value |
+-----------------------+-------+
| replicate_server_mode | ON    |
+-----------------------+-------+
set global read_only=0;
```

主机房主库与同城机房 2 主库建立主从关系：

```
change master to master_host='192.169.0.23',master_user='repl',master_password='xxxxxx',MASTER_LOG_FILE='mysql-bin.000003',MASTER_LOG_POS=254 for channel 'dr_center2_s0_23';
start slave for channel 'dr_center2_s0_23';
set global read_only=1;
```

与异地机房 3 主库建立主从关系：

```
change master to master_host='232.17.0.33',master_user='repl',master_password='xxxxxx',MASTER_LOG_FILE='mysql-bin.000004',MASTER_LOG_POS=4 for channel 'dr_center3_s0_33';
start slave for channel 'dr_center3_s0_33';
set global read_only=1;
```

此时，查看主从状态会显示 3 个通道各自的同步状态：

```
mysql> show slave status\G;
*************************** 1. row ***************************
               Slave_IO_State: Waiting for master to send event
                  Master_Host: 192.168.0.16
                  Master_User: repl
                  Master_Port: 3306
                Connect_Retry: 60
              Master_Log_File: mysql-bin.001880
          Read_Master_Log_Pos: 641525432
               Relay_Log_File: relay-bin-dr_center1_s0_16.002560
                Relay_Log_Pos: 367
        Relay_Master_Log_File: mysql-bin.001880
             Slave_IO_Running: Yes
            Slave_SQL_Running: Yes
              Replicate_Do_DB:
          Replicate_Ignore_DB:
           Replicate_Do_Table:
       Replicate_Ignore_Table:
      Replicate_Wild_Do_Table:
  Replicate_Wild_Ignore_Table:
```

```
                  Last_Errno: 0
                  Last_Error:
                Skip_Counter: 0
         Exec_Master_Log_Pos: 641525432
             Relay_Log_Space: 633
             Until_Condition: None
              Until_Log_File:
               Until_Log_Pos: 0
          Seconds_Behind_Master: 0
Master_SSL_Verify_Server_Cert: No
               Last_IO_Errno: 0
               Last_IO_Error:
              Last_SQL_Errno: 0
              Last_SQL_Error:
  Replicate_Ignore_Server_Ids:
             Master_Server_Id: 16
                  Master_UUID: b9104d4d-ef19-11e9-a66d-fa163e8df3f7
             Master_Info_File: mysql.slave_master_info
                    SQL_Delay: 0
          SQL_Remaining_Delay: NULL
      Slave_SQL_Running_State: Slave has read all relay log; waiting for more updates
           Master_Retry_Count: 86400
                  Master_Bind:
      Last_IO_Error_Timestamp:
     Last_SQL_Error_Timestamp:
                Auto_Position: 0
         Replicate_Rewrite_DB:
                 Channel_Name: dr_center1_s0_16
           Master_TLS_Version:
*************************** 2. row ***************************
               Slave_IO_State: Waiting for master to send event
                  Master_Host: 192.169.0.23
                  Master_User: repl
                  Master_Port: 3306
                Connect_Retry: 60
              Master_Log_File: mysql-bin.001825
          Read_Master_Log_Pos: 180546381
               Relay_Log_File: relay-bin-dr_center2_s0_23.002482
                Relay_Log_Pos: 367
        Relay_Master_Log_File: mysql-bin.001825
             Slave_IO_Running: Yes
            Slave_SQL_Running: Yes
              Replicate_Do_DB:
          Replicate_Ignore_DB:
           Replicate_Do_Table:
       Replicate_Ignore_Table:
      Replicate_Wild_Do_Table:
  Replicate_Wild_Ignore_Table:
```

```
                 Last_Errno: 0
                 Last_Error:
               Skip_Counter: 0
        Exec_Master_Log_Pos: 180546381
            Relay_Log_Space: 633
            Until_Condition: None
             Until_Log_File:
              Until_Log_Pos: 0
        Seconds_Behind_Master: 0
Master_SSL_Verify_Server_Cert: No
              Last_IO_Errno: 0
              Last_IO_Error:
             Last_SQL_Errno: 0
             Last_SQL_Error:
  Replicate_Ignore_Server_Ids:
           Master_Server_Id: 23
                Master_UUID: 9e3b537f-f178-11ea-ad55-fa163efd4f2a
           Master_Info_File: mysql.slave_master_info
                  SQL_Delay: 0
        SQL_Remaining_Delay: NULL
      Slave_SQL_Running_State: Slave has read all relay log; waiting for more updates
         Master_Retry_Count: 86400
                Master_Bind:
    Last_IO_Error_Timestamp:
   Last_SQL_Error_Timestamp:
              Auto_Position: 0
        Replicate_Rewrite_DB:
               Channel_Name: dr_center2_s0_23
         Master_TLS_Version:
*************************** 3. row ***************************
             Slave_IO_State: Waiting for master to send event
                Master_Host: 232.17.0.33
                Master_User: repl
                Master_Port: 3306
              Connect_Retry: 60
            Master_Log_File: mysql-bin.001265
        Read_Master_Log_Pos: 180546381
             Relay_Log_File: relay-bin-dr_center3_s0_33.002312
              Relay_Log_Pos: 367
      Relay_Master_Log_File: mysql-bin.001265
           Slave_IO_Running: Yes
          Slave_SQL_Running: Yes
            Replicate_Do_DB:
        Replicate_Ignore_DB:
         Replicate_Do_Table:
     Replicate_Ignore_Table:
     Replicate_Wild_Do_Table:
  Replicate_Wild_Ignore_Table:
```

```
                Last_Errno: 0
                Last_Error: 
              Skip_Counter: 0
       Exec_Master_Log_Pos: 180546381
           Relay_Log_Space: 633
           Until_Condition: None
            Until_Log_File: 
             Until_Log_Pos: 0
        Seconds_Behind_Master: 0
Master_SSL_Verify_Server_Cert: No
             Last_IO_Errno: 0
             Last_IO_Error: 
            Last_SQL_Errno: 0
            Last_SQL_Error: 
  Replicate_Ignore_Server_Ids: 
          Master_Server_Id: 33
               Master_UUID: 9e3b537f-f178-12ae-ad75-fa163efd4f2a
          Master_Info_File: mysql.slave_master_info
                 SQL_Delay: 0
       SQL_Remaining_Delay: NULL
   Slave_SQL_Running_State: Slave has read all relay log; waiting for more updates
        Master_Retry_Count: 86400
               Master_Bind: 
   Last_IO_Error_Timestamp: 
  Last_SQL_Error_Timestamp: 
             Auto_Position: 0
      Replicate_Rewrite_DB: 
              Channel_Name: dr_center3_s0_33
        Master_TLS_Version: 
3 rows in set (0.00 sec)
```

18.4 容灾切换演练

18.4.1 主机房高可用演练

18.4.1.1 以 keepalived 高可用方案为例

在主机房边界节点主库数据库 A（192.168.0.15）上，运行命令 ip a，查看 VIP 是否在主库上：

```
[xuliang@mysql01 ~]$ ip a
1: eth0: <BROADCAST,MULTICAST,UP,LOWER_UP> mtu 1500 qdisc mq state UP qlen 1000
    link/ether fa:16:3e:72:e7:87 brd ff:ff:ff:ff:ff:ff
    inet 192.168.0.15/24 brd 192.168.0.255 scope global eth0
       valid_lft forever preferred_lft forever
    inet 192.168.0.217/24 scope global secondary eth0
       valid_lft forever preferred_lft forever
```

```
        inet6 34ef::f816:3eff:fe72:c787/64 scope link
           valid_lft forever preferred_lft forever
```

确认后停掉 keepavlied，VIP 漂移到备库：

```
[xuliang@mysql01 ~]$ service keepalived stop
```

查看 VIP，发现其已经下线：

```
[xuliang@mysql01 ~]$ ip a
1: eth0: <BROADCAST,MULTICAST,UP,LOWER_UP> mtu 1500 qdisc mq state UP qlen 1000
    link/ether fa:16:3e:72:e7:87 brd ff:ff:ff:ff:ff:ff
    inet 192.168.0.15/24 brd 192.168.0.255 scope global eth0
       valid_lft forever preferred_lft forever
    inet6 34ef::f816:3eff:fe72:c787/64 scope link
       valid_lft forever preferred_lft forever
```

18.4.1.2 主机房数据库 B（192.168.0.16）

运行命令 ip a，查看 VIP 是否绑定：

```
[xuliang@mysql02 ~]$ ip a
1: eth0: <BROADCAST,MULTICAST,UP,LOWER_UP> mtu 1500 qdisc mq state UP qlen 1000
    link/ether fa:78:3e:72:e7:87 brd ff:ff:ff:ff:ff:ff
    inet 192.168.0.16/24 brd 192.168.0.255 scope global eth0
       valid_lft forever preferred_lft forever
    inet 192.168.0.217/24 scope global secondary eth0
       valid_lft forever preferred_lft forever
    inet6 fe80::4568:3eff:fe72:c787/64 scope link
       valid_lft forever preferred_lft forever
```

18.4.2 机房间容灾演练

业务通过内部域名 DNS 系统访问数据库时，通过修改域名指向的 IP 地址来实现机房切换，无需重启业务模块，大幅缩短切换时间（见表 10）。

(1) 在同城容灾数据库 C（192.169.0.23）上查看主从状态，延迟太大，不建议进行此操作。

```
    [xuliang@mysql03 ~]$ mysql -pxxxxxx -e "show slave status" 2>&1 |grep -v
" Using a password on the command line interface can be insecure."|grep -E "_
Running:|Seconds_Behind_Master|Master_Host|Error:";
```

(2) 设置主机房数据库 A（192.168.0.15）为只读。

```
[xuliang@mysql01 ~]$ mysql -pxxxxxx -e "set global read_only=1";
```

(3) 等待 10s，降低潜在大事务和主从延迟风险。

```
[xuliang@mysql01 ~]$ sleep 10
```

(4) 设置同城容灾数据库 C（192.169.0.23）为可写。

```
[xuliang@mysql03 ~]$ mysql -pxxxxxx -e "set global read_only=0;"
```

(5) 修改 DNS 域名指向的 IP 地址。域名由主机房 VIP（192.168.0.217）更改为同城机房（192.169.0.23），见表 10。

表 10

业务域名	原 IP 地址	目标 IP 地址
province.andlink-mysql.svc.cluster.local	192.168.0.217	192.169.0.23

(6) kill 主机房数据库 A（192.168.0.15）的数据库连接。

```
[xuliang@mysql01 ~]$ mysql -pxxxxxx -e "select id from information_schema.processlist where user not in (repl','system user') and (locate('information_schema.processlist',INFO) is null or locate('information_schema.processlist',INFO)=0);" 2>&1 |grep -v \" Using a password on the command line interface can be insecure.\"|grep -v id|awk '{print "kill "$1";"}'>/tmp/kill.sql
[xuliang@mysql01 ~]$ mysql -pxxxxxx -f < /tmp/kill.sql
```

(7) 查看切换后的流量情况。

```
[xuliang@mysql01 ~]$ mysql -pxxxxxx -e ""select id,user,host,db,command,time,state,substr(info,1,100) from information_schema.processlist  where user not in ('system user','repl','monitor','backup') and (info is null or info not like '%information_schema.processlist%') order by TIME limit 20;"
```

(8) 原库如果有残留流量，等待 10s，第二次执行第 6 步。

为了规避 Java 程序 DNS 缓存问题，可以多次执行 kill 操作。

18.5 总结

通过尝试融合多源复制和主主复制这两种技术，实现多通道主主复制，使得主库和备库、容灾机房之间分别建立主主复制关系，尽可能减少切换演练中的操作步骤。这种方式无需更改数据库主从拓扑，减少了网络、连接、权限设置、增量数据补偿等额外操作，降低了容灾演练出错的风险。同时，只需对主机房边界节点进行配置升级，运维较为简单，减少因误操作而导致数据不一致的可能性。

演练回切时间由传统的约 1 小时缩短到 5 分钟以内，效率大幅提升。这一改进不仅提升了系统的稳定性和可靠性，同时也让运维团队有机会专注于更加重要的任务，而不是陷入烦琐的手动操作之中。

02 MySQL 篇
—— 故障分析

在数据库管理的过程中，故障分析无疑是每位 DBA 的必备技能。本篇将通过一系列真实的现场故障案例，带领读者深入故障现场，一探究竟。我们的作者团队将从故障的背景出发，逐步复现和分析故障原因，并分享他们在定位和解决故障过程中的宝贵经验。这些案例不仅展示了一线 DBA 的真实工作状态，更是将相关知识点融会贯通，为读者提供了一场知识的盛宴。

我们精心挑选了 18 个最具代表性的案例，涵盖了 MySQL 的各个方面，从数据页损坏到权限丢失，从复制异常到性能瓶颈。每一个案例都是对 DBA 技能的一次挑战，也是对问题解决能力的一次提升。

1 生产环境遇到 MySQL 数据页损坏问题如何解决？

作者：徐文梁

1.1 问题背景

笔者遇到一次实例异常 crash 的问题，当时数据库自动重启，未对生产造成影响，因此未做处理，但还是记录了错误信息。错误日志中有如下信息：

```
InnoDB: End of page dump
InnoDB: Page may be an index page where index id is 8196
2023-04-11T07:57:42.508371+08:00 0 [ERROR] [FATAL] InnoDB: Apparent corruption of an index page [page id: space=3859, page number=842530] to be written to data file. We intentionally crash the server to prevent corrupt data from ending up in data files.
2023-04-11 07:57:42 0x7fe4d42cf080 InnoDB: Assertion failure in thread 140620788985984 in file ut0ut.cc line 921
InnoDB: We intentionally generate a memory trap.
```

因为当时自动恢复了，所以笔者并未重视这个问题，但两个月后实例又 crash 了。查看报错信息，报错信息如下：

```
2023-06-23T04:32:36.538380+08:00 0 [ERROR] InnoDB: Probable data corruption on page 673268. Original record on that page;
(compact record)2023-06-23T04:32:36.538426+08:00 0 [ERROR] InnoDB: Cannot find the dir slot for this record on that page;
(compact record)2023-06-23 04:32:36 0x7fe2bf68f080 InnoDB: Assertion failure in thread 140611850662016 in file page0page.cc line 153
InnoDB: We intentionally generate a memory trap.
```

两次的报错信息很相似，出现一次是偶然，出现两次就值得重视了。虽然之前很幸运，未对生产造成影响，但是如果后面哪一天异常导致实例无法启动，那不就是生产故障吗？作为 DBA 要有忧患意识，必须提前准备好应对之策。针对此类问题，该如何排查以及解决？通过查阅资料和向前辈请教，也算有所收获，想着如果有其他同学遇到类似问题也可作为参考，于是有了此文。

1.2 问题分析

一般来说，数据页损坏，错误日志中都会显示具体的 page number，其他情况暂不考虑。在此前提下，根据实例状态可以将数据页损坏分为两种场景：实例能正常启动和实例无法正常启动。场景不同，处理方法也略有不同，下面分别展开详细分析。

1.2.1 场景一：实例能正常启动

此时通过错误日志中的信息，可以查询元数据表获取数据页所属信息。考虑生产环境信息安全，在测试环境中建立测试表进行展示。

测试环境表结构如下：

```
mysql> use test;
Reading table information for completion of table and column names
You can turn off this feature to get a quicker startup with -A

Database changed
mysql> show create table t_user\G;
*************************** 1. row ***************************
Table: t_user
Create Table: CREATE TABLE `t_user` (
 `id` bigint(20) NOT NULL AUTO_INCREMENT,
 `name` varchar(255) DEFAULT NULL,
 `age` tinyint(4) DEFAULT NULL,
 `create_time` datetime DEFAULT NULL,
 `update_time` datetime DEFAULT NULL,
 PRIMARY KEY (`id`),
 KEY `idx_name` (`name`)
) ENGINE=InnoDB AUTO_INCREMENT=178120 DEFAULT CHARSET=utf8
1 row in set (0.00 sec)

ERROR:
No query specified
```

根据错误信息中提示的 page number 信息来查看数据页信息，查询方式如下：

```
mysql> use information_schema;
Reading table information for completion of table and column names
You can turn off this feature to get a quicker startup with -A

Database changed
mysql> select * from  INNODB_BUFFER_PAGE where PAGE_NUMBER=1156 limit 10;
+---------+--------+--------+-------------+----------+------------+--------
----+------------+----------------+--------------+--------------+---------
--------+------------+-----------------+
```

```
        | POOL_ID | BLOCK_ID | SPACE   | PAGE_NUMBER | PAGE_TYPE | FLUSH_TYPE | FIX_
COUNT | IS_HASHED | NEWEST_MODIFICATION | OLDEST_MODIFICATION  | ACCESS_TIME  |
TABLE_NAME           | INDEX_NAME  | NUMBER_RECORDS | DATA_SIZE | COMPRESSED_SIZE |
PAGE_STATE | IO_FIX  | IS_OLD | FREE_PAGE_CLOCK |
        +---------+----------+---------+-------------+-----------+------------+-------
----+-----------+---------------------+----------------------+--------------+--------
---------+-------------+----------------+-----------+-----------------+
------------+---------+--------+-----------------+
        |       0 |       64 |     126 |        1156 | INDEX     |          0
|       0 | NO        |                   0 |                    0 |            0
| `test`.`t_user` | idx_name    |            515 |     15965 |               0 |
FILE_PAGE  | IO_NONE | NO     |               0 |
        +---------+----------+---------+-------------+-----------+------------+-------
----+-----------+---------------------+----------------------+--------------+--------
---------+-------------+----------------+-----------+-----------------+
------------+---------+--------+-----------------+
        1 row in set (0.18 sec)
```

查询 INNODB_BUFFER_PAGE 系统表会对性能有影响，因此不建议随意在生产环境执行。

另外，如果错误日志中有 space id 和 index id 相关信息，则也可以通过如下方式（涉及 INNODB_SYS_INDEXES 和 INNODB_SYS_TABLES 系统表）进行查询：

```
    mysql> select b.INDEX_ID, a.NAME as TableName, a.SPACE as Space,b.NAME as
IndexName from INNODB_SYS_TABLES a,INNODB_SYS_INDEXES b where a.SPACE =b.SPACE and
a.SPACE=126 and b.INDEX_ID=225;
    +----------+-------------+-------+-----------+
    | INDEX_ID | TableName   | Space | IndexName |
    +----------+-------------+-------+-----------+
    |      225 | test/t_user |   126 | idx_name  |
    +----------+-------------+-------+-----------+
    1 row in set (0.01 sec)
```

根据上面的查询结果，确定损坏的页是属于主键还是辅助索引，如果属于主键索引，因为在 MySQL 中索引即数据，则可能会导致数据丢失；如果是辅助索引，删除索引重建即可。

1.2.2 场景二：实例无法正常启动

此时可以通过两种方式尝试拉起实例。

1.2.2.1 方法一

使用 innodb_force_recovery 参数强制拉起 MySQL 实例。

正常情况下，innodb_force_force_recovery 值应该设置为 0。在紧急情况下，当实例无法正常启动时，可以尝试将其设置为大于 0 的值，强制拉起实例，然后将数据逻辑备份导出进行恢复。innodb_force_recovery 值最大可设置为 6，但是值为 4 或更大可能会永久损坏数据文件。因此当强制 InnoDB 恢复时，应始终以 innodb_force_recovery=1 开头，并仅在必要时递增该值。

1.2.2.2 方法二

使用 inno_space 工具进行数据文件修复。

inno_space 是一个可以直接访问 InnoDB 内部文件的命令行工具，可以通过该工具查看 MySQL 数据文件的具体结构，修复 corrupt page。

如果 InnoDB 表文件中的 page 损坏，导致实例无法启动，可以尝试通过该工具进行修复，如果损坏的只是 leaf page，inno_space 可以将 corrupt page 跳过，从而保证实例能够启动，并且将绝大部分的数据找回。以下为示例：

```
# 假设 MySQL 错误日志中有类似报错
  [ERROR] [MY-030043] [InnoDB] InnoDB: Corrupt page resides in file: .test/t_user.ibd, offset: 163840, len: 16384
  [ERROR] [MY-011906] [InnoDB] Database page corruption on disk or a failed file read of page [page id: space=126, page number=1158]. You may have to recover from a backup.

# 通过如下方式进行修复：
# 删除损坏的数据页中损坏部分
./inno -f /opt/mysql/data/3307/test/t_user.ibd   -d 10

# 更新损坏的数据页中 checksum 值
./inno -f /opt/mysql/data/3307/test/t_user.ibd   -u 10
```

启动 MySQL 服务。

1.3 问题总结

通过前面分析，我们了解了数据页损坏场景的处理方式。哪怕在极端场景下，也可以做到从容不迫，尽可能少丢数据甚至不丢数据。但是如果是生产环境，尤其是金融行业，是无法容忍丢失一条数据的，可能这一条数据就涉及几个"小目标"呢。因此，重要的事情说三遍，一定要备份！一定要备份！一定要备份！

2 一则 MySQL 从节点无响应问题分析

作者：李锡超

近期，我发现一个 MySQL 从节点提示同步异常，执行 show replica status 时系统被挂起。相关重要信息如表 1 所示。

表 1

MySQL 版本	8.0.27
架构	主从模式
binlog_transaction_dependency_tracking	WRITESET
replica_parallel_workers	16

2.1 初步分析

2.1.1 连接情况

连接情况如图 1 所示。

Id	User	Host	db	Command	Time	State	Info
5	event_scheduler	localhost	NULL	Daemon	7240932	Waiting on empty queue	NULL
73	system user	connecting host	NULL	Connect	7239045	Queueing source event to the relay log	NULL
74	system user		NULL	Query	138020	Waiting for replica workers to process their queues	NULL
75	system user		zabbixdb	Query	137272	Applying batch of row changes (update)	NULL
76	system user		zabbixdb	Query	137273	Applying batch of row changes (delete)	NULL
77	system user		NULL	Query	137272	Waiting for preceding transaction to commit	NULL
78	system user		NULL	Query	137272	Waiting for preceding transaction to commit	NULL
79	system user		NULL	Query	137272	Waiting for preceding transaction to commit	NULL
80	system user		zabbixdb	Query	137272	Applying batch of row changes (write)	NULL
81	system user		zabbixdb	Query	137272	Applying batch of row changes (delete)	NULL
82	system user		zabbixdb	Query	137272	Applying batch of row changes (write)	NULL
83	system user		zabbixdb	Query	137273	Applying batch of row changes (delete)	NULL
84	system user		zabbixdb	Query	137273	Applying batch of row changes (delete)	NULL
85	system user		NULL	Query	137273	Executing event	NULL
86	system user		NULL	Query	137273	Waiting for preceding transaction to commit	NULL
87	system user		zabbixdb	Query	137273	Applying batch of row changes (delete)	NULL
88	system user		zabbixdb	Query	137273	Applying batch of row changes (delete)	NULL
89	system user		zabbixdb	Query	137272	Applying batch of row changes (delete)	NULL
90	system user		zabbixdb	Query	137272	Applying batch of row changes (delete)	NULL
256793	monitor	127.0.0.1:48204	performance_schema	Query	137990	starting	show replica status
256795	root	127.0.0.1:52188	NULL	Query	95432	starting	flush logs
256797	root	127.0.0.1:55242	NULL	Query	62930	starting	flush logs
256799	root	127.0.0.1:55954	NULL	Query	55549	starting	flush logs
256801	root	127.0.0.1:32890	zabbixdb	Query	324	starting	show slave status
256808	root	127.0.0.1:32910	zabbixdb	Query	238	starting	show slave status
256811	root	127.0.0.1:32922	zabbixdb	Sleep	171		NULL
256812	root	127.0.0.1:32924	NULL	Query	173	starting	show replica status
256817	root	127.0.0.1:32946	NULL	Sleep	58		NULL
256818	root	127.0.0.1:32952	zabbixdb	Query	0	init	show processlist

图 1

当前连接中运行着 16 个 worker 线程。其中：

(1)4 个状态为 Waiting for preceding transaction to commit。

(2)11 个状态为 Applying batch of row changes。

(3) 1 个状态为 Executing event。

线程等待时间为 137272 秒（38 小时左右）。同时，看到执行的 show replica status、flush logs 命令都被挂起。

2.1.2 InnoDB Status 输出

InnoDB Status 输出如图 2 所示。

```
------------------
ROW OPERATIONS
------------------
0 queries inside InnoDB, 0 queries in queue
0 read views open inside InnoDB
Process ID=106175, Main thread ID=140079150397184 , state=doing insert buffer merge
Number of rows inserted 22762046787, updated 1355386122, deleted 22658498269, read 28846531036
0.00 inserts/s, 0.00 updates/s, 0.00 deletes/s, 0.00 reads/s
Number of system rows inserted 62377072, updated 6978087504, deleted 62377020, read 7041786024
0.00 inserts/s, 0.00 updates/s, 0.00 deletes/s, 0.00 reads/s
------------------

------------------
FILE I/O
------------------
I/O thread 0 state: waiting for completed aio requests (insert buffer thread)
I/O thread 1 state: waiting for completed aio requests (log thread)
I/O thread 2 state: waiting for completed aio requests (read thread)
I/O thread 3 state: waiting for completed aio requests (read thread)
I/O thread 4 state: waiting for completed aio requests (read thread)
I/O thread 5 state: waiting for completed aio requests (read thread)
I/O thread 6 state: waiting for completed aio requests (read thread)
I/O thread 7 state: waiting for completed aio requests (read thread)
I/O thread 8 state: waiting for completed aio requests (read thread)
I/O thread 9 state: waiting for completed aio requests (read thread)
I/O thread 10 state: waiting for completed aio requests (write thread)
I/O thread 11 state: waiting for completed aio requests (write thread)
I/O thread 12 state: waiting for completed aio requests (write thread)
I/O thread 13 state: waiting for completed aio requests (write thread)
I/O thread 14 state: waiting for completed aio requests (write thread)
I/O thread 15 state: waiting for completed aio requests (write thread)
I/O thread 16 state: waiting for completed aio requests (write thread)
I/O thread 17 state: waiting for completed aio requests (write thread)
Pending normal aio reads: [0, 0, 0, 0, 0, 0, 0, 0] , aio writes: [0, 0, 0, 0, 0, 0, 0, 0] ,
 ibuf aio reads:, log i/o's:, sync i/o's:
Pending flushes (fsync) log: 0; buffer pool: 75869
280107507 OS file reads, 12940403442 OS file writes, 1429178917 OS fsyncs
0.00 reads/s, 0 avg bytes/read, 0.00 writes/s, 0.00 fsyncs/s
------------------
```

图 2

根据 Innodb Status 输出结果，检查 ROW OPERATIONS、FILE I/O，未见几个典型的问题，比如 innodb_thread_concurrency 配置不合理导致无法进入 InnoDB。

2.1.3 负载情况

负载情况如图 3 所示。

```
top - 17:30:46 up 505 days, 19:44,  3 users,  load average: 2.06, 2.06, 2.05
Threads:  96 total,    2 running,   94 sleeping,    0 stopped,    0 zombie
%Cpu(s):  7.8 us,  5.1 sy,  0.0 ni, 87.1 id,  0.0 wa,  0.0 hi,  0.0 si,  0.0 st
KiB Mem :  80.9/32765892 [
KiB Swap:  37.4/8388604  [

   PID USER      PR  NI    VIRT    RES    SHR S  %CPU %MEM     TIME+ COMMAND
106175 mysql     20   0 33.985g 0.024t   8564 S   0.0 77.8   0:09.75 mysqld
106181 mysql     20   0 33.985g 0.024t   8564 S   0.0 77.8   3:04.55 ib_io_ibuf
106182 mysql     20   0 33.985g 0.024t   8564 S   0.0 77.8   3:02.91 ib_io_log
106183 mysql     20   0 33.985g 0.024t   8564 S   0.0 77.8 195:01.07 ib_io_rd-1
106184 mysql     20   0 33.985g 0.024t   8564 S   0.0 77.8 194:36.37 ib_io_rd-2
106185 mysql     20   0 33.985g 0.024t   8564 S   0.0 77.8 195:08.55 ib_io_rd-3
106186 mysql     20   0 33.985g 0.024t   8564 S   0.0 77.8 194:37.77 ib_io_rd-4
106187 mysql     20   0 33.985g 0.024t   8564 S   0.0 77.8 195:02.27 ib_io_rd-5
106188 mysql     20   0 33.985g 0.024t   8564 S   0.0 77.8 194:58.94 ib_io_rd-6
106189 mysql     20   0 33.985g 0.024t   8564 S   0.0 77.8 194:54.11 ib_io_rd-7
106190 mysql     20   0 33.985g 0.024t   8564 S   0.0 77.8 194:51.15 ib_io_rd-8
106191 mysql     20   0 33.985g 0.024t   8564 S   0.0 77.8 168:47.09 ib_io_wr-1
106192 mysql     20   0 33.985g 0.024t   8564 S   0.0 77.8 168:23.63 ib_io_wr-2
106193 mysql     20   0 33.985g 0.024t   8564 S   0.0 77.8 167:31.07 ib_io_wr-3
106194 mysql     20   0 33.985g 0.024t   8564 S   0.0 77.8 166:48.45 ib_io_wr-4
106195 mysql     20   0 33.985g 0.024t   8564 S   0.0 77.8 167:10.82 ib_io_wr-5
106196 mysql     20   0 33.985g 0.024t   8564 S   0.0 77.8 166:27.39 ib_io_wr-6
106197 mysql     20   0 33.985g 0.024t   8564 S   0.0 77.8 166:03.80 ib_io_wr-7
106198 mysql     20   0 33.985g 0.024t   8564 S   0.0 77.8 165:27.43 ib_io_wr-8
106199 mysql     20   0 33.985g 0.024t   8564 S   0.0 77.8   1182:50 ib_pg_flush_co
106200 mysql     20   0 33.985g 0.024t   8564 S   0.0 77.8   1155:04 ib_pg_flush-1
106201 mysql     20   0 33.985g 0.024t   8564 S   0.0 77.8   1154:15 ib_pg_flush-2
106202 mysql     20   0 33.985g 0.024t   8564 S   0.0 77.8   1154:30 ib_pg_flush-3
106233 mysql     20   0 33.985g 0.024t   8564 R  99.9 77.8   2771:21 ib_log_checkpt
106234 mysql     20   0 33.985g 0.024t   8564 S   0.0 77.8  71:05.95 ib_log_fl_notif
106235 mysql     20   0 33.985g 0.024t   8564 S   0.0 77.8  73:43.49 ib_log_flush
106236 mysql     20   0 33.985g 0.024t   8564 S   0.0 77.8 672:03.36 ib_log_wr_notif
106237 mysql     20   0 33.985g 0.024t   8564 S   0.0 77.8   1084:36 ib_log_writer
106242 mysql     20   0 33.985g 0.024t   8564 S   0.0 77.8   1:44.19 ib_srv_lock_to
106243 mysql     20   0 33.985g 0.024t   8564 S   0.0 77.8   2:01.46 ib_srv_err_mon
106244 mysql     20   0 33.985g 0.024t   8564 S   0.0 77.8   0:19.25 ib_srv_mon
106248 mysql     20   0 33.985g 0.024t   8564 S   0.0 77.8   0:00.00 ib_buf_resize
106249 mysql     20   0 33.985g 0.024t   8564 S   0.0 77.8  72:13.59 ib_src_main
106250 mysql     20   0 33.985g 0.024t   8564 S   0.0 77.8   1:20.14 ib_dict_stats
106251 mysql     20   0 33.985g 0.024t   8564 S   0.0 77.8   0:22.61 ib_fts_opt
106252 mysql     20   0 33.985g 0.024t   8564 S   0.0 77.8   0:01.64 xpl_worker-1
106253 mysql     20   0 33.985g 0.024t   8564 S   0.0 77.8   0:01.66 xpl_worker-2
106254 mysql     20   0 33.985g 0.024t   8564 S   0.0 77.8   0:00.00 xpl_accept-1
106258 mysql     20   0 33.985g 0.024t   8564 S   0.0 77.8   0:00.00 ib_buf_dump
106259 mysql     20   0 33.985g 0.024t   8564 S   0.0 77.8 151:30.30 ib_clone_gtid
106260 mysql     20   0 33.985g 0.024t   8564 S  11.0 77.8   1699:39 ib_srv_purge
106261 mysql     20   0 33.985g 0.024t   8564 S   0.0 77.8   6485:06 ib_srv_wkr-1
106262 mysql     20   0 33.985g 0.024t   8564 S   0.0 77.8  871:23.00 ib_srv_wkr-2
```

图 3

检查系统负载、线程模式查看所有线程的状态，发现 MySQL 的 ib_log_checkpt 线程 CPU 使用率一直处于 100% 的状态，其他线程都处于空闲状态。

2.1.4 错误日志

错误日志如图 4 所示。

```
2023-10-24T03:06:58.438427+08:00 216128 [Warning] [MY-011975] [InnoDB] Waiting for prepared transaction to exit
2023-11-08T09:54:19.793393Z mysqld_safe Number of processes running now: 0
```

图 4

检查错误日志，未见相关的错误日志记录。

2.1.5 慢查询

慢查询日志见图 5。

```
Time                 Id Command    Argument
# Time: 2023-11-08T18:07:01.923132+08:00
# User@Host: monitor[monitor] @  [127.0.0.1]  Id:    38
# Query_time: 0.113738  Lock_time: 0.000000 Rows_sent: 0  Rows_examined: 0
use performance_schema;
SET timestamp=1699438021;
show replica status;
```

图 5

检查慢查询日志，未见相关的慢查询记录。

2.1.6 分析小结

我们知道当前 MySQL 版本存在一些已知缺陷，根据主从线程的状态，首先想到可能是由于多线程复制（MTS）的 Bug（#103636：MySQL 8.0 MTS 定时炸弹）导致。

但根据之前的案例，如果是该 Bug 导致，Bug 触发了这么久，worker 线程应该都是处于 Waiting for preceding transaction to commit 状态，与此处现象不相符。

2.2 结合源码进一步分析

结合源码对 MTS 相关的逻辑进行梳理，关键逻辑如下：

```
    // STEP-1：启动时，根据配置参数，分配 commit_order_mngr
    |-handle_slave_sql (./sql/rpl_replica.cc:6812)
      |-Commit_order_manager *commit_order_mngr = nullptr
      |-if (opt_replica_preserve_commit_order && !rli->is_parallel_exec() && rli-
>opt_replica_parallel_workers > 1)
        |-commit_order_mngr = new Commit_order_manager(rli->opt_replica_parallel_
workers)
      |-rli->set_commit_order_manager(commit_order_mngr) // 设置 commit_order_mngr
    // STEP-2：执行 event 时，分配提交请求的序列号
    |-handle_slave_sql (./sql/rpl_replica.cc:7076)
      |-exec_relay_log_event (./sql/rpl_replica.cc:4905)
        |-apply_event_and_update_pos (./sql/rpl_replica.cc:4388)
          |-Log_event::apply_event (./sql/log_event.cc:3307)
            |-Log_event::get_slave_worker (./sql/log_event.cc:2770)
              |-Mts_submode_logical_clock::get_least_occupied_worker (./sql/rpl_mta_submode.
cc:976)
                |-Commit_order_manager::register_trx (./sql/rpl_replica_commit_order_manager.
cc:67)
                  |-this->m_workers[worker->id].m_stage = cs::apply::Commit_order_queue::enum_
worker_stage::REGISTERED
                  |-this->m_workers.push(worker->id)
                  // 此节点的worker正在处理的提交请求的序列号，其取值为每次从m_commit_sequence_
generator(提交序列号计数器)加1后所得
```

```
    // m_commit_sequence_generator 应该是全局的
    // 每次 start slave 之后会重新初始化
    |-this->m_workers[index].m_commit_sequence_nr->store(this->m_commit_sequence_
generator->fetch_add(1))
    // 将 worker id 加入 m_commit_queue 的尾部
    |-this->m_commit_queue << index
    |-...
    |-handle_slave_worker (./sql/rpl_replica.cc:5891)
    |-slave_worker_exec_job_group (./sql/rpl_rli_pdb.cc:2549)
    |-Slave_worker::slave_worker_exec_event (./sql/rpl_rli_pdb.cc:1760)
    |-Xid_apply_log_event::do_apply_event_worker (./sql/log_event.cc:6179)
    |-Xid_log_event::do_commit (./sql/log_event.cc:6084)
    |-trans_commit (./sql/transaction.cc:246)
    |-ha_commit_trans (./sql/handler.cc:1765)
    |-MYSQL_BIN_LOG::commit (./sql/binlog.cc:8170)
    // STEP-3: 提交事务前，判断 worker 队列状态和位置，使用 MDL 锁，等待释放事务锁
    |-MYSQL_BIN_LOG::ordered_commit (./sql/binlog.cc:8749)
    // 在该函数中会控制，只有 slave 的 worker 线程由于持有 commit_order_manager，才会在此处进
行 mngr->wait
    |-Commit_order_manager::wait (./sql/rpl_replica_commit_order_manager.cc:375)
    |-if (has_commit_order_manager(thd)): // 确认有 commit_order_manager 对象，该对象
只要在开启并行符合和从库提交顺序一致后，才不为 nullptr
    |-Commit_order_manager::wait (./sql/rpl_replica_commit_order_manager.cc:153)
    |-Commit_order_manager::wait_on_graph (./sql/rpl_replica_commit_order_manager.
cc:71)
    |-this->m_workers[worker->id].m_stage = cs::apply::Commit_order_queue::enum_
worker_stage::FINISHED_APPLYING
    // Worker is first in the queue
    |-if (this->m_workers.front() != worker->id) {}
    // Worker is NOT first in the queue
    |-if (this->m_workers.front() != worker->id):
    |-this->m_workers[worker->id].m_stage = cs::apply::Commit_order_queue::enum_
worker_stage::REQUESTED_GRANT
    |-set_timespec(&abs_timeout, LONG_TIMEOUT); // Wait for a year
    // 等待 MDL 锁
    |-auto wait_status = worker_thd->mdl_context.m_wait.timed_wait(worker_thd,
&abs_timeout,...)
    // STEP-4: 提交事务后，自增获取下一个 worker、sequence_number，然后释放对应的锁。
    //         释放了之后，其他 worker 就会在 STEP-3 中获取到锁，否则就会一直在 STEP-3 等待
    |-MYSQL_BIN_LOG::ordered_commit (./sql/binlog.cc:8763)
    |-MYSQL_BIN_LOG::change_stage (./sql/binlog.cc:8483)
    |-Commit_stage_manager::enroll_for (./sql/rpl_commit_stage_manager.cc:209)
    |-Commit_order_manager::finish_one (./sql/rpl_replica_commit_order_manager.
cc:454)
    |-Commit_order_manager::finish_one (./sql/rpl_replica_commit_order_manager.
cc:300)
    // 获取 commit_queue 的第一个 worker 线程 :next_worker
    |-auto next_worker = this->m_workers.front()
```

```
        // 确认next_worker线程的状态,并比对m_commit_sequence_nr。
        // 如果满足,则设置mdl锁状态为GRANTED,即唤醒next_worker执行ordered_commit的后续操作
        |-if (this->m_workers[next_worker].m_stage in (FINISHED_APPLYING,REQUESTED_
GRANT) and
            /* 对比 next_worker 的 m_commit_sequence_nr 与 next_seq_nr:
            如果 m_commit_sequence_nr == next_seq_nr,则返回true,然后设置m_commit_sequence_
nr = SEQUENCE_NR_FROZEN(1)
            如果 m_commit_sequence_nr != next_seq_nr,则返回false
            */
            this->m_workers[next_worker].freeze_commit_sequence_nr(next_seq_nr))
            // 如果if条件满足,设置next_worker为授权状态
            |-this->m_workers[next_worker].m_mdl_context->m_wait.set_status(MDL_
wait::GRANTED)
            /* 对比 next_worker 的 m_commit_sequence_nr 与 next_seq_nr
            如果m_commit_sequence_nr == 1,则返回true,然后设置m_commit_sequence_nr = next_
seq_nr
            如果 m_commit_sequence_nr != 1,则返回false
            */
            |-this->m_workers[next_worker].unfreeze_commit_sequence_nr(next_seq_nr)
            |-this->m_workers[this_worker].m_mdl_context->m_wait.reset_status()
            // 标记当前状态为终态(FINISHED)
            |-this->m_workers[this_worker].m_stage = cs::apply::Commit_order_queue::enum_
worker_stage::FINISHED
```

为便于下文说明,对于MTS的16个worker线程,分类如下:

(1)A类:4个worker线程的状态为Waiting for preceding transaction to commit。

(2)B类:11个worker线程的状态为Applying batch of row changes。

(3)C类:1个worker线程的状态为Executing event。

2.2.1 论点1

假设是Bug 103636导致了该问题,那么worker线程应该会在执行提交操作的时候,在ordered_commit函数执行的开始位置,由于无法获取MDL锁而等待,通过show processlist查看worker线程的状态应该为Waiting for preceding transaction to commit。

而且通过debug验证发现,一旦该问题被触发,那么所有的worker(即使只有一个worker在执行事务),都会进入ordered_commit函数,由于无法获取MDL锁,而在相同的位置等待,即所有worker线程应该处于Waiting for preceding transaction to commit状态。

因此,初步判断该问题现象和Bug 103636不相符合。

2.2.2 论点2

从因果关系来看:如果bug 103636是问题的原因,结合论点1的分析,无法完全解释问题现象。

更可能是由于其他原因，导致 B 类和 C 类的 worker 线程无法顺利执行事务。这样一来，A 类 worker 线程在提交的时候，由于具有较小的 commit_sequence 事务的 worker 线程还未提交执行操作，因此其等待，导致其线程状态为 Waiting for preceding transaction tocommit。如此更能合理地解释当前数据库的状态。

那又是什么原因导致 B 类、C 类线程无法提交？根据采集的数据，我发现一个奇怪的症状：

在问题时段，ib_log_checkpt 的 CPU 使用率一直是处于 100% 的。

2. 3 根因分析

2.3.1 ib_log_checkpt 堆栈分析

根据 top 的线程 ID、堆栈以及 perf report 信息（见图 6），结合 log_checkpoint 的实现，我们看到该线程主要是执行 dirty page flush 操作。具体流程如下：

分析从 buf_pool_get_oldest_modification_approx 的实现，该过程一直在循环遍历，以找到 dirty page 的最小 LSN。然后，根据该 LSN 进行异步 IO 刷 dirty page 操作。

由于 buf_pool_get_oldest_modification_approx 一直在跑，猜测可能是因为异步 IO 慢，导致检查点无法完成，一直在寻找最小的 LSN。为此，需要进一步分析系统 IO 压力。

```
106233 mysql     20   0 33.985g 0.024t   8564 R 99.9 77.8   2771:21 ib_log_checkpt

Thread 74 (Thread 0x7f6700e38700 (LWP 106233)):
#0  0x00000000023fbb28 in log_checkpointer(log_t*) ()
#1  0x0000000002407ec0 in std::thread::_State_impl<std::thread::_Invoker<std::tuple<Detached_thread, void (*)(log_t*), log_t*>
>::_M_run() ()
#2  0x0000000002b33280 in execute_native_thread_routine ()
#3  0x00007f6d490fcdc5 in start_thread () from /lib64/libpthread.so.0
#4  0x00007f6d4779973d in clone () from /lib64/libc.so.6

Samples: 37K of event 'cpu-clock', Event count (approx.): 37429000000
  Children      Self  Command          Shared Object        Symbol
-   48.43%     0.00%  ib_log_checkpt   mysqld               [.] std::thread::_State_impl<std::thread::_Invoker<s
   + std::thread::_State_impl<std::thread::_Invoker<std::tuple<Detached_thread, void (*)(log_t*), log_t*> > >::
-   48.43%     0.00%  ib_log_checkpt   mysqld               [.] execute_native_thread_routine
     execute_native_thread_routine
   + std::thread::_State_impl<std::thread::_Invoker<std::tuple<Detached_thread, void (*)(log_t*), log_t*> > >::
-   47.89%     5.35%  ib_log_checkpt   mysqld               [.] log_checkpointer
   - 42.54% log_checkpointer
      - 19.72% log_sync_flush_lsn
         - 16.29% log_update_available_for_checkpoint_lsn
            - 13.17% buf_pool_get_oldest_modification_lwm
                 12.31% buf_pool_get_oldest_modification_approx
      - 16.63% log_update_available_for_checkpoint_lsn
         - 13.28% buf_pool_get_oldest_modification_lwm
            - 12.41% buf_pool_get_oldest_modification_approx
                 0.55% fsp_is_global_temporary
               0.51% fsp_is_system_temporary
        2.98% log_should_checkpoint
        1.53% pthread_mutex_unlock
        1.13% pthread_mutex_lock
   + 5.35% execute_native_thread_routine
```

图 6

2.3.2 系统 IO 负载情况

系统 IO 负载情况如图 7 所示。

```
|05:00:01 PM   dev8-176      0.00     0.00    0.00    0.00    0.00    0.00    0.00    0.00
|05:10:01 PM   dev8-0        0.47     0.00    5.79   12.31    0.00    0.97    0.71    0.03
|05:10:01 PM   dev8-16       0.00     0.00    0.00    0.00    0.00    0.00    0.00    0.00
|05:10:01 PM   dev8-32       0.00     0.00    0.00    0.00    0.00    0.00    0.00    0.00
|05:10:01 PM   dev11-0       0.00     0.00    0.00    0.00    0.00    0.00    0.00    0.00
|05:10:01 PM   dev253-0      0.10     0.00    0.80    8.00    0.00    1.28    1.00    0.01
|05:10:01 PM   dev253-1      0.00     0.00    0.00    0.00    0.00    0.00    0.00    0.00
|05:10:01 PM   dev253-2      0.00     0.00    0.00    0.00    0.00    0.00    0.00    0.00
|05:10:01 PM   dev253-3      0.00     0.00    0.00    0.00    0.00    0.00    0.00    0.00
|05:10:01 PM   dev253-4      0.00     0.00    0.00    0.00    0.00    0.00    0.00    0.00
|05:10:01 PM   dev253-5      0.27     0.00    2.18    8.20    0.00    1.33    0.46    0.01
|05:10:01 PM   dev253-6      0.39     0.00    3.16    8.07    0.00    0.91    0.29    0.01
|05:10:01 PM   dev8-48       0.00     0.00    0.00    0.00    0.00    0.00    0.00    0.00
|05:10:01 PM   dev8-64       0.00     0.00    0.00    0.00    0.00    0.00    0.00    0.00
|05:10:01 PM   dev8-80       0.00     0.00    0.00    0.00    0.00    0.00    0.00    0.00
|05:10:01 PM   dev8-96       0.00     0.00    0.00    0.00    0.00    0.00    0.00    0.00
|05:10:01 PM   dev8-112      0.00     0.00    0.00    0.00    0.00    0.00    0.00    0.00
|05:10:01 PM   dev8-128      0.04     0.00    0.35    8.00    0.00    1.42    0.54    0.00
|05:10:01 PM   dev8-144      0.00     0.00    0.00    0.00    0.00    0.00    0.00    0.00
|05:10:01 PM   dev8-160      0.00     0.00    0.00    0.00    0.00    0.00    0.00    0.00
|05:10:01 PM   dev8-176      0.00     0.00    0.00    0.00    0.00    0.00    0.00    0.00
|05:20:01 PM   dev8-0        0.48     0.00    5.93   12.45    0.00    2.66    0.84    0.04
|05:20:01 PM   dev8-16       0.00     0.00    0.00    0.00    0.00    0.00    0.00    0.00
|05:20:01 PM   dev8-32       0.00     0.00    0.00    0.00    0.00    0.00    0.00    0.00
|05:20:01 PM   dev11-0       0.00     0.00    0.00    0.00    0.00    0.00    0.00    0.00
|05:20:01 PM   dev253-0      0.10     0.00    0.80    8.00    0.00    0.83    0.78    0.01
|05:20:01 PM   dev253-1      0.00     0.00    0.00    0.00    0.00    0.00    0.00    0.00
|05:20:01 PM   dev253-2      0.00     0.00    0.00    0.00    0.00    0.00    0.00    0.00
|05:20:01 PM   dev253-3      0.00     0.00    0.00    0.00    0.00    0.00    0.00    0.00
|05:20:01 PM   dev253-4      0.00     0.00    0.00    0.00    0.00    0.00    0.00    0.00
|05:20:01 PM   dev253-5      0.28     0.00    2.27    8.14    0.00    1.16    0.46    0.01
|05:20:01 PM   dev253-6      0.40     0.00    3.24    8.03    0.00    3.16    0.49    0.02
|05:20:01 PM   dev8-48       0.00     0.00    0.00    0.00    0.00    0.00    0.00    0.00
|05:20:01 PM   dev8-64       0.00     0.00    0.00    0.00    0.00    0.00    0.00    0.00
|05:20:01 PM   dev8-80       0.00     0.00    0.00    0.00    0.00    0.00    0.00    0.00
|05:20:01 PM   dev8-96       0.00     0.00    0.00    0.00    0.00    0.00    0.00    0.00
|05:20:01 PM   dev8-112      0.00     0.00    0.00    0.00    0.00    0.00    0.00    0.00
|05:20:01 PM   dev8-128      0.05     0.00    0.37    8.00    0.00    0.36    0.21    0.00
|05:20:01 PM   dev8-144      0.00     0.00    0.00    0.00    0.00    0.00    0.00    0.00
|05:20:01 PM   dev8-160      0.00     0.00    0.00    0.00    0.00    0.00    0.00    0.00
|05:20:01 PM   dev8-176      0.00     0.00    0.00    0.00    0.00    0.00    0.00    0.00
|05:30:01 PM   dev8-0       18.66   456.10   14.78   25.24    0.02    1.34    0.09    0.17
```

图 7

根据以上 IO 负载情况，发现问题时段服务器 IO 压力分析与上述猜想不相符！那是因为什么原因导致该问题呢？难道是无法找到最小 LSN，或者寻找比较慢？

2.3.3 再次分析 InnoDB Status

根据以上分析结果，再次分析相关采集数据。发现 Innodb Status 输出包含如下信息：

```
---
LOG
---
Log sequence number              54848990703166
Log buffer assigned up to        54848990703166
Log buffer completed up to       54848990703166
Log written up to                54848990703166
Log flushed up to                54848990703166
Added dirty pages up to          54848990703166
Pages flushed up to              54846113541560
Last checkpoint at               54846113541560
5166640859 log i/o's done, 0.00 log i/o's/second
---------------------
```

当前 InnoDB 中的 Redo 日志使用量 = 最新 lsn - checkpoint lsn=54848990703166-54846113541560=2877161606=2.68（GB）。

而此时数据库的 InnoDB Redo Log 配置为：

```
innodb_log_file_size                           | 1073741824
innodb_log_files_in_group                      | 3
```

即总的日志大小 =1073741824×3=3（GB）。

根据 MySQL 官方说明，需要确保 Redo 日志使用量不超过 Redo 总大小的 75%，否则就会导致数据库出现性能问题。但在问题时段，我们的日志使用率 =2.68/3=89%，超过 Redo 日志使用率的建议值 14%。

2.4 问题总结与建议

2.4.1 问题总结

综合以上分析过程，导致此次故障的根本原因还是在于数据库的 Redo 配置参数过小，在问题时段从节点的压力下，Redo 的使用率过高，导致 InnoDB 无法完成检查点。这进一步导致从节点的 worker 线程在执行事务时，检查 Redo Log 是否存在有剩余 Log 文件时，而发生等待。当前 worker 线程执行事务挂起后，由于从节点采用 MTS，且 slave_preserve_commit_order=on，因此其他 worker 线程需要等待之前的事务提交，最终导致所有 worker 线程挂起。

2.4.2 解决建议

（1）增加 Redo Log 文件的大小。

（2）升级到 MySQL 8.0 的最新版本，解决 Bug 103636 等关键 Bug。

（3）增加 innodb_buffer_pool_size 内存大小。

2.4.3 针对复杂问题的数据采集建议

针对上述问题数据的采集，提供以下用于 MySQL 复杂问题的问题采集命令：

```
su - mysql
currdt=`date +%Y%m%d_%H%M%S`
echo "$currdt" > /tmp/currdt_tmp.txt
mkdir /tmp/diag_info_`hostname -i`_$currdt
cd /tmp/diag_info_`hostname -i`_$currdt

-- 1. mysql 进程负载
su -
cd /tmp/diag_info_`hostname -i`_`cat /tmp/currdt_tmp.txt`

ps -ef | grep -w mysqld
mpid=`pidof mysqld`
echo $mpid
```

```
-- b: 批量模式；  n: 制定采集测试；  d: 间隔时间；  H: 线程模式；  p: 指定进程号
echo $mpid
top -b -n 120 -d 1 -H -p $mpid > mysqld_top_`date +%Y%m%d_%H%M%S`.txt

-- 2. 信息采集步骤 --- 以下窗口，建议启动额外的窗口执行
mysql -uroot -h127.1 -p

tee var-1.txt
show global variables;

tee stat-1.txt
show global status;

tee proclist-1.txt
show full processlist\G
show full processlist;

tee slave_stat-1.txt
show slave status\G

tee threads-1.txt
select * from performance_schema.threads \G

tee innodb_trx-1.txt
select * from information_schema.innodb_trx \G

tee innodb_stat-1.txt
show engine innodb status\G

tee innodb_mutex-1.txt
SHOW ENGINE INNODB MUTEX;

-- 锁与等待信息
tee data_locks-1.txt

-- mysql8.0
select * from performance_schema.data_locks\G
select * from performance_schema.data_lock_waits\G

-- mysql5.7
select * from information_schema.innodb_lock_waits \G
select * from information_schema.innodb_locks\G

-- 3. 堆栈信息
su -
cd /tmp/diag_info_`hostname -i`_`cat /tmp/currdt_tmp.txt`

ps -ef | grep -w mysqld
```

```
mpid=`pidof mysqld`
echo $mpid

-- 堆栈信息
echo $mpid
pstack $mpid > mysqld_stack_`date +%Y%m%d_%H%M%S`.txt

-- 线程压力
echo $mpid

perf top
echo $mpid
perf top -p $mpid

perf record -a -g -F 1000 -p $mpid -o pdata_1.dat
perf report -i pdata_1.dat

-- 4. 等待30秒
SELECT SLEEP(60);

-- 5. 信息采集步骤 --- 以下窗口,建议启动额外的窗口执行
mysql -uroot -h127.1 -p

tee var-2.txt
show global variables;

tee stat-2.txt
show global status;

tee proclist-2.txt
show full processlist\G
show full processlist;

tee slave_stat-2.txt
show slave status\G

tee threads-2.txt
select * from performance_schema.threads \G

tee innodb_trx-2.txt
select * from information_schema.innodb_trx \G

tee innodb_stat-2.txt
show engine innodb status\G

tee innodb_mutex-2.txt
SHOW ENGINE INNODB MUTEX;
```

```
-- 锁与等待信息
tee data_locks-2.txt

-- mysql8.0
select * from performance_schema.data_locks\G
select * from performance_schema.data_lock_waits\G

-- mysql5.7
select * from information_schema.innodb_lock_waits \G
select * from information_schema.innodb_locks\G

-- 6. 堆栈信息
su -
cd /tmp/diag_info_`hostname -i`_`cat /tmp/currdt_tmp.txt`

ps -ef | grep -w mysqld
mpid=`pidof mysqld`
echo $mpid

-- 堆栈信息
echo $mpid
pstack $mpid > mysqld_stack_`date +%Y%m%d_%H%M%S`.txt

-- 线程压力
echo $mpid

perf top
echo $mpid
perf top -p $mpid

perf record -a -g -F 1000 -p $mpid -o pdata_2.dat
perf report -i pdata_2.dat


--- END --
```

3 如何通过 blktrace 排查磁盘异常？

作者：张昊

3.1 背景描述

客户反馈业务经常出现运行效率低的情况，希望我们从数据库层面进行排查。之前已经定位到磁盘问题，但是当时没有继续深入排查。

3.2 日志分析

3.2.1 慢日志分析

在 MySQL 慢日志中发现，慢日志是一批一批地被记录的，并不是实时记录。

```
[root@big hao]# less slow_back.log |grep "# Time: 2022-08-26T" | awk -F "." '{print $1}'|termsql -0 "select COL2 time,COUNT(COL2) sql_count from tbl group by COL2"
time|sql_count
2022-08-26T00:19:29|23
2022-08-26T00:34:58|14
2022-08-26T00:41:18|22
2022-08-26T00:56:41|20
...
...
2022-08-26T23:24:32|22
2022-08-26T23:32:53|19
2022-08-26T23:46:16|29
2022-08-26T23:54:41|28
```

3.2.2 错误日志分析

查看 mysql-error 日志发现，平均每 22 分钟左右会输出一次 page_cleaner 信息。

```
[root@dqynj142113 was]# grep "2022-08-26T" mysql-error.log |grep flush
2022-08-26T00:19:29.317569+08:00 0 [Note] [MY-011953] [InnoDB] Page cleaner took 10614ms to flush 9 and evict 0 pages
2022-08-26T00:41:18.053724+08:00 0 [Note] [MY-011953] [InnoDB] Page cleaner took 11653ms to flush 97 and evict 0 pages
2022-08-26T01:03:05.205966+08:00 0 [Note] [MY-
```

011953] [InnoDB] Page cleaner took 10077ms to flush 8 and evict 0 pages
　　2022-08-26T01:24:51.147325+08:00　0　[Note]　[MY-011953] [InnoDB] Page cleaner took 10871ms to flush 10 and evict 0 pages
......
......
　　2022-08-26T22:49:24.264033+08:00　0　[Note]　[MY-011953] [InnoDB] Page cleaner took 11051ms to flush 10 and evict 0 pages
　　2022-08-26T23:11:08.958039+08:00　0　[Note]　[MY-011953] [InnoDB] Page cleaner took 9625ms to flush 8 and evict 0 pages
　　2022-08-26T23:32:53.922992+08:00　0　[Note]　[MY-011953] [InnoDB] Page cleaner took 9886ms to flush 8 and evict 0 pages
　　2022-08-26T23:54:41.143097+08:00　0　[Note]　[MY-011953] [InnoDB] Page cleaner took 12151ms to flush 19 and evict 0 pages

3.2.3 page_cleaner 分析

(1) page_cleaner 输出信息解释。

```
2022-08-26T23:54:41.143097+08:00 0 [Note] [MY-011953] [InnoDB] Page cleaner took 12151ms to flush 19 and evict 0 pages
```

(2) page cleaner 耗时 12151ms。

意义：耗时为 12.121s。

(3) flush 19 and evict 0 pages。

意义：InnoDB 尝试刷新 19 个脏页。

3.2.3.1 打印机制代码如图 8 所示。

```
3207            if (curr_time > next_loop_time + 3000) {
3208                if (warn_count == 0) {
3209                    ib::info() << "page_cleaner: 1000ms"
3210                                " intended loop took "
3211                             << 1000 + curr_time
3212                                - next_loop_time
3213                             << "ms. The settings might not"
3214                                " be optimal. (flushed="
3215                             << n_flushed_last
3216                             << " and evicted="
3217                             << n_evicted
3218                             << ", during the time.)";
3219                    if (warn_interval > 300) {
3220                        warn_interval = 600;
3221                    } else {
3222                        warn_interval *= 2;
3223                    }
3224
3225                    warn_count = warn_interval;
3226                } else {
3227                    --warn_count;
3228                }
3229            } else {
3230                /* reset counter */
3231                warn_interval = 1;
3232                warn_count = 0;
3233            }
```

图 8

3.2.3.2 代码解释

每一轮刷脏时间都超过 4s 的情况下，第一轮刷脏会被记录在 mysql-error 日志中，第二轮、第三轮刷脏不会被记录，第四轮刷脏会被记录在 mysql-error 中。

3.2.3.3 分析

连续刷脏慢的情况下，两个 mysql-error 的 page_cleaner 记录之间最大可能相差 600 轮。

3.3 持续观测磁盘 IO

通过 iostat 命令看到磁盘确实会出现一段时间的 IO 异常（此时磁盘 IO 使用几乎为 0，但是磁盘使用率为 100%）（见图 9）。

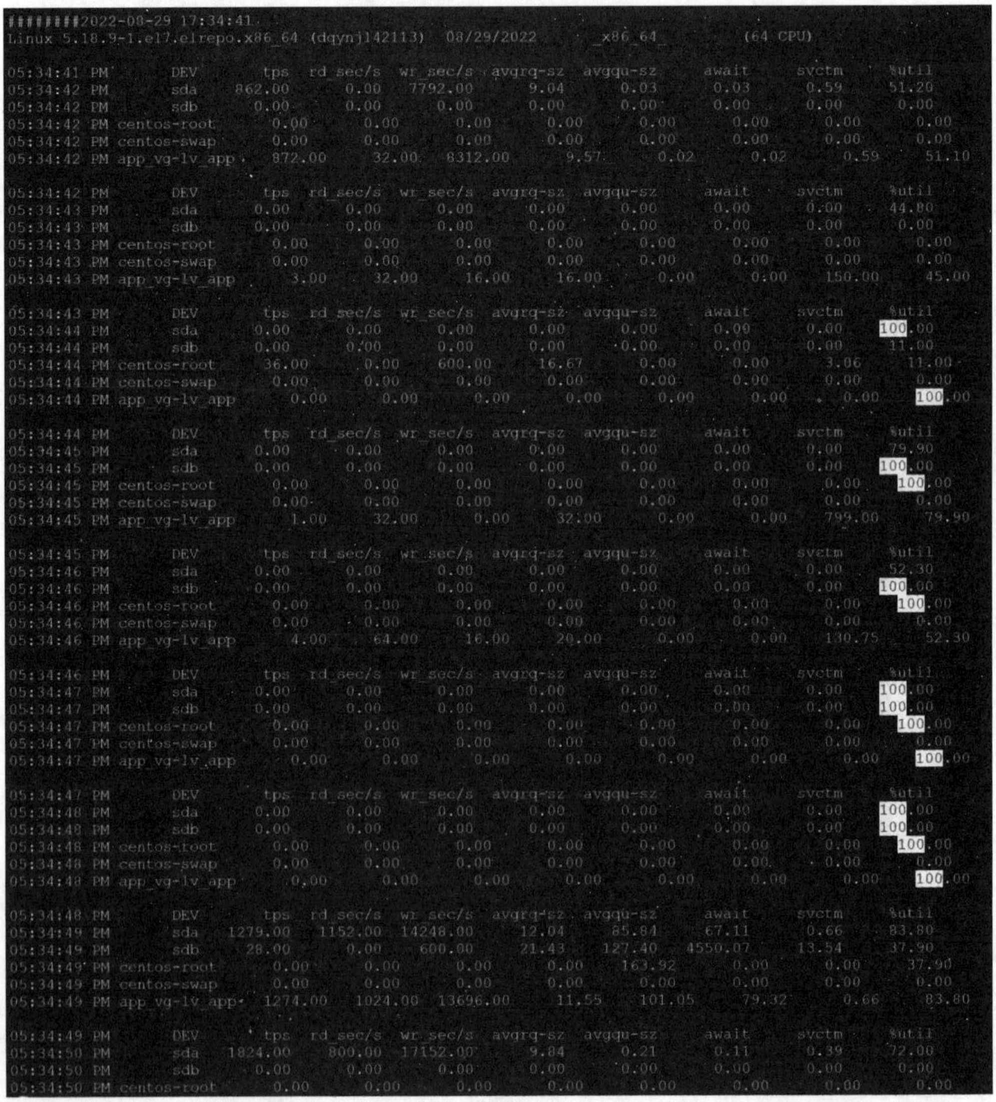

```
########2022-08-29 17:34:41
Linux 5.18.9-1.el7.elrepo.x86_64 (dqynj142113)  08/29/2022   _x86_64_   (64 CPU)

05:34:41 PM       DEV      tps  rd_sec/s  wr_sec/s  avgrq-sz  avgqu-sz     await     svctm     %util
05:34:42 PM       sda   862.00      0.00   7792.00      9.04      0.03      0.03      0.59     51.20
05:34:42 PM       sdb     0.00      0.00      0.00      0.00      0.00      0.00      0.00      0.00
05:34:42 PM centos-root      0.00      0.00      0.00      0.00      0.00      0.00      0.00      0.00
05:34:42 PM centos-swap      0.00      0.00      0.00      0.00      0.00      0.00      0.00      0.00
05:34:42 PM app_vg-lv_app    872.00     32.00   8312.00      9.57      0.02      0.02      0.59     51.10

05:34:42 PM       DEV      tps  rd_sec/s  wr_sec/s  avgrq-sz  avgqu-sz     await     svctm     %util
05:34:43 PM       sda     0.00      0.00      0.00      0.00      0.00      0.00      0.00     44.80
05:34:43 PM       sdb     0.00      0.00      0.00      0.00      0.00      0.00      0.00      0.00
05:34:43 PM centos-root      0.00      0.00      0.00      0.00      0.00      0.00      0.00      0.00
05:34:43 PM centos-swap      0.00      0.00      0.00      0.00      0.00      0.00      0.00      0.00
05:34:43 PM app_vg-lv_app      3.00     32.00     16.00     16.00      0.00      0.00    150.00     45.00

05:34:43 PM       DEV      tps  rd_sec/s  wr_sec/s  avgrq-sz  avgqu-sz     await     svctm     %util
05:34:44 PM       sda     0.00      0.00      0.00      0.00      0.00      0.09      0.00    100.00
05:34:44 PM       sdb     0.00      0.00      0.00      0.00      0.00      0.00      0.00     11.00
05:34:44 PM centos-root     36.00      0.00    600.00     16.67      0.00      0.00      3.06     11.00
05:34:44 PM centos-swap      0.00      0.00      0.00      0.00      0.00      0.00      0.00      0.00
05:34:44 PM app_vg-lv_app      0.00      0.00      0.00      0.00      0.00      0.00      0.00    100.00

05:34:44 PM       DEV      tps  rd_sec/s  wr_sec/s  avgrq-sz  avgqu-sz     await     svctm     %util
05:34:45 PM       sda     0.00      0.00      0.00      0.00      0.00      0.00      0.00     79.90
05:34:45 PM       sdb     0.00      0.00      0.00      0.00      0.00      0.00      0.00    100.00
05:34:45 PM centos-root      0.00      0.00      0.00      0.00      0.00      0.00      0.00    100.00
05:34:45 PM centos-swap      0.00      0.00      0.00      0.00      0.00      0.00      0.00      0.00
05:34:45 PM app_vg-lv_app      1.00     32.00      0.00     32.00      0.00      0.00    799.00     79.90

05:34:45 PM       DEV      tps  rd_sec/s  wr_sec/s  avgrq-sz  avgqu-sz     await     svctm     %util
05:34:46 PM       sda     0.00      0.00      0.00      0.00      0.00      0.00      0.00     52.30
05:34:46 PM       sdb     0.00      0.00      0.00      0.00      0.00      0.00      0.00    100.00
05:34:46 PM centos-root      0.00      0.00      0.00      0.00      0.00      0.00      0.00    100.00
05:34:46 PM centos-swap      0.00      0.00      0.00      0.00      0.00      0.00      0.00      0.00
05:34:46 PM app_vg-lv_app      4.00     64.00     16.00     20.00      0.00      0.00    130.75     52.30

05:34:46 PM       DEV      tps  rd_sec/s  wr_sec/s  avgrq-sz  avgqu-sz     await     svctm     %util
05:34:47 PM       sda     0.00      0.00      0.00      0.00      0.00      0.00      0.00    100.00
05:34:47 PM       sdb     0.00      0.00      0.00      0.00      0.00      0.00      0.00    100.00
05:34:47 PM centos-root      0.00      0.00      0.00      0.00      0.00      0.00      0.00    100.00
05:34:47 PM centos-swap      0.00      0.00      0.00      0.00      0.00      0.00      0.00      0.00
05:34:47 PM app_vg-lv_app      0.00      0.00      0.00      0.00      0.00      0.00      0.00    100.00

05:34:47 PM       DEV      tps  rd_sec/s  wr_sec/s  avgrq-sz  avgqu-sz     await     svctm     %util
05:34:48 PM       sda     0.00      0.00      0.00      0.00      0.00      0.00      0.00    100.00
05:34:48 PM       sdb     0.00      0.00      0.00      0.00      0.00      0.00      0.00    100.00
05:34:48 PM centos-root      0.00      0.00      0.00      0.00      0.00      0.00      0.00    100.00
05:34:48 PM centos-swap      0.00      0.00      0.00      0.00      0.00      0.00      0.00      0.00
05:34:48 PM app_vg-lv_app      0.00      0.00      0.00      0.00      0.00      0.00      0.00    100.00

05:34:48 PM       DEV      tps  rd_sec/s  wr_sec/s  avgrq-sz  avgqu-sz     await     svctm     %util
05:34:49 PM       sda  1279.00   1152.00  14248.00     12.04     85.84     67.11      0.66     83.80
05:34:49 PM       sdb    28.00      0.00    600.00     21.43    127.40   4550.07     13.54     37.90
05:34:49 PM centos-root      0.00      0.00      0.00      0.00    163.92      0.00      0.00     37.90
05:34:49 PM centos-swap      0.00      0.00      0.00      0.00      0.00      0.00      0.00      0.00
05:34:49 PM app_vg-lv_app   1274.00   1024.00  13696.00     11.55    101.05     79.32      0.66     83.80

05:34:49 PM       DEV      tps  rd_sec/s  wr_sec/s  avgrq-sz  avgqu-sz     await     svctm     %util
05:34:50 PM       sda  1824.00    800.00  17152.00      9.84      0.21      0.11      0.39     72.00
05:34:50 PM       sdb     0.00      0.00      0.00      0.00      0.00      0.00      0.00      0.00
05:34:50 PM centos-root      0.00      0.00      0.00      0.00      0.00      0.00      0.00      0.00
```

图 9

3.4 客户诉求

目前已知是磁盘 IO 的问题影响了日常业务运行，客户还是希望我们能协助继续排查一下是 IO 的哪个环节出现了问题。

3.5 blktrace 工具

3.5.1 工具简述

blktrace 工具可以更好地追踪 IO 的过程，统计一个 IO 是在调度队列停留的时间长还是在硬件上消耗的时间长，这个工具可以协助分析和优化问题。

3.5.2 工具使用

3.5.2.1 blktrace 采集命令

根据磁盘 IO 异常规律，使用 blktrace 工具采集磁盘异常期间 25s 的数据。

```
// blktrace 采集命令
blktrace -w 25 -d /dev/sda1  -o sda1$(date "+%Y%m%d%H%M%S")
```

3.5.2.2 统计分析

```
// blkparse 合并成一个二进制文件
blkparse -i sda120220915112930 -d blkpares2930.out


// btt 命令查看每个阶段的耗时
btt -i blkpares2930.out
```

不同阶段的 IO 解释如下：

Q2G：生成 IO 请求所消耗的时间，包括 remap 和 split 的时间。

G2I：IO 请求进入 IO Scheduler 所消耗的时间，包括 merge 的时间。

I2D：IO 请求在 IO Scheduler 中等待的时间，可以作为 IO Scheduler 性能的指标。

D2C：IO 请求在 Driver 和硬件上所消耗的时间，可以作为硬件性能的指标。

Q2C：整个 IO 请求所消耗的时间 (Q2G+G2I+I2D+D2C=Q2C)，相当于 iostat 的 await。

Q2Q：2 个 IO 请求的间隔时间。

从分析结果来看 IO 卡在了 D2C 阶段（见图 10）。

```
======================== All Devices ========================

          ALL           MIN           AVG           MAX            N
---------- ------------- ------------- ------------- -------------
Q2Q         0.000000060   0.005471628   3.251351051         4555
Q2A         0.000203429   3.791985163  13.836781063           10
Q2G         0.000000451   0.009956189  13.836266102         4559
G2I         0.000000177   0.000001926   0.000392639         4557
Q2M         0.000000518   0.000001625   0.000003476           23
I2D         0.000001576   0.000008036   0.000058495         4525
M2D         0.000009325   0.000061893   0.000315968           23
D2C         0.000019770   0.030146471  10.949388461         4548
Q2C         0.000023592   0.035966391  10.949398271         4548

==================== Device Overhead ====================

       DEV |       Q2G       G2I       Q2M       I2D       D2C
----------- | --------- --------- --------- --------- ---------
 (  8,  1) |  27.7489%   0.0054%   0.0000%   0.0222%  83.8184%
----------- | --------- --------- --------- --------- ---------
   Overall |  27.7489%   0.0054%   0.0000%   0.0222%  83.8184%

================= Device Merge Information =================

       DEV |        #Q        #D   Ratio |   BLKmin   BLKavg   BLKmax    Total
----------- | --------- --------- ------- | -------- -------- -------- --------
 (  8,  1) |      4559      4559    1.0  |        8       17      512    78928

============== Device Q2Q Seek Information ==============

       DEV |      NSEEKS           MEAN         MEDIAN | MODE
----------- | ---------- -------------- -------------- | ------
 (  8,  1) |       4556   3441144117.3              0 | 0(198)
----------- | ---------- -------------- -------------- | ------
   Overall |      NSEEKS           MEAN         MEDIAN | MODE
   Average |        4556   3441144117.3              0 | 0(198)

============== Device D2D Seek Information ==============

       DEV |      NSEEKS           MEAN         MEDIAN | MODE
----------- | ---------- -------------- -------------- | ------
 (  8,  1) |       4559   3489903833.1              0 | 0(195)
----------- | ---------- -------------- -------------- | ------
   Overall |      NSEEKS           MEAN         MEDIAN | MODE
   Average |        4559   3489903833.1              0 | 0(195)
```

图 10

3.5.2.3 通过 blkparse 命令分析

```
// blkparse 直接进行分析
blkparse -i sda120220915112930 |less

// 图 11 中第六 gq 列表示 IO 事件，代表 IO 请求到了哪一阶段
```

A：remap，对于栈式设备，进来的 IO 将被重新映射到 IO 栈中的具体设备。

X：split，对于做了 Raid 或进行了 device mapper(dm) 的设备，进来的 IO 可能需要切割，然后发送给不同的设备。

Q：queued，IO 进入 block layer，将要被 request 代码处理（即将生成 IO 请求）。

G：get request，IO 请求（request）生成，为 IO 分配一个 request 结构体。

M：back merge，若之前已经存在的 IO 请求的终止 block 号和该 IO 的起始 block 号一致，就会合并，也就是向后合并。

F：front merge，若之前已经存在的 IO 请求的起始 block 号和该 IO 的终止 block 号一致，就会合并，也就是向前合并。

I：inserted，IO 请求被插入到 IO scheduler 队列。

S: sleep，没有可用的request结构体，也就是IO满了，只能等待request结构体被释放。

P：plug，当一个 IO 入队一个空队列时，Linux 会锁住这个队列，不处理该 IO，这样做是为了等待一会，看有没有新的 IO 进来，以便合并。

U：unplug，当队列中已经有 IO request 时，会放开这个队列，准备向磁盘驱动发送该 IO。这个动作的触发条件是超时（plug 的时候，会设置超时时间），或者是有一些 IO 在队列中（多于 1 个 IO）。

D：issued，IO 将会被传送给磁盘驱动程序处理。

C：complete，IO 被磁盘处理完。

图 11 中第五列表示开始执行 blktrace 采集之后的时间（本次 blktrace 一共采集了 25s），第八列表示起始 block number；图 11 中被标记的这一片 IO，有不同时间发出的 D，都在同一时间完成 C。也就是说设备看上去是卡在某一个时间点后就突然活过来了，D 表示已经开始下发到 driver → RAID → Disk 路径，跟 OS 层已经没有关系了。

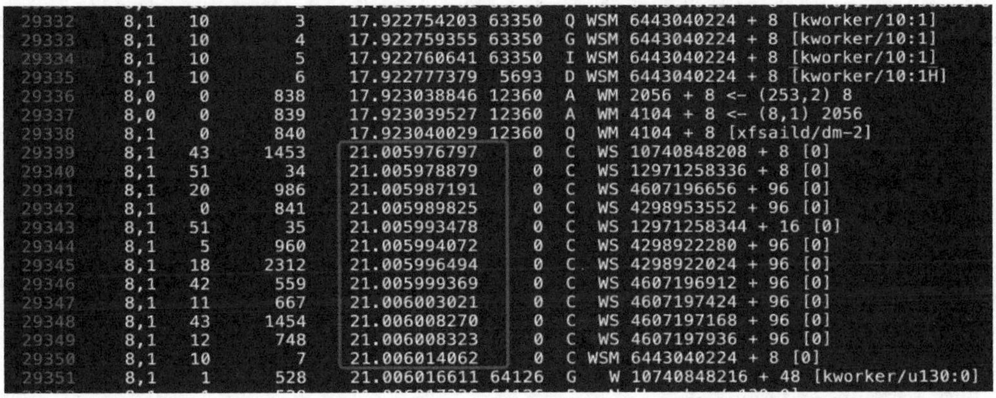

```
29332   8,1   10      3   17.922754203 63350  Q WSM 6443040224 + 8 [kworker/10:1]
29333   8,1   10      4   17.922759355 63350  G WSM 6443040224 + 8 [kworker/10:1]
29334   8,1   10      5   17.922760641 63350  I WSM 6443040224 + 8 [kworker/10:1]
29335   8,1   10      6   17.922777379  5693  D WSM 6443040224 + 8 [kworker/10:1H]
29336   8,0    0    838   17.923038846 12360  A  WM 2056 + 8 <- (253,2) 8
29337   8,0    0    839   17.923039527 12360  A  WM 4104 + 8 <- (8,1) 2056
29338   8,1    0    840   17.923040029 12360  Q  WM 4104 + 8 [xfsaild/dm-2]
29339   8,1   43   1453   21.005976797     0  C  WS 10740848208 + 8 [0]
29340   8,1   51     34   21.005978879     0  C  WS 12971258336 + 8 [0]
29341   8,1   20    986   21.005987191     0  C  WS 4607196656 + 96 [0]
29342   8,1    0    841   21.005989825     0  C  WS 4298953552 + 96 [0]
29343   8,1   51     35   21.005993478     0  C  WS 12971258344 + 16 [0]
29344   8,1    5    960   21.005994072     0  C  WS 4298922280 + 96 [0]
29345   8,1   18   2312   21.005996494     0  C  WS 4298922024 + 96 [0]
29346   8,1   42    559   21.005999369     0  C  WS 4607196912 + 96 [0]
29347   8,1   11    667   21.006003021     0  C  WS 4607197424 + 96 [0]
29348   8,1   43   1454   21.006008270     0  C  WS 4607197168 + 96 [0]
29349   8,1   12    748   21.006008323     0  C  WS 4607197936 + 96 [0]
29350   8,1   10      7   21.006014062     0  C WSM 6443040224 + 8 [0]
29351   8,1    1    528   21.006016611 64126  G   W 10740848216 + 48 [kworker/u130:0]
```

图 11

3.5.2.4 对比试验

客户环境使用 SATA 盘做的 RAID5（这里使用了 DELL 的 RAID 控制器固件），客户找了一台相同配置的机器，直接用 SATA 盘做数据盘，没有发现磁盘异常，因此初步定位故障点在 RAID 设备层面。

3.6 最终排查

3.6.1 问题原因

客户环境使用的 RAID 控制器固件版本为 25.5.6.0009，通过官网查询产品信息发现这个版本存在 Bug，会短暂影响磁盘 IO，修复版本为 25.5.8.0001（见图 12）。

图 12

3.6.2 问题解决

升级 RAID 控制器固件版本前，如图 13 所示。

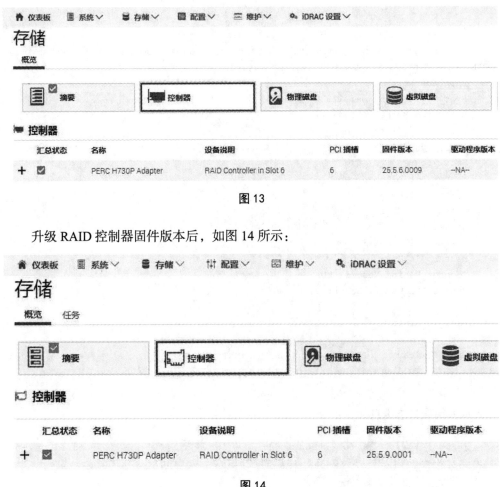

图 13

升级 RAID 控制器固件版本后，如图 14 所示：

图 14

4 MySQL 主从复制遇到 1590 报错

作者：王祥

4.1 故障描述

DMP（数据库管理平台）收到告警：从库的 SQL 线程停止工作。MySQL 版本为 5.7.32。登录到从库查看复制信息，报错如下：

```
mysql> show slave status\G
*************************** 1. row ***************************
Slave_IO_State: Waiting for master to send event
```

```
    ......
    Last_Errno: 1590
    Last_Error: The incident LOST_EVENTS occured on the master. Message: REVOKE/GRANT failed while granting/revoking privileges in databases.
    Skip_Counter: 0
    Exec_Master_Log_Pos: 12531
    Relay_Log_Space: 69304
    Until_Condition: None
    Until_Log_File:
    Until_Log_Pos: 0
    Master_SSL_Allowed: No
    Master_SSL_CA_File:
    Master_SSL_CA_Path:
    Master_SSL_Cert:
    Master_SSL_Cipher:
    Master_SSL_Key:
    Seconds_Behind_Master: NULL
    Master_SSL_Verify_Server_Cert: No
    Last_IO_Errno: 0
    Last_IO_Error:
    Last_SQL_Errno: 1590
    Last_SQL_Error: The incident LOST_EVENTS occured on the master. Message: REVOKE/GRANT failed while granting/revoking privileges in databases.
    ......
```

从库错误日志信息如下:

```
    [ERROR] Slave SQL for channel '': The incident LOST_EVENTS occured on the master. Message: REVOKE/GRANT failed while granting/revoking privileges in databases. Error_code: 1590

    [ERROR] Error running query, slave SQL thread aborted. Fix the problem, and restart the slave SQL thread with "SLAVE START". We stopped at log 'mysql-bin.000003' position 12531.
```

解析对应 Binlog 信息如下:

```
    # Incident: LOST_EVENTS
    RELOAD DATABASE; # Shall generate syntax error
```

主库错误信息如下:

```
    [ERROR] REVOKE/GRANT failed while granting/revoking privileges in databases. An incident event has been written to the binary log which will stop the slaves.
```

客户反馈在执行一些授权操作后复制就出现报错,执行的语句如下:

```
    mysql> create user test@'%',app@'%' identified by 'Root@123';
    ERROR 1819 (HY000): Your password does not satisfy the current policy
```

```
requirements
    mysql> grant all on test.* to test@'%',app@'%';
    ERROR 1819 (HY000): Your password does not satisfy the current policy
requirements
```

4.2 故障分析

根据以上报错信息可知，权限变更时发生了错误，主库在 binlog 中写一条 INCIDENT_EVENT，备库在解析到 INCIDENT_EVENT 就直接报错。

那在什么情况下执行授权语句会导致主库在 binlog 中写 INCIDENT_EVENT 呢？

当权限变更操作只处理了一部分就发生错误时，主库会在 binlog 中写一条 INCIDENT_EVENT。

那什么情况下会发生权限变更只处理了一部分而还有一部分没处理完呢？

下面举例说明两种相关场景。

4.2.1 MySQL 5.7 的问题

在 MySQL 5.7 里使用 GRANT 语句新建用户，其中有部分权限有问题。

使用 GRANT 创建 test 用户（MySQL 8.0 版本已经不支持使用 GRANT 创建用户）：

```
mysql> grant select,insert,file on test.* to test@'%' identified by 'Q1w2e3E$';
ERROR 1221 (HY000): Incorrect usage of DB GRANT and GLOBAL PRIVILEGES
mysql> select user,host from mysql.user where user='test' and host='%';
+------+------+
| user | host |
+------+------+
| test | %    |
+------+------+
1 row in set (0.00 sec)

mysql> show grants for test@'%';
+---------------------------------------+
| Grants for test@%                     |
+---------------------------------------+
| GRANT USAGE ON *.* TO 'test'@'%'      |
+---------------------------------------+
1 row in set (0.00 sec)
```

在创建用户时，对 test 库授予 SELECT、INSERT、FILE 权限，因 FILE 权限不能授予某个数据库而导致语句执行失败。最终结果是：用户 test@'%' 创建成功，授权部分失败。从上面的测试可知，使用 GRANT 创建用户其实分为两个步骤：创建用户和授权。权限有问题并不影响用户的创建，上述语句会导致主库在 binlog 中写 INCIDENT_EVENT，从而导致主从复制报错。

4.2.2 GRANT 对两个用户同时授权

使用一条 GRANT 语句，同时给 test@'10.186.63.5' 与 test@'10.186.63.29' 用户授权，其中 test@'10.186.63.5' 用户存在，而 test@'10.186.63.29' 不存在。

```
mysql> create user test@'10.186.63.5' identified by '123';
Query OK, 0 rows affected (0.00 sec)

mysql> grant all on test.* to test@'10.186.63.5',test@'10.186.63.29';
ERROR 1133 (42000): Can't find any matching row in the user table
mysql> show grants for test@'10.186.63.5';
+---------------------------------------------------------+
| Grants for test@10.186.63.5                             |
+---------------------------------------------------------+
| GRANT USAGE ON *.* TO 'test'@'10.186.63.5'              |
| GRANT ALL PRIVILEGES ON `test`.* TO 'test'@'10.186.63.5'|
+---------------------------------------------------------+
2 rows in set (0.00 sec)
```

根据上面的实验可知：test@'10.186.63.5' 用户存在，故授权成功；而 test@'10.186.63.29' 用户不存在，故授权失败。上述语句也会导致主库在 binlog 中写 INCIDENT_EVENT，从而导致主从复制报错。

但以上两种情况似乎都不符合客户执行语句的情况。从报错来看是因为密码复杂度不够而导致创建用户失败了，那到底是什么原因导致从库出现 1590 错误呢？下面我们来看看在使用了密码复杂度插件后使用 create 语句同时创建两个用户会有什么问题。

```
mysql> show global variables like '%validate%';
+--------------------------------------+--------+
| Variable_name                        | Value  |
+--------------------------------------+--------+
| query_cache_wlock_invalidate         | OFF    |
| validate_password_check_user_name    | OFF    |
| validate_password_dictionary_file    |        |
| validate_password_length             | 8      |
| validate_password_mixed_case_count   | 1      |
| validate_password_number_count       | 1      |
| validate_password_policy             | MEDIUM |
| validate_password_special_char_count | 1      |
+--------------------------------------+--------+

mysql> select user,host from mysql.user;
+---------------+-----------+
| user          | host      |
+---------------+-----------+
| universe_op   | %         |
| root          | 127.0.0.1 |
```

```
| mysql.session | localhost |
| mysql.sys     | localhost |
| root          | localhost |
+---------------+-----------+
5 rows in set (0.00 sec)

mysql> create user test@'%',app@'%' identified by 'Root@123';
ERROR 1819 (HY000): Your password does not satisfy the current policy requirements

mysql> select user,host from mysql.user; (app@'%' 创建成功, test@'%' 创建失败)
+---------------+-----------+
| user          | host      |
+---------------+-----------+
| app           | %         |
| universe_op   | %         |
| root          | 127.0.0.1 |
| mysql.session | localhost |
| mysql.sys     | localhost |
| root          | localhost |
+---------------+-----------+
6 rows in set (0.00 sec)
```

上述测试使用 CREATE USER 同时创建 test@'%'、app@'%'，但因为密码复杂度不符合要求而失败报错（多次测试后发现并不是密码复杂度不够，只要同时创建两个用户都会报错说密码复杂度不符合要求。在未使用密码复杂度插件时可以同时创建两个用户），正常来说这两个用户应该都会创建失败，但实际上 app@'%' 用户创建成功了。

到这里，我们就明白文章开头描述的故障触发原因：数据库实例开启了密码复杂度插件，使用 CREATE USER 同时创建两个用户，因为密码复杂度不符合要求而报错，但 app@'%' 用户已经创建，test@'%' 用户未创建，紧接着又执行了 GRANT 语句给两个用户同时授权，此时因为 test@'%' 用户不存在，而导致 GRANT 语句部分执行的问题，最终导致了主从复制报错。

4.3 故障解决

当主从复制出现 1590 报错时，可以先解析 binlog，找到 INCIDENT_EVENT（搜索关键字 LOST_EVENTS）和对应的 GTID，然后通过跳过这个 GTID 的方式来恢复主从复制。在跳过 GTID 之前还需要先将数据补全，因为主库有一个用户已经授权成功而从库这部分授权是没有执行的。具体操作如下（在从库执行）：

```
mysql>set global super_read_only=0;
```

```
mysql>set sql_log_bin=0;
mysql>grant all on test.* to test@'10.186.63.5';
```

跳过报错对应的 GTID，具体操作如下：

(1) 解析 binlog 找到 INCIDENT_EVENT（见图 15）。

```
mysqlbinlog --no-defaults -vv --base64-output=decode-rows mysql-bin.000016 > /root/bin.log
```

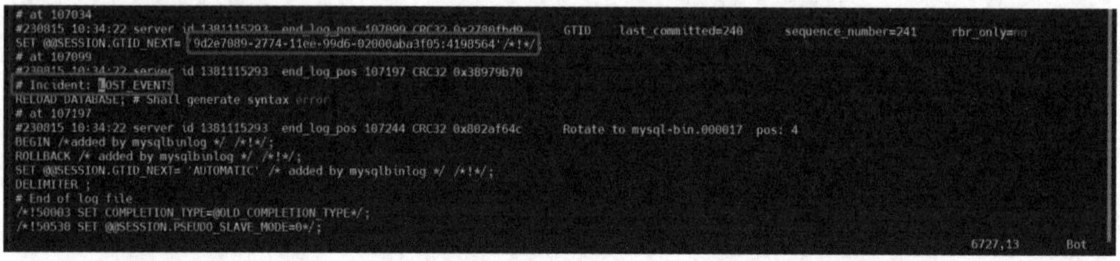

图 15

(2) 跳过 LOST_EVENTS 对应的 GTID（在从库执行）。

```
stop slave sql_thread;
set gtid_next='9d2e7089-2774-11ee-99d6-02000aba3f05:4198564';
begin;commit;
set gtid_next='automatic';
start slave sql_thread;
```

4.4 总结

(1) 权限变更操作只处理了一部分就发生错误时，会导致在 binlog 中写一条 INCIDENT_EVENT，从而导致主从复制报错。

(2) 在使用密码复杂度插件时，使用 CREATE 语句同时创建两个用户，会出现一个用户创建成功而另一个用户创建失败的情况。

4.5 建议

(1) 若使用了密码复杂度插件，在创建用户时一条 CREATE 语句只创建一个用户。

(2) 授权时一条 GRANT 语句只对一个用户授权，防止因权限错误导致部分授权成功的问题。

5 MySQL 的隐式转换导致诡异现象的案例一则

作者：刘晨

5.1 背景

同事问了个 MySQL 的问题，现象看起来确实诡异。具体情况是 SELECT 表的数据，WHERE 条件是 "a=0"，其中 a 字段是 VARCHAR 类型，该字段存在 NULL 以及包含字符的记录，但是并无 "0" 的记录，然而执行 SQL 后返回的记录恰恰就是所有包含中文字符的记录。

明明没有 "0" 值记录，却可以返回，而且有规律。这是什么现象？

```
select * from test where a = 0;
```

5.2 问题分析

为了对比说明，我们分别用 MySQL、Oracle 和 SQL Server 进行模拟。

5.2.1 准备测试表

三种数据库建表和插入数据的语句分别如下。

5.2.1.1 MySQL

```
create table test (id int, a varchar(3000), b varchar(2000));
insert into test values(1, '测试a', '测试b'),(2, NULL, '测试');
```

5.2.1.2 Oracle

```
create table test (id NUMBER(1), a varchar2(3000), b varchar2(2000));
insert into test values(1, '测试a', '测试b');
insert into test values(2, NULL, '测试');
```

5.2.1.3 SQL Server

```
create table test (id numeric(1,0), a varchar(3000), b varchar(2000));
insert into test values(1, '测试a', '测试b');
insert into test values(2, NULL, '测试');
```

5.2.2 对比查询结果

预期 test 表返回的记录都应该是表 2 这样的。

表 2

id	a	b
1	测试 a	测试 b
2	NULL	测试

我们看一下在三种数据库中，都执行如下语句，得到的是什么。

```
select * from test where a = 0;
```

5.2.2.1 MySQL

执行返回如下带字符的记录（见表 3），但实际逻辑上肯定是错的。

表 3

id	a	b
1	测试 a	测试 b

执行时，还会抛出以下警告：

```
warning:Truncated incorrect DOUBLE value: '测试a'
```

5.2.2.2 Oracle

执行直接报错，提示"无效数字"。因为 a 是 VARCHAR2，0 是数字，因此报错是针对字段 a 的，需要将 a 转成数字，但字符是无法转成数字的，所以提示"无效数字"是合情合理的。

```
ORA-01722: 无效数字
```

5.2.2.3 SQL Server

执行直接报错，但是提示信息更加清晰明了，说的就是字段 a 的值"测试 a"不能转成 INT 数值型。

```
SQL 错误 [245] [S0001]: 在将 varchar 值'测试a' 转换成数据类型 int 时失败。
```

5.2.2.4 小结

通过以上对比，可以知道 Oracle 和 SQL Server 对"字符型=数值型"的条件，会自动将字符型转成数值型，如果因为值的问题不能转成数值型，就会提示错误。而 SQL Server 给出的提示，比 Oracle 更具体。

相比之下，MySQL 针对"字符型=数值型"的条件，不仅能执行，而且执行是错的，这就很"拉垮"了。毕竟对产品来说，避免错误可能比表面上能执行更加重要，但就这个问题，Oracle 和 SQL Server 可以说更胜一筹。

5.2.3 问题分析

MySQL 为什么在这里会给出错误的结果？

从官方文档的这几段内容，我们可以得到一些线索（见图16）。

The following examples illustrate conversion of strings to numbers for comparison operations:

```
mysql> SELECT 1 > '6x';
        -> 0
mysql> SELECT 7 > '6x';
        -> 1
mysql> SELECT 0 > 'x6';
        -> 0
mysql> SELECT 0 = 'x6';
        -> 1
```

For comparisons of a string column with a number, MySQL cannot use an index on the column to look up the value quickly. If *str_col* is an indexed string column, the index cannot be used when performing the lookup in the following statement:

```
SELECT * FROM tbl_name WHERE str_col=1;
```

The reason for this is that there are many different strings that may convert to the value 1, such as '1', ' 1', or '1a'.

图 16

MySQL 中将 VARCHAR 转成 INT，会自动截断字符串，例如 "1 测试 a" 会被截成 "1"。通过如下例子，可以证明。

```
bisal@mysqldb 23:26:  [test]> select 1="1 测试 a";
+--------------+
| 1="1 测试 a" |
+--------------+
|            1 |
+--------------+
1 row in set, 1 warning (0.00 sec)
```

在上述例子中 "测试 a" 会被截成 ""，因此 a=0，返回字段才不为空。

```
bisal@mysqldb 23:27:  [test]> select 0=" 测试 a";
+--------------+
| 0=" 测试 a"  |
+--------------+
|            1 |
+--------------+
1 row in set, 1 warning (0.00 sec)
```

通过把 0 和 "" 进行比较，则可以进一步证明这个问题。

```
bisal@mysqldb 23:29:  [test]> select 0="";
+------+
| 0="" |
```

159

```
+------+
|  1   |
+------+
1 row in set (0.00 sec)
```

因此，正是因为 MySQL 对字符串进行隐式转换时会截断再转，而不是像 Oracle、SQL Server 这些数据库针对这种问题直接报错，才出现了这个诡异的问题。

5.3 总结

不知道这种设计是出于什么考虑，但这种"容错性"不可取，毕竟返回了错误的结果集。

当然，这个问题也和数据类型的使用有关，SQL 条件中"a=0"实际上是"varchar=int"。两边类型不一致，所以导致了数据库的隐式转换。

有可能是数据库设计的问题，比如，字段应该是 INT，但是定义成了 VARCHAR；还可能是开发人员的问题（SQL 条件右值应该用字符类型，例如"0"，但实际上用了 INT 数值类型的 0）。

总之，按照数据库设计开发规范的要求，"="号两边的数据类型保持一致，就不会引发数据库的隐式转换。

6 主从数据不一致竟然不报错？

作者：孙绪宗

背景须知：

(1) Retrieved_Gtid_Set：从库已经从主库接收到的事务编号（从库的 IO 线程已经接收到了）。

(2) Executed_Gtid_Set：从库已经执行的事务编号（从库的执行 SQL 线程已经执行了的 SQL）。

6.1 故障现象

主从数据不一致，但是复制是正常状态（双 Yes）。此时主库执行本应导致从库报错错误代码（1062 或者 1032）的 SQL，从库复制线程还是双 Yes 状态，没有报错。

MySQL 版本为 5.7.35。

6.2 故障复现

查看主库状态:

```
MySQL [xuzong]> show master status;
+------------------+----------+--------------+------------------+-------------------------------------------------+
| File             | Position | Binlog_Do_DB | Binlog_Ignore_DB | Executed_Gtid_Set                               |
+------------------+----------+--------------+------------------+-------------------------------------------------+
| mysql-bin.000008 | 39349641 |              |                  | c233aec0-58d3-11ec-a74a-a0a33ba8b3ff:1-104345   |
+------------------+----------+--------------+------------------+-------------------------------------------------+
```

在从库上只需要设置 GTID 值的 gno 大于主库的值即可复现。

```
mysql> stop slave;
Query OK, 0 rows affected (0.00 sec)

mysql> reset slave all;
Query OK, 0 rows affected (0.01 sec)

mysql> reset master;
Query OK, 0 rows affected (0.01 sec)

mysql> set @@GLOBAL.GTID_PURGED='c233aec0-58d3-11ec-a74a-a0a33ba8b3ff:1-1000000'; #1000000>104345
Query OK, 0 rows affected (0.00 sec)
```

然后用 POSITION MODE 点位复制,启动从库,执行 show slave status \G 就会看到如下信息:

```
Slave_IO_Running: Yes
Slave_SQL_Running: Yes
......
Last_IO_Error_Timestamp:
Last_SQL_Error_Timestamp:
Master_SSL_Crl:
Master_SSL_Crlpath:
Retrieved_Gtid_Set: c233aec0-58d3-11ec-a74a-a0a33ba8b3ff:104221-104345
Executed_Gtid_Set: c233aec0-58d3-11ec-a74a-a0a33ba8b3ff:1-1000000
Auto_Position: 0
Replicate_Rewrite_DB:
```

双 Yes,看上去一点问题都没有,但是这时在主库执行任何操作,都不会被复制,因为在复制时,从库校验 GTID 时显示这些操作已经执行过了。

主库：

```
MySQL [xuzong]> insert into test(passtext) values('test');
Query OK, 1 row affected (0.00 sec)
```

从库：

```
mysql> select * from xuzong.test;
+----+----------------------------------+
| id | passtext                         |
+----+----------------------------------+
|  1 | 7e5a44af63552be3f2f819cebbe0832a |
|  2 | 7e5a44af63552be3f2f819cebbe0832a |
|  3 | 7e5a44af63552be3f2f819cebbe0832a |
|  4 | d13baf7019431e0c75dee85bc923a91b |
|  5 | 11                               |
+----+----------------------------------+
5 rows in set (0.00 sec)

# 从库能看出来 Retrieved_Gtid_Set 在变化
mysql> show slave status \G
Retrieved_Gtid_Set: c233aec0-58d3-11ec-a74a-a0a33ba8b3ff:104221-104367
Executed_Gtid_Set: c233aec0-58d3-11ec-a74a-a0a33ba8b3ff:1-1000000
```

问题在于，这种情况不应该报错吗？从库的 GTID 大于主库的 GTID？

笔者猜测根源在于没有设置 Auto_Position=1，那么接下来验证一下这个猜测。

```
    Last_IO_Error: Got fatal error 1236 from master when reading data from binary
log: 'Slave has more GTIDs than the master has, using the master's SERVER_UUID.
This may indicate that the end of the binary log was truncated or that the last
binary log file was lost, e.g., after a power or disk failure when sync_binlog
!= 1. The master may or may not have rolled back transactions that were already
replicated to the slave. Suggest to replicate any transactions that master has
rolled back from slave to master, and/or commit empty transactions on master to
account for transactions that have been'
    Last_SQL_Errno: 0
    Last_SQL_Error:
    Replicate_Ignore_Server_Ids:
    Master_Server_Id: 1362229455
    Master_UUID: c233aec0-58d3-11ec-a74a-a0a33ba8b3ff
    Master_Info_File: mysql.slave_master_info
    SQL_Delay: 0
    SQL_Remaining_Delay: NULL
    Slave_SQL_Running_State: Slave has read all relay log; waiting for more
updates
    Master_Retry_Count: 86400
    Master_Bind:
    Last_IO_Error_Timestamp: 230817 11:11:44
```

```
Last_SQL_Error_Timestamp:
Master_SSL_Crl:
Master_SSL_Crlpath:
Retrieved_Gtid_Set:
Executed_Gtid_Set: c233aec0-58d3-11ec-a74a-a0a33ba8b3ff:1-1000000
Auto_Position: 1
```

果不其然报错了，猜测被验证了。

6.3 问题处理

这种情况目前看只能重做从库。因为案例中复制进程为双 Yes，无法被监控捕获，而且无法得知数据不一致出现的时间点，所以无法通过分析 binlog 恢复。

本案例中的实例还进行过主从切换，只能联系业务方做一次全量的数据对比，得到一个完整数据的实例，然后重新构建从库。

6.4 总结

最后总结一下 Auto_Position 的作用。

(1) 它会立即清理原来的 Relay Log。

(2) 它会根据从库的 Executed_Gtid_Set 和 Retrieved_Gtid_Set 的并集定位 Binlog，此时 MASTER_LOG_FILE 和 MASTER_LOG_POS 不再存储实例的值。

在生产环境中，如果实例开了 GTID，主从复制建议用 AUTO_POSITION MODE 的方式，即 Auto_Position=1。避免一些未知操作导致从库复制线程未产生预期的报错，进而导致从库未正确复制，数据丢失。

6.5 拓展：源码解读

```
rel_slave.cc
--> request_dump

# 部分源码：
......
enum_server_command command= mi->is_auto_position() ?
COM_BINLOG_DUMP_GTID : COM_BINLOG_DUMP;
......
if (command == COM_BINLOG_DUMP_GTID)
{
if (gtid_executed.add_gtid_set(mi->rli->get_gtid_set()) != RETURN_STATUS_OK ||
// 加 Retrieved_Gtid_Set
    gtid_executed.add_gtid_set(gtid_state->get_executed_gtids()) !=
RETURN_STATUS_OK) // 加 Executed_Gtid_Set
```

```
......
int2store(ptr_buffer, binlog_flags);
ptr_buffer+= ::BINLOG_FLAGS_INFO_SIZE;
int4store(ptr_buffer, server_id);
ptr_buffer+= ::BINLOG_SERVER_ID_INFO_SIZE;
int4store(ptr_buffer, static_cast<uint32>(BINLOG_NAME_INFO_SIZE));
ptr_buffer+= ::BINLOG_NAME_SIZE_INFO_SIZE;
memset(ptr_buffer, 0, BINLOG_NAME_INFO_SIZE); // 设置 MASTER_LOG_FILE 为 0
ptr_buffer+= BINLOG_NAME_INFO_SIZE;
int8store(ptr_buffer, 4LL); // 设置 MASTER_LOG_POS 为 4
ptr_buffer+= ::BINLOG_POS_INFO_SIZE; // 存 gtid_set
```

如果 relay_log_recovery 是打开的，则忽略 Retrieved_Gtid_Set 值。

```
/*
In the init_gtid_set below we pass the mi->transaction_parser.
This will be useful to ensure that we only add a GTID to
the Retrieved_Gtid_Set for fully retrieved transactions. Also, it will
be useful to ensure the Retrieved_Gtid_Set behavior when auto
positioning is disabled (we could have transactions spanning multiple
relay log files in this case).
We will skip this initialization if relay_log_recovery is set in order
to save time, as neither the GTIDs nor the transaction_parser state
would be useful when the relay log will be cleaned up later when calling
init_recovery.
*/ // 注释解释得很清楚，relay_log_recovery=1 会跳过初始化 gtid_retrieved_initialized
if (!is_relay_log_recovery &&
!gtid_retrieved_initialized &&
relay_log.init_gtid_sets(&gtid_set, NULL,
opt_slave_sql_verify_checksum,
true/*true=need lock*/,
&mi->transaction_parser, &gtid_partial_trx))
{
sql_print_error("Failed in init_gtid_sets() called from Relay_log_info::rli_init_info().");
DBUG_RETURN(1);
}
gtid_retrieved_initialized= true;
#ifndef NDEBUG
global_sid_lock->wrlock();
gtid_set.dbug_print("set of GTIDs in relay log after initialization");
global_sid_lock->unlock();
#endif
```

7 从一个死锁问题分析优化器特性

作者：李锡超

7.1 问题现象

自从发布了 INSERT 并发死锁问题的文章，收到了多次关于死锁问题的咨询。其中一个具体案例如下：研发人员反馈应用发生死锁，并收集了如下诊断内容。

```
------------------------
LATEST DETECTED DEADLOCK
------------------------
2023-07-04 06:02:40 0x7fc07dd0e700
*** (1) TRANSACTION:
TRANSACTION 182396268, ACTIVE 0 sec fetching rows
mysql tables in use 1, locked 1
LOCK WAIT 21 lock struct(s), heap size 3520, 2 row lock(s), undo log entries 1
MySQL thread id 59269692, OS thread handle 140471135803136, query id 373851495
3 192.168.0.215 user1 updating
delete from ltb2 where c = 'CCRSFD07E' and j = 'Y15' and b >= '20230717' and d != '1' and e != '1'
*** (1) WAITING FOR THIS LOCK TO BE GRANTED:
RECORD LOCKS space id 603 page no 86 n bits 248 index PRIMARY of table `testdb`.`ltb2` trx id 182396268 lock_mode X locks rec but not gap waiting
*** (2) TRANSACTION:
TRANSACTION 182396266, ACTIVE 0 sec fetching rows, thread declared inside InnoDB 1729
mysql tables in use 1, locked 1
28 lock struct(s), heap size 3520, 2 row lock(s), undo log entries 1
MySQL thread id 59261188, OS thread handle 140464721291008, query id 373851496
4 192.168.0.214 user1 updating
update ltb2 set f = '0', g = '0', is_value_date = '0', h = '0', i = '0' where c = '22115001B' and j = 'Y4' and b >= '20230717'
*** (2) HOLDS THE LOCK(S):
RECORD LOCKS space id 603 page no 86 n bits 248 index PRIMARY of table `testdb`.`ltb2` trx id 182396266 lock_mode X locks rec but not gap
*** (2) WAITING FOR THIS LOCK TO BE GRANTED:
RECORD LOCKS space id 603 page no 86 n bits 248 index PRIMARY of table `testdb`.`ltb2` trx id 182396266 lock_mode X locks rec but not gap waiting
*** WE ROLL BACK TRANSACTION (1)
------------
```

其中的 "space id 603 page no 86 n bits 248"，space id 表示表空间 ID；page no 表示

记录锁在表空间内的哪一页；n bits 是锁位图中的位数，而不是页面偏移量。记录的页偏移量一般以 heap no 的形式输出，但此例并未输出该信息。

7.1.1 基本环境信息

确认如下问题相关信息：

(1) 数据库版本：Percona MySQL 5.7。

(2) 事务隔离级别：Read-Commited。

(3) 表结构和索引如下。

```
CREATE TABLE `ltb2` (
  `ID` bigint(20) unsigned NOT NULL AUTO_INCREMENT COMMENT 'ID',
  `j` varchar(16) DEFAULT NULL COMMENT '',
  `c` varchar(32) NOT NULL DEFAULT '' COMMENT '',
  `b` date NOT NULL DEFAULT '2019-01-01' COMMENT '',
  `f` varchar(1) NOT NULL DEFAULT '' COMMENT '',
  `g` varchar(1) NOT NULL DEFAULT '' COMMENT '',
  `d` varchar(1) NOT NULL DEFAULT '' COMMENT '',
  `e` varchar(1) NOT NULL DEFAULT '' COMMENT '',
  `h` varchar(1) NOT NULL DEFAULT '' COMMENT '',
  `i` varchar(1) DEFAULT NULL COMMENT '',
  `LAST_UPDATE_TIME` timestamp NOT NULL DEFAULT CURRENT_TIMESTAMP ON UPDATE CURRENT_TIMESTAMP COMMENT '修改时间',
  PRIMARY KEY (`ID`),
  UNIQUE KEY `uidx_1` (`b`,`c`)
) ENGINE=InnoDB AUTO_INCREMENT=270983 DEFAULT CHARSET=utf8mb4 COMMENT='';
```

7.1.2 关键信息梳理

关键信息如表 4 所示。

表 4

	事务 T1
语句	delete from ltb2 where c = 'code001' and j = 'Y15' and b >= '20230717' and d != '1' and e != '1'
关联对象及记录	space id 603 page no 86 n bits 248 index PRIMARY of table testdb.ltb2
持有的锁	未知
等待的锁	lock_mode X locks rec but not gap waiting
	事务 T2
语句	update ltb2 set f = '0', g = '0', is_value_date = '0', h = '0', i = '0' where c = '22115001B' and j = 'Y4' and b >= '20230717'
关联对象及记录	space id 603 page no 86 n bits 248 index PRIMARY of table testdb.ltb2
持有的锁	lock_mode X locks rec but not gap
等待的锁	lock_mode X locks rec but not gap waiting

可以看到在主键索引上发生了死锁，但是在查询的条件中并未使用主键列。

那为什么会在主键列出现死锁？在分析死锁根因问题前，需要先弄清楚 SQL 的执行情况。

7.2 SQL 执行情况

7.2.1 执行计划

表 4 中的两个 SQL 都有列 b、c 作为条件，且这两列构成了唯一索引 uidx_1。简化 SQL，将其改为查询语句，并确认其执行计划：

```
mysql> desc select * from ltb2 where b >= '20230717' and c = 'code001';

# 部分结果
+------+---------------+------+-------------+
| type | possible_keys | key  | Extra       |
+------+---------------+------+-------------+
| ALL  | uidx_1        | NULL | Using where |
+------+---------------+------+-------------+
```

注意：自 MySQL 5.6 开始，可以直接查看 UPDATE/DELETE/INSERT 等语句的执行计划。因个人习惯、避免误操作等原因，笔者还是习惯将语句改为 SELECT 查看执行计划。

在执行计划中，可能的索引有 uidx_1(b,c)，但实际并未使用该索引，而是采用全表扫描方式执行。

根据经验，由于列 b 为索引的最左列，但查询的条件为 b>= '20230717'，即该条件不是等值查询，因此数据库可能只能"使用"到 b 列。为进一步确认不使用 b 列索引的原因，查询数据分布：

```
mysql> select count(1) from ltb2;

+----------+
| count(1) |
+----------+
|     4509 |
+----------+

mysql> select count(1) from ltb2 where b >= '20230717';

+----------+
| count(1) |
+----------+
|     1275 |
+----------+
```

计算满足 b 列条件的数据占比为 1275/4509 = 28%，差不多达到了 1/3。此时也的确不应使用该索引。

难道已经是 MySQL 5.7 的数据库，优化器还是这么简单？

7.2.2 ICP 特性

带着问题，将条件设置为一个更大的值（但小于该列的最大值），再次执行验证查询语句：

```
mysql> desc select * from ltb2 where b >= '20990717';
# 部分结果
+---------+-------+----------------------+
| key_len | rows  | Extra                |
+---------+-------+----------------------+
| 3       | 64    | Using Index condition |
+---------+-------+----------------------+
```

优化器预估返回 64 行，数据占比为 64/4509 = 1.4%，因此可以使用索引。但从执行计划的 Extra 列看到 Using index condition 提示，该提示说明使用了索引条件下推（Index Condition Pushdown，ICP）。针对该特性，参考官方简要说明如下。

使用 ICP 时，扫描将像这样进行：

(1) 获取下一行的索引元组（但不是完整的表行）。

(2) 测试 WHERE 条件中应用于此表的部分，并且只能使用索引列的进行检查。如果不满足条件，则继续到下一行的索引元组。

(3) 如果满足条件，则使用索引元组定位并读取整个表行。

(4) 测试适用于此表的 WHERE 条件的其余部分。根据测试结果接受或拒绝该行。

既然可以使用 ICP 特性，进一步执行如下验证语句：

```
mysql> desc select * from ltb2 where b >= '20990717' and c = 'code001';

# 部分结果
+---------+-------+----------------------+
| key_len | rows  | Extra                |
+---------+-------+----------------------+
| 133     | 64    | Using Index condition |
+---------+-------+----------------------+
```

发现当新增 c 列作为条件后，根据 key_len（索引里使用的字节数）可以判断，的确用到了 uidx_1 索引中的 c 列。但 rows 的结果与实际返回的结果差异较大（实际执行时仅返回 0 行）。

更重要的是，既然具有 ICP 特性，原始的 SQL 查询为什么不能利用 ICP 特性使用索引呢？

```
mysql> select * from ltb2 where b >= '20230717' and c = 'code001'
```

7.2.3 执行计划跟踪

继续带着问题，通过 MySQL 提供的 OPTIMIZER TRACE，跟踪执行计划生成过程。命令如下：

```
SET OPTIMIZER_TRACE="enabled=on",END_MARKERS_IN_JSON=on;
SET OPTIMIZER_TRACE_MAX_MEM_SIZE=1000000;
-- sql-1:
select * from ltb2 where b >= '20990717' and c = 'code001';
-- sql-2:
select * from ltb2 where b >= '20990717';
-- sql-3
select * from ltb2 where b >= '20230717' and c = 'code001';

SELECT * FROM INFORMATION_SCHEMA.OPTIMIZER_TRACE\G
SET optimizer_trace="enabled=off";
```

由于分析结果较长，仅截取 SQL-1 和 SQL-2 的部分结果（rows_estimation 和 considered_execution_plans），具体内容如下。

7.2.3.1 SQL-1

```
select * from ltb2 where b >= '20990717' and c = 'code001'

# 分析结果
"analyzing_range_alternatives":{
  "range_scan_alternatives":[
    {
      "index":"uidx_1",
      "ranges":[
        "0xe76610 <= b"
      ] /* ranges */,
      "index_dives_for_eq_ranges": true,
      "rowid_ordered": false,
      "using_mrr": false,
      "index_only": false,
      "rows":64,
      "cost": 77.81,
      "chosen": true
    }
  ] /* range_scan alternatives */
}

"best_access_path":{
    "considered_access_paths":[
      "rows_to_scan": 64,
      "access_type":"range",
```

```
      "range_details":{
        "used index";"uidx_1"
      } /* range_details */,
      "resulting_rows": 64,
      "cost": 90.61,
      "chosen": true
    }
  ] /* considered access_paths */
} /* best access_path */,
```

7.2.3.2 SQL-2

```
select * from ltb2 where b >= '20990717'

# 分析结果
"analyzing_range_alternatives":{
  "range_scan_alternatives":[
    {
      "index":"uidx_1",
      "ranges":[
        "0xe76610 <= b"
      ] /* ranges */,
      "index_dives_for_eq_ranges": true,
      "rowid_ordered": false,
      "using_mrr": false,
      "index_only": false,
      "rows":64,
      "cost": 77.81,
      "chosen": true
    }
  ] /* range_scan_alternatives */
}

"considered_access_paths":[
  {
    "rows_to_scan": 64,
    "access_type":"range",
    "range_details":{
      "used index":"uidx_1"
    } /* range_details */,
    "resulting_rows": 64,
    "cost": 90.61,
    "chosen": true
  }
] /* considered access_paths */,
```

根据以上信息，两个 SQL 的 cost 部分是完全相同的，而且在优化器分析阶段只能识别到 b 的条件。在分析阶段，只能根据优化器认为可用的列来计算 cost。ICP 特性应该是在执行阶段用到的特性。

同时，根据 SQL-3 的执行跟踪结果，对比全表扫描和索引扫描的 cost，截取部分结果如下。

7.2.3.3 SQL-3

```
select * from ltb2 where b >= '20230717' and c = 'code001';

# 全表扫描结果
"range_analysis": {
  "table_scan": {
    "rows": 4669,
    "cost": 1018.9
  } /* table_scan */,

# 索引扫描评估结果
"analyzing_range_alternatives": {
  "range_scan_alternatives": [
    {
      "index":"uidx_1",
      "ranges":[
        "@xe7ce0f] <= b"
      ] /* ranges */,
      "index dives_for_eq_ranges": true,
      "rowid_ordered": false,
      "using_mrr": false,
      "index_only": false,
      " rows": 1273,
      "cost": 1528.6,
      "chosen": false,
      "cause":"cost"
    }
  ] /* range scan_alternatives */,

# 最优执行计划
"best_access_path": {
  "considered access_paths":[
    {
      "rows_to_scan": 4669,
      "access_type":"scan",
      "resulting_rows": 4669,
      "cost": 1016.8,
      "chosen": true
    }
  ] /* considered access_paths *//* best access_path */
}
```

由于优化器阶段使用列 b，使用索引的成本高于全表扫描，最终数据库就会选择使用全表扫描。除非应用使用 hint 强制索引：

```
mysql> desc select * from ltb2 FORCE INDEX (uidx_1) where b >= '20230717' and c = 'code001';

# 部分结果
+---------+------+---------------------+
| key_len | rows | Extra               |
+---------+------+---------------------+
| 133     | 1273 | Using Index condition |
+---------+------+---------------------+
```

同时，根据执行计划的输出结果，rows 列应该是优化器阶段的输出，key_len/Extra 则包括了执行阶段的输出。

7.2.4 小结

综上所述，对于问题 SQL 和索引结构，由于列 b 为索引的最左列，且查询时的条件为 b>= '20230717'（非等值条件），数据库优化器只能"使用" b 列，并基于"使用"的列，评估扫描的行数和 cost。

如果优化器评估后，发现使用索引的成本更低，则可以使用该索引，并利用 ICP 特性进一步提高查询性能；

如果优化器评估后，发现使用全表扫描的成本更低，数据库就会选择使用全表扫描。

7.3 SQL 优化方案

根据 7.2 节明确了问题的原因后，通过调整索引可以解决最左列尾范围查询的问题。具体操作如下：

```
alter table ltb2 drop index uidx_1;
alter table ltb2 add index uidx_1(c,b);
alter table ltb2 add index idx_(b);
```

7.4 死锁为何发生

自此，分析和解决了 SQL 执行计划问题。但直接的问题是死锁，因查询语句无法使用索引，正常情况下就应该使用全表扫描。但是全表扫描为什么会出现死锁呢？

在此，对死锁过程进行如下大胆猜想。

7.4.1 T1 时刻

trx-2 执行了 UPDATE，在处理行时，row_search_mvcc 函数查询到数据，并获取了对应行的 LOCK_X,LOCK_REC_NOT_GAP 锁。

7.4.2 T2 时刻

trx-1 执行了 DELETE，在处理行时，row_search_mvcc 函数查询到数据，尝试获取行的 LOCK_X 和 LOCK_REC_NOT_GAP。但由于 trx-1 已经持有了该锁，因此该操作被堵塞，并会创建一个锁（以指示锁等待）。

7.4.3 T3 时刻

trx-2 继续执行 UPDATE 操作。该操作除了在 T1 时刻的操作外，在其他位置还需要获取锁（lock_mode X locks rec but not gap）。但由于在 T2 时刻，trx-1 尝试获取该锁而被堵塞，并且也增加了一个锁。

假如此时，此处的实现机制和 INSERT 死锁案例一样，也没有先进行冲突检查，而只是看记录上是否存在锁的话，那么此时也会看到该记录上有 trx-1 事务的锁，从而导致 trx-2 第二次获取锁时被堵塞。

死锁发生！

以上仅为根据经验进行的猜想，真正的原因还需要进一步分析和验证。有兴趣的读者可结合如下几个问题进一步研究。

(1) 以上各步骤获取锁的位置是否正确？

(2)T3 时刻，UPDATE 操作在其他的什么位置再次获取了锁？

(3)T3 时刻，发起的假设是否成立？如成立，具体逻辑是什么？如不成立，那正确的逻辑是什么？

(4)T3 时刻，如果假设不成立，那死锁的原因又是什么？

(5) 以上都是针对唯一索引/主键索引的执行逻辑分析的，结合该案例，全表扫描和索引查询的执行逻辑是否存在差异？差异在哪里？

(6) 除了调整索引，还能通过什么方式避免该问题发生？

8 MySQL 全文索引触发 OOM 一例

作者：付祥

MySQL 版本为 5.7.34。

8.1 故障现象

某业务监控报警提示内存不足，发现 mysqld 进程由于内存不足被 kill 而自动重启了。

```
[root@xxxxxx ~]# ps -ef|grep mysqld
root     17117 62542  0 20:26 pts/1    00:00:00 grep --color=auto mysqld
mysql    27799     1  7 09:54 ?        00:48:32 /usr/sbin/mysqld --daemonize --pid-file=/var/run/mysqld/mysqld.pid
[root@xxxxxx ~]#

# 操作系统日志记录 MySQL 被 OOM
    Dec  8 09:54:42 xxxxxx kernel: Out of memory: Kill process 22554 (mysqld) score 934 or sacrifice child
    Dec  8 09:54:42 xxxxxx kernel: Killed process 22554 (mysqld), UID 27, total-vm:11223284kB, anon-rss:7444620kB, file-rss:0kB, shmem-rss:0kB
```

8.2 故障分析

机器总内存 8GB，还有其他应用占用了少许内存。尽管 MySQL 重启了，但使用的内存依然很高，内存监控数据如图 17 所示。

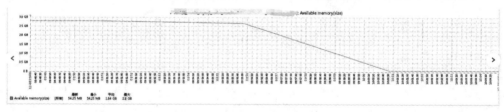

图 17

有效内存并不是一点一点地减少，而是突然下降的。内存监控数据每 5 分钟采集一次。MySQL 在 09：54 重启后，09：55：54 的有效内存是 2869899264 字节，10：00：54 采集降低至 56885248 字节。

```
2023-12-08 10:20:54
963796992
```

```
2023-12-08 10:15:54
93224960
2023-12-08 10:10:54
111407104
2023-12-08 10:05:54
113987584
2023-12-08 10:00:54
56885248
2023-12-08 09:55:54
2869899264
```

InnoDB Buffer 分配的内存为 1GB：

```
| innodb_buffer_pool_size    | 1073741824    |
```

通过 top 查看，MySQL 当前使用了 6GB 内存：

```
27799 mysql      20   0 8888376   6.1g   6120 S  26.2 80.5  30:19.01 mysqld
```

通过 gdb 调用 malloc_stats() 函数查看内存分配情况：

```
gdb -ex "call (void) malloc_stats()" --batch -p $(pidof mysqld)
```

查看 MySQL 日志中的内存使用情况：

```
MALLOC:      853070392 (  813.6 MiB) Bytes in use by application
MALLOC: +   6369394688 ( 6074.3 MiB) Bytes in page heap freelist
MALLOC: +      9771872 (    9.3 MiB) Bytes in central cache freelist
MALLOC: +       863232 (    0.8 MiB) Bytes in transfer cache freelist
MALLOC: +     25216616 (   24.0 MiB) Bytes in thread cache freelists
MALLOC: +     25559040 (   24.4 MiB) Bytes in malloc metadata
MALLOC:   ------------
MALLOC: =   7283875840 ( 6946.4 MiB) Actual memory used (physical + swap)
MALLOC: +    329924608 (  314.6 MiB) Bytes released to OS (aka unmapped)
MALLOC:   ------------
MALLOC: =   7613800448 ( 7261.1 MiB) Virtual address space used
MALLOC:
MALLOC:          12564              Spans in use
MALLOC:             52              Thread heaps in use
MALLOC:           8192              Tcmalloc page size
------------------------------------------------
Call ReleaseFreeMemory() to release freelist memory to the OS (via madvise()).
Bytes released to the OS take up virtual address space but no physical memory.
```

MySQL 当前使用 Tcmalloc 内存分配器，Bytes in page heap freelist 使用了将近 6GB 内存，猜测是有操作比较"吃"内存，操作完后 MySQL 释放了内存，但是 Tcmalloc 并没有将内存释放给操作系统。那到底是什么操作比较"吃"内存呢？分析相关时间段的慢 SQL，发现有一个使用全文索引 SQL 比较可疑：

```
# Time: 2023-12-08T01:52:23.084854Z
# User@Host: xxxxxx @  [x.x.x.x]  Id: 259892877
# Query_time: 1.436714  Lock_time: 0.000049 Rows_sent: 1  Rows_examined: 0
SET timestamp=1702000343;
SELECT count(*) FROM `xx` inner JOIN (select id from  xx_content  where
MATCH(content) AGAINST('\"Elasticsearch Cluster in 1 state\"' IN BOOLEAN MODE)) al
ON xx.id = al.id WHERE (xx.handle_status in ('pending','processing','completed'))
AND `xx`.`sub_type` = 1;
# Time: 2023-12-08T01:52:24.540847Z
# User@Host: xxxxxx @  [x.x.x.x]  Id: 259892879
# Query_time: 1.454352  Lock_time: 0.000052 Rows_sent: 0  Rows_examined: 0
SET timestamp=1702000344;
SELECT xx.*,SUBSTRING(xx.content, 1, 1024) as content,SUBSTRING(xx.sub_
content, 1, 1024) as sub_content FROM `xx` inner JOIN (select id from xx_content
where  MATCH(content) AGAINST('\"Elasticsearch Cluster in 1 state\"' IN BOOLEAN
MODE)) al ON xx.id = al.id WHERE (xx.handle_status in ('pending','processing','com
pleted')) AND `xx`.`sub_type` = 1 ORDER BY xx.sub_time DESC LIMIT 50;
# Time: 2023-12-08T01:53:26.546353Z
# User@Host: xxxxxx @  [x.x.x.x]  Id: 259893335
# Query_time: 44.198100  Lock_time: 0.000041 Rows_sent: 1  Rows_examined: 48
SET timestamp=1702000406;
SELECT count(*) FROM `xx` inner JOIN (select id from xx_content where
MATCH(content) AGAINST('\"Elasticsearch Cluster in \"' IN BOOLEAN MODE)) al ON
xx.id = al.id WHERE (xx.handle_status in ('pending','processing','completed')) AND
`xx`.`sub_type` = 1;
73
# Time: 2023-12-08T01:56:35.790820Z
# User@Host: xxxxxx @  [x.x.x.x]  Id:  1671
# Query_time: 29.259303  Lock_time: 0.000042 Rows_sent: 1  Rows_examined: 48
SET timestamp=1702000595;
SELECT count(*) FROM `xx` inner JOIN (select id from xx_content where
MATCH(content) AGAINST('\"Elasticsearch Cluster in \"' IN BOOLEAN MODE)) al ON
xx.id = al.id WHERE (xx.handle_status = 'pending') AND `xx`.`sub_type` = 1;
# Time: 2023-12-08T01:56:36.350983Z
# User@Host: xxxxxx @  [x.x.x.x]  Id:  1672
# Query_time: 28.870504  Lock_time: 0.000050 Rows_sent: 1  Rows_examined: 48
SET timestamp=1702000596;
SELECT count(*) FROM `xx` inner JOIN (select id from xx_content where
MATCH(content) AGAINST('\"Elasticsearch Cluster in \"' IN BOOLEAN MODE)) al ON
xx.id = al.id WHERE (xx.handle_status in ('pending','processing','completed')) AND
`xx`.`sub_type` = 1;
```

表结构及数据量如下:

```
root@3306 xxxxxx> show create table xx_content\G
*************************** 1. row ***************************
Table: xx_content
Create Table: CREATE TABLE `xx_content` (
  `id` bigint(20) NOT NULL AUTO_INCREMENT,
```

```
`content` longtext,
PRIMARY KEY (`id`),
FULLTEXT KEY `ngram_content` (`content`) /*!50100 WITH PARSER `ngram` */
) ENGINE=InnoDB AUTO_INCREMENT=100377976 DEFAULT CHARSET=utf8mb4
1 row in set (0.00 sec)

root@3306 xxxxxx> select count(*) from xx_content;
+----------+
| count(*) |
+----------+
|   360215 |
+----------+
1 row in set (0.11 sec)
```

全文索引相关参数均采用默认配置:

```
root@3306 (none)> show variables like '%ft%';
+---------------------------------+------------------+
| Variable_name                   | Value            |
+---------------------------------+------------------+
| ft_boolean_syntax               | + -><()~*:""&|   |
| ft_max_word_len                 | 84               |
| ft_min_word_len                 | 4                |
| ft_query_expansion_limit        | 20               |
| ft_stopword_file                | (built-in)       |
| innodb_ft_aux_table             |                  |
| innodb_ft_cache_size            | 8000000          |
| innodb_ft_enable_diag_print     | OFF              |
| innodb_ft_enable_stopword       | ON               |
| innodb_ft_max_token_size        | 84               |
| innodb_ft_min_token_size        | 3                |
| innodb_ft_num_word_optimize     | 2000             |
| innodb_ft_result_cache_limit    | 2000000000       |
| innodb_ft_server_stopword_table |                  |
| innodb_ft_sort_pll_degree       | 2                |
| innodb_ft_total_cache_size      | 640000000        |
| innodb_ft_user_stopword_table   |                  |
+---------------------------------+------------------+
17 rows in set (0.01 sec)
```

SQL 分别在 9:53 和 9:56 执行，正好在 MySQL 自动重启前后，和内存监控数据比较吻合（9:53 执行，9:54 机器内存不足，MySQL 被 OOM；9:56 执行后，10:00:54 采集降低至 56885248 字节）。这个环境还有一个从库，从库未承载业务，将 SQL 拿到从库执行，观察 MySQL 内存使用情况的变化。开两个窗口，一个窗口执行 SQL：

```
root@3306 xxxxxx> SELECT count(*) FROM `xx` inner JOIN (select id from xx_
content where MATCH(content) AGAINST('\"Elasticsearch Cluster in 1 state\"' IN
BOOLEAN MODE)) a1 ON xx.id = a1.id WHERE (xx.handle_status in ('pending','processi
```

```
ng','completed')) AND `xx`.`sub_type` = 1;
+----------+
| count(*) |
+----------+
|        3 |
+----------+
1 row in set (1 min 9.31 sec)
```

另一个窗口观察 mysqld 进程内存使用情况：

```
[root@xxxxxx ~]# ps aux|grep mysqld|grep -v grep|awk '{print $6}'
3453980
[root@xxxxxx ~]# while true;do ps aux|grep mysqld|grep -v grep|awk '{print $6}';sleep 1;done;

3453980
3453980
3453980
3459064
3617600
3822828
3969212
4128056
4533612
4677028
4756868
4844452
5011176
5070292
5123844
5188556
5263880
5410368
5410368
5412200
5412200
5412200
5412200
5412200
```

可以观察到在 SQL 执行过程中，使用的内存不断上涨，SQL 执行完后使用的内存从 3453980KB 涨到 5412200KB，但是 Tcmalloc 并没有将内存释放给操作系统。

到目前为止，总算定位了问题，MySQL 并不擅长全文索引，可以交给 ElasticSearch 等数据库去做。在业务不调整情况下，怎么解决问题呢？不妨换 Jemalloc 内存分配器试试：

```
[root@xxxxxx ~]# yum install -y jemalloc
[root@xxxxxx ~]# cat /etc/sysconfig/mysql
#LD_PRELOAD=/usr/lib64/libtcmalloc.so
```

```
LD_PRELOAD=/usr/lib64/libjemalloc.so.1
[root@xxxxxx ~]# systemctl restart mysqld
```

果然有惊喜，SQL 执行完后会释放内存，使用的内存从 822948KB 涨到 2738040KB，最终回落到 916400KB：

```
[root@xxxxxx ~]# while true;do ps aux|grep mysqld|grep -v grep|awk '{print $6}';sleep 1;done;
822948
822948
822948
874216
1057240
1273684
1443820
1662924
1873304
2177760
2502488
2738040
1296604
899580
900636
902412
903680
904384
......
914492
914492
915020
915284
915736
916524
916524
916524
916524
916524
916400
916400
```

8.3 总结

在线上环境中，MySQL 都是使用 Tcmalloc 内存分配器的，一直很稳定，并未出现服务器内存不足问题。本次出现服务器内存不足，是因为使用了全文索引这种极少使用的场景，换成 Jemalloc 后，内存的使用整体上得到了控制。

9 MySQL 扩展 VARCHAR 长度遭遇问题的总结

作者：莫善

9.1 背景介绍

最近，业务反馈有个扩展 VARCHAR 改表需求失败多次，需要干预处理一下。

经过排查分析发现，这是由于改表系统解析改表需求时得出错误的改表方案导致，即这类改表操作可以快速改表（直接使用 ALTER TABLE），理论上任务下发后能马上改完，但是工单结果是执行触发 10 秒超时，最终工单失败。

原则上，扩展 VARCHAR 类型是可以快速改表的，我们的改表工单对这类需求也是支持的，但是实际结果与预期不符，这到底是工单系统的 Bug，还是 MySQL 的坑呢？

本文就来总结一下扩展 VARCHAR 长度可能会遇到的一些问题，并给出解决方案，仅供参考。

> 本文仅讨论 MySQL 5.7 及以后的版本。

9.2 MySQL Online DDL

表 5

Operation	Extending VARCHAR column size
In Place	Yes
Rebuilds Table	No
Permits Concurrent DML	Yes
Only Modifies Metadata	Yes

表 5 摘自 MySQL 官方文档中关于 Online DDL 章节的一部分。可以看到，关于 VARCHAR 类型的字段的扩展是可以原地改表的且仅仅改元数据，理论上敲完回车就执行结束。

当然，针对这种场景，还是有一些条件的，以下是相关内容的原文：

> The number of length bytes required by a VARCHAR column must remain the same. For VARCHAR columns of 0 to 255 bytes in size, one length byte is required to encode the value. For VARCHAR columns of 256 bytes in size or more, two length bytes are required.

As a result, in-place ALTER TABLE only supports increasing VARCHAR column size from 0 to 255 bytes, or from 256 bytes to a greater size. In-place ALTER TABLE does not support increasing the size of a VARCHAR column from less than 256 bytes to a size equal to or greater than 256 bytes. In this case, the number of required length bytes changes from 1 to 2, which is only supported by a table copy (ALGORITHM=COPY).

VARCHAR 是变长类型，实际存储的内容不固定，需要 1 或 2 字节来表示实际长度，所以修改前和修改后，这个字节数必须一致。

有了这个技术基础，我们的改表系统就针对这类需求做了优化，可以支持直接使用 ALTER TABLE 改表，如果是大表可以节省很多时间，提升效率，也因此遇到了很多问题，才有了这篇文章。

9.3 问题汇总

首先简单介绍一下我们的改表系统的处理逻辑。系统会根据业务的改表需求去选择最优的改表方案：

（1）若满足快速改表条件，就直接使用 ALTER TABLE 进行操作。比如，删除索引、修改表名 / 列名、修改默认值 / 注释、扩展 VARCHAR 长度、小表添加唯一索引以及 MySQL8.0 快速加列等等。

（2）若不满足快速改表条件，就优先选择 gh-ost 进行改表。如果 binlog format 不为 ROW 格式，则不能使用 gh-ost。添加唯一索引必须使用 gh-ost。

（3）若不满足 gh-ost 的使用条件，都会选择 pt-osc 进行改表。但在此模式下添加唯一索引会直接失败。

那么问题来了，我们如何判断业务改表需求是不是扩展 VARCHAR？

其实思路也很简单，就是检查改表前后的 information_schema.columns 记录，用到的 SQL 如下：

```
select * from information_schema.columns where table_schema = 'db' and table_name = 'table' and column_name = 'col';

# 样例数据
*************************** 1. row ***************************
TABLE_CATALOG: def
TABLE_SCHEMA: information_schema
TABLE_NAME: CHARACTER_SETS
COLUMN_NAME: CHARACTER_SET_NAME
ORDINAL_POSITION: 1
COLUMN_DEFAULT:
```

```
            IS_NULLABLE: NO
             DATA_TYPE: varchar
CHARACTER_MAXIMUM_LENGTH: 32
  CHARACTER_OCTET_LENGTH: 96
       NUMERIC_PRECISION: NULL
           NUMERIC_SCALE: NULL
      DATETIME_PRECISION: NULL
      CHARACTER_SET_NAME: utf8
          COLLATION_NAME: utf8_general_ci
             COLUMN_TYPE: varchar(32)
              COLUMN_KEY:
                   EXTRA:
              PRIVILEGES: select
          COLUMN_COMMENT:
     GENERATION_EXPRESSION:
```

(1) DATA_TYPE 值是 VARCHAR。

(2) CHARACTER_MAXIMUM_LENGTH 的值在改表后要大于或等于改表前的值。

(3) CHARACTER_OCTET_LENGTH 的值在改表前后应位于同一区间，即这个值要么都小于或等于 255，要么都大于 255。

(4) 除 DATA_TYPE/COLUMN_TYPE/CHARACTER_MAXIMUM_LENGTH/CHARACTER_OCTET_LENGTH 字段外的其余字段，在改表前后应保持一致。

9.3.1 问题一：默认值问题

我们发现，如果在改表时还要改字段名、注释、默认值这种元数据信息，依旧可以快速改表。为了优化，不再比较 COLUMN_NAME、COLUMN_COMMENT、COLUMN_DEFAULT 这三个属性。

关于默认值，问题有点复杂。最开始也是思路跑偏了，认为判断 COLUMN_DEFAULT 的值就行，比较这个值前后要么都是 null，要么都不是 null。在都不是 null 的情况下可以是任意值，比如可以用下面的逻辑判断改表前后是否一致。

```
if(COLUMN_DEFAULT is null ,null,"")
```

但是有个问题，如果一个字段从允许为 null 且默认值为 1 变成不允许为 null 且默认值也是 1，该值改表前后也是一致的。具体的测试如下：

```
CREATE TABLE `tb_test` (
`id` bigint(20) NOT NULL AUTO_INCREMENT,
`rshost` varchar(20) DEFAULT '1' COMMENT '主机地址',
`cpu_info` json DEFAULT NULL COMMENT 'cpu 信息 json 串',
`mem_info` json DEFAULT NULL COMMENT 'mem 信息 json 串,单位是 GB',
`io_info` json DEFAULT NULL COMMENT '磁盘 io 使用情况,单位是 KB',
`net` json DEFAULT NULL COMMENT '网络使用情况,单位是 KB(speed 单位是 MB/S)',
```

```
    `a_time` datetime NOT NULL DEFAULT '2022-01-01 00:00:00',
    PRIMARY KEY (`id`),
    KEY `idx_a_time` (`a_time`),
    KEY `idx_rshost` (`rshost`)
) ENGINE=InnoDB AUTO_INCREMENT=623506728 DEFAULT CHARSET=utf8mb4
>select * from information_schema.columns where table_name = 'tb_test' and column_name = 'rshost'\G
*************************** 1. row ***************************
       TABLE_CATALOG: def
        TABLE_SCHEMA: dbzz_monitor
          TABLE_NAME: tb_test
         COLUMN_NAME: rshost
    ORDINAL_POSITION: 2
      COLUMN_DEFAULT: 1
         IS_NULLABLE: YES
           DATA_TYPE: varchar
CHARACTER_MAXIMUM_LENGTH: 30
  CHARACTER_OCTET_LENGTH: 120
       NUMERIC_PRECISION: NULL
           NUMERIC_SCALE: NULL
       DATETIME_PRECISION: NULL
      CHARACTER_SET_NAME: utf8mb4
          COLLATION_NAME: utf8mb4_general_ci
             COLUMN_TYPE: varchar(30)
              COLUMN_KEY:
                   EXTRA:
              PRIVILEGES: select,insert,update,references
          COLUMN_COMMENT: 主机地址
   GENERATION_EXPRESSION:
1 row in set (0.00 sec)

>alter table tb_test modify `rshost` varchar(30) not null DEFAULT '1' COMMENT '主机地址';
Query OK, 1000000 rows affected (13.68 sec)
Records: 1000000  Duplicates: 0  Warnings: 0

>select * from information_schema.columns where table_name = 'tb_test' and column_name = 'rshost'\G
*************************** 1. row ***************************
       TABLE_CATALOG: def
        TABLE_SCHEMA: dbzz_monitor
          TABLE_NAME: tb_test
         COLUMN_NAME: rshost
    ORDINAL_POSITION: 2
      COLUMN_DEFAULT: 1
         IS_NULLABLE: NO
           DATA_TYPE: varchar
CHARACTER_MAXIMUM_LENGTH: 30
  CHARACTER_OCTET_LENGTH: 120
```

```
    NUMERIC_PRECISION: NULL
      NUMERIC_SCALE: NULL
 DATETIME_PRECISION: NULL
 CHARACTER_SET_NAME: utf8mb4
     COLLATION_NAME: utf8mb4_general_ci
        COLUMN_TYPE: varchar(30)
         COLUMN_KEY:
              EXTRA:
         PRIVILEGES: select,insert,update,references
     COLUMN_COMMENT: 主机地址
GENERATION_EXPRESSION:
1 row in set (0.00 sec)
```

可以看到 COLUMN_DEFAULT 这个列的值不是 null 且不变，按照上面的判断逻辑是可以快速改表的，但是我们知道实际上这个需求是需要 copy 数据的。

其实，关于默认值问题，使用 IS_NULLABLE 的值就可以完美解决，如果是 null 到 not null，这个值会从 yes 变成 no；如果是 not null 到 null，这个值会从 no 变成 yes。

所以最终解决方案是仅比较 IS_NULLABLE 即可，只要改表前后一致就认为默认值这个属性满足快速改表条件。

在测试这个问题的时候发现一个现象：从 not null 到 null 可以使用 inplace 算法，但是需要 copy 数据，从 null 到 not null 不能使用 inplace。请看下面的用例：

```
    -- not null --> null 可以使用 inplace
    >alter table tb_test modify `rshost` varchar(30) DEFAULT '1' COMMENT '主机地址' ,ALGORITHM=INPLACE, LOCK=NONE;
    Query OK, 0 rows affected (3.45 sec)
    Records: 0  Duplicates: 0  Warnings: 0

    -- null --> not null 不可以使用 inplace
    >alter table tb_test modify `rshost` varchar(30) not null DEFAULT '1' COMMENT '主机地址' ,ALGORITHM=INPLACE, LOCK=NONE;
    ERROR 1846 (0A000): ALGORITHM=INPLACE is not supported. Reason: cannot silently convert NULL values, as required in this SQL_MODE. Try ALGORITHM=COPY.
    >

    -- 可以使用下面的操作查看改表进度及拷贝数据的情况，第一次使用时需要开启此功能
    -- UPDATE performance_schema.setup_instruments SET ENABLED = 'YES' WHERE NAME LIKE 'stage/innodb/alter%';
    -- UPDATE performance_schema.setup_consumers SET ENABLED = 'YES' WHERE NAME LIKE '%stages%';

    >SELECT EVENT_NAME, WORK_COMPLETED, WORK_ESTIMATED FROM performance_schema.events_stages_current;
    +------------------------------+----------------+----------------+
    | EVENT_NAME                   | WORK_COMPLETED | WORK_ESTIMATED |
```

```
+-------------------------------+----------------+----------------+
| stage/sql/copy to tmp table   |        272289  |        978903  |
+-------------------------------+----------------+----------------+
1 row in set (0.00 sec)
```

```
-- 为了避免测试干扰，检查 events_stages_history 表之前可以先清空，切记不要对线上环境做此操作
-- TRUNCATE TABLE performance_schema.events_stages_history;
>SELECT EVENT_NAME, WORK_COMPLETED, WORK_ESTIMATED FROM performance_schema.events_stages_history;
+-------------------------------+----------------+----------------+
| EVENT_NAME                    | WORK_COMPLETED | WORK_ESTIMATED |
+-------------------------------+----------------+----------------+
| stage/sql/copy to tmp table   |       1000000  |        978903  |
+-------------------------------+----------------+----------------+
1 row in set (0.00 sec)
```

9.3.2 问题二：索引字段问题

过了一段时间又发现第二个问题，部分工单会触发执行 10 秒超时失败。

工单系统判断用户的改表需求，满足直接使用 ALTER TABLE 进行操作会有一个 10 秒超时的兜底策略，来避免因为解析错误导致方案选择错误最终影响主从延迟。

另外，建议带上 ALGORITHM=INPLACE, LOCK=NONE，避免因为不是使用 inplace 导致 DML 阻塞。

这个问题排查了很久都没什么眉目，我们反反复复地查阅文档及测试，始终认为这个需求一定是满足快速改表的方案，实在想不明白到底是哪里的问题，还一度认为是 MySQL 的 Bug。

下面是一张 100 万记录表的测试用例：

```
> show create table tb_test\G
*************************** 1. row ***************************
Table: tb_test
Create Table: CREATE TABLE `tb_test` (
  `id` bigint(20) NOT NULL AUTO_INCREMENT,
  `rshost` varchar(30) NOT NULL DEFAULT '1' COMMENT '主机地址',
  `cpu_info` json DEFAULT NULL COMMENT 'cpu 信息 json 串',
  `mem_info` json DEFAULT NULL COMMENT 'mem 信息 json 串，单位是 GB',
  `io_info` json DEFAULT NULL COMMENT '磁盘 io 使用情况，单位是 KB',
  `net` json DEFAULT NULL COMMENT '网络使用情况，单位是 KB(speed 单位是 MB/S)',
  `a_time` datetime NOT NULL DEFAULT '2022-01-01 00:00:00',
  PRIMARY KEY (`id`),
  KEY `idx_a_time` (`a_time`),
  KEY `idx_rshost` (`rshost`)
) ENGINE=InnoDB AUTO_INCREMENT=623506728 DEFAULT CHARSET=utf8mb4
1 row in set (0.00 sec)
```

```
> select count(*) from tb_test;
+----------+
| count(*) |
+----------+
| 1000000  |
+----------+
1 row in set (0.15 sec)

> alter table tb_test modify `rshost` varchar(31) NOT NULL DEFAULT '1' COMMENT '主机地址';
Query OK, 0 rows affected (3.61 sec)
Records: 0  Duplicates: 0  Warnings: 0

> alter table tb_test modify `rshost` varchar(32) NOT NULL DEFAULT '1' COMMENT '主机地址', ALGORITHM=INPLACE, LOCK=NONE;
Query OK, 0 rows affected (3.66 sec)
Records: 0  Duplicates: 0  Warnings: 0

> alter table tb_test drop index idx_rshost;
Query OK, 0 rows affected (0.00 sec)
Records: 0  Duplicates: 0  Warnings: 0

> alter table tb_test modify `rshost` varchar(33) NOT NULL DEFAULT '1' COMMENT '主机地址', ALGORITHM=INPLACE, LOCK=NONE;
Query OK, 0 rows affected (0.00 sec)
Records: 0  Duplicates: 0  Warnings: 0

> alter table tb_test modify `rshost` varchar(34) NOT NULL DEFAULT '1' COMMENT '主机地址', ALGORITHM=INPLACE, LOCK=NONE;
Query OK, 0 rows affected (0.00 sec)
Records: 0  Duplicates: 0  Warnings: 0

>
```

可以看到 rshost 字段有一个索引，在扩展字段的时候虽然支持 inplace，但实际执行速度很慢，推测内部应该是重建索引了。后来将索引删除后，扩展操作几乎瞬间完成。

针对这个场景，我们的解决方案是使用 gh-ost/pt-osc 进行改表。那么问题来了，我们应该怎么判断目标字段是否是被索引了呢？

请看下面的 SQL，information_schema.STATISTICS 记录了一个表的所有索引字段信息，可以很方便地判断某个字段是否被索引。

```
select * from information_schema.STATISTICS where table_schema = 'db' and table_name = 'table' and column_name = 'col';
```

9.3.3 其他问题

这个问题也是执行触发 10 秒超时，也就是文章开头提到的业务反馈的问题，跟问题二差不多同期发生，但在问题二解决后还是一直找不到原因及解决方案。

这个问题甚至都没法复现，不像问题二可以方便复现，当时在业务的线上库做操作能 100% 复现，但是将他们的表及数据单独导出来放在测试环境就不行。

在业务库上测试是选了一个从库，不记录 binlog（set sql_log_bin = 0）。虽然不建议这么做，但是实属迫不得已，因为在测试环境复现不出来。

后来实在找不到原因，就跳过快速改表的方案，使用改表工具进行处理，这个事情就算不了了之了。直到前几天业务突然找我，说之前的那个表能快速改表了。我赶紧去查看了工单详情，发现确实如业务所述。这回我就更加郁闷了，难不成是见鬼了？这玩意还自带歇业窗口的吗？

本着严谨的态度，又去测了一下，发现确实是可以快速改表了，但还是找不到原因，这感觉真的很难受。

最后，静下来认真梳理了一下，发现了一些猫腻。下面是我的测试思路。

9.3.3.1 将线上的表导出并导入到测试环境

因为表本身就几个 GB，不算大，就使用了 mysqldump 进行导出导入。这个操作并非 100% 复原线上的环境，有个隐藏的变量被修改了，那就是这个表被重建了，这跟之前业务用改表工具进行修改后的操作有点类似。所以我就猜想，会不会是因为这个表本身存在空洞导致的呢？

最后通过拉历史备份，还原一个环境进行测试，果不其然不能快速改表。为了印证自己的想法，就去查了一下这个表的空洞。十分遗憾，这个表并没有空洞（空洞只有几 MB）。这回又郁闷了，还以为要破案了。但是不管怎么样既然怀疑是重建表能解决，那就开干。

9.3.3.2 重建前的状态

业务从 varchar(300) 扩展到 varchar(500)，其他属性没变更。

```
|  1170930999 | dba        | 192.168.1.100:47522   | dbzz_dbreport | Query
|    45 | altering table                                          | ALTER TABLE
t_recycle_express  MODIFY  address VARCHAR(500) NOT NULL DEFAULT '' COMMENT '地址',
ALGORITHM=INPLACE, LOCK=NONE;       |
```

9.3.3.3 重建后的状态

```
>ALTER TABLE t_recycle_express engine = innodb;
Query OK, 0 rows affected (18 min 52.60 sec)
```

```
Records: 0  Duplicates: 0  Warnings: 0

>ALTER TABLE
->    t_recycle_express
-> MODIFY
->    address VARCHAR(500) NOT NULL DEFAULT '' COMMENT '地址';
Query OK, 0 rows affected (0.07 sec)
Records: 0  Duplicates: 0  Warnings: 0
```

还真的是重建表后就能解决了！虽然很郁闷，但终究是有一个解决方案了。后期我们决定对此做个优化，将满足快速改表的工单且触发 10 秒超时的改为使用 gh-ost/pt-osc 重新执行，以此避免业务反复提交工单，应该能大幅提升好感度。

这个问题虽然知道解决方案，但是依旧不知道原因，笔者猜测可能是跟统计信息不准确有关系（或者约束）。

9.4 总结

MySQL Online DDL 特性给 DBA 带来了很多的便利，提升了工作效率，我们可以以官方的理论作为指导去优化我们的系统。但是实际情况是理论知识很简单，线上环境十分复杂，可能会遇到各种意料之外的事情，任何线上的操作都要给自己留好后路，做好兜底，这是十分必要的。

我们的系统，如果没有添加 10 秒超时的兜底，那势必会因为解析错误导致选了错误的改表方案，然后导致从库延迟，可能会影响线上业务，想想都有点心慌。

这里有个注意事项，针对执行超时不能简单地使用 timeout 等属性进行控制，还需要添加检查逻辑，要到数据库里面去查一下任务是否真的已经终止了，避免因为 timeout 异常导致终止信号没有给到 MySQL。这种情况可能会引发一系列问题，切记切记。

10 MySQL 执行 Online DDL 操作报错空间不足？

作者：徐文梁

10.1 问题背景

客户反馈对某张表执行 alter table table_name engine=innodb; 时，报错显示空间不足。通过登录数据库查看客户的 tmpdir 设置的路径，发现是 /tmp。该目录磁盘空间本身较小，调整 tmpdir 的路径使其与数据目录相同，重新执行 ALTER 操作则成功。

问题到此结束了，但是故事并没有结束。通过查看官网信息，我们可以从这个小小的报错中深挖出更多信息。

10.2 信息解读

从官网的论述中，我们可以了解到，在进行 Online DDL 操作时，需要保证以下三个方面的空间充足，否则可能会导致空间不足而报错。

10.2.1 临时日志文件

当进行 Online DDL 操作创建索引或者更改表时，临时日志文件会记录此期间的并发 DML 操作，临时日志文件最大值由 innodb_online_alter_log_max_size 参数控制，如果 Online DDL 操作耗时较长（如果表数据量较大，这是很有可能的），并且期间并发 DML 对表中的记录修改较多，则可能导致临时日志文件大小超过 innodb_online_alter_log_max_size 值，从而引发 DB_ONLINE_LOG_TOO_BIG 错误，并回滚未提交的并发 DML 操作。

10.2.2 临时排序文件

对于会重建表的 Online DDL 操作，在创建索引期间，会将临时排序文件写入到 MySQL 的临时目录。仅考虑 UNIX 系统，对应的参数为 tmpdir，如果 /tmp 目录比较小，请设置该参数为其他目录，否则可能会因为无法容纳排序文件而导致 Online DDL 失败。

10.2.3 中间表文件

对于会重建表的 Online DDL 操作，会在与原始表相同的目录中创建一个临时中间表文件，中间表文件可能需要与原始表大小相等的空间。中间表文件名以 #sql-ib 前缀开头，仅在 Online DDL 操作期间短暂出现。

10.3 前置准备

针对官网的论述，我们可以进行实际测试，这里对临时排序文件和中间表文件场景进行测试。为了故事更好地发展，先做一些准备工作。

10.3.1 创建一个测试库

数据目录对应为 /opt/mysql/data/3310/my_test。

```
create database my_test;
```

10.3.2 限制数据目录大小

```
# 创建一个指定大小的磁盘镜像文件，这里为 600M
dd if=/dev/zero of=/root/test.img bs=60M count=10
```

```
# 挂载设备
losetup /dev/loop0 /root/test.img

# 格式化设备
mkfs.ext3 /dev/loop0

# 挂载为文件夹，则限制其文件夹空间大小为 600M
mount -t ext3 /dev/loop0 /opt/mysql/data/3310/my_test

# 修改属组为 MySQL 服务对应用户
chown -R mysql.mysql /opt/mysql/data/3310/my_test
```

10.3.3 创建一张测试表

```
CREATE TABLE `student` (
`id` int(11) NOT NULL,
`name` varchar(50) NOT NULL,
`score` int(11) DEFAULT NULL,
`age` int(11) DEFAULT NULL,
`sex` varchar(10) DEFAULT NULL,
PRIMARY KEY (`id`)
) ENGINE=InnoDB DEFAULT CHARSET=utf8;
```

10.3.4 插入一些数据

> 注意：数据量不要太大，小于 /opt/mysql/data/3310/my_test 目录的一半，建议 30% 左右。

```
./mysql_random_data_load -h127.0.0.1 -P3310 -uuniverse_op -p'xxx' --max-threads=8 my_test student 1500000
```

10.3.5 修改 /tmp 大小

这里 tmpdir 目录为 /tmp，修改 /tmp 大小为一个较小值。

```
mount -o remount,size=1M tmpfs /tmp
```

10.3.6 修改其他参数

修改 tmp_table_size 和 max_heap_table_size 值为较小值，这里仅仅为了便于生成磁盘临时文件，在生产环境中不建议这样做，会严重影响性能。

```
set sort_buffer_size=128*1024;
set tmp_table_size=128*1024;
```

10.4 场景测试

登录数据库执行如下操作，可以观察到添加索引失败，报错信息如下：

```
mysql> alter table student add  idx_name index(name);
ERROR 1878 (HY000): Temporary file write failure.
```

执行如下操作，修改 /tmp 目录大小，再次执行 ALTER 操作则成功。

```
[root@localhost ~]# mount -o remount,size=500M tmpfs /tmp
mysql> alter table student add index(name);
Query OK, 0 rows affected (4.92 sec)
Records: 0 Duplicates: 0 Warnings: 0
```

观察 /opt/mysql/data/3310/my_test 目录已使用的空间，如果使用率较低，建议继续插入数据直到磁盘空间使用率超过 50%

执行如下操作，会报如下错误：

```
mysql> alter table student engine=innodb;
ERROR 1114 (HY000): The table 'student' is full
```

10.5 问题总结

好了，最后总结一下。为了我们的 Online DDL 操作顺利进行，需要注意在进行操作前，先检查 innodb_online_alter_log_max_size 值，预估是否需要修改。

(1) 可以直接修改为一个较大值，但是没有百分百的好事，如果业务在 DDL 操作期间并发 DML 修改记录较多，Online DDL 结束时锁定表以应用记录的 DML 时间会增加。所以，选择好时机很重要，在对的时间做对的事，当然是在业务低峰期来做，或者利用工具（pt-osc 或 ghost）进行操作。

(2) 在安装实例时即设置 tmpdir 为合理的值。温馨提示，该值不支持动态修改，真出现问题就晚了，毕竟在生产环境中不允许随便重启服务。

(3) 及时关注磁盘空间。不要等到磁盘空间快满了才想着通过 Online DDL 操作进行碎片空间清理，例如 optimize table table_name;、alter table table_name engine=innodb; 等操作，这些操作本身也是需要额外的空间的，否则，等待你的可能是 FAILURE。

11 server_id 引发的级联复制同步异常

作者：蒋士峰

11.1 业务场景

在日常运维的某个系统下，由于之前数据库主机所用硬盘是传统机械硬盘，容量小、传输速度低，并且数据库服务器整体性能不高，随着业务访问量的增加，现有数据库服务器无法满足需求，所以需要搭建一台高性能的数据库服务器，并且所用硬盘是 SSD。

由于原先数据库采用的是主从复制架构，所以新搭建的数据库也要采用主从架构。跟旧数据库集群组成一套级联复制的 MySQL 数据库集群（旧集群的主库作为主，新集群的主库为旧集群主库的从，新集群从库还继续为新集群主库的从），先数据同步一段时间，再找时间点进行业务割接。

由此，从旧集群主库→新集群主库→新集群从库，形成了一条类似于链条式的同步关系，具体关系图如图 18 所示。

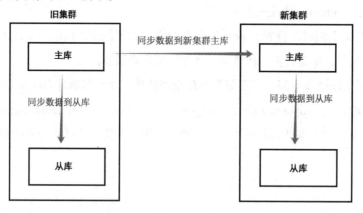

图 18

11.2 问题的发现

搭建完新集群，做级联复制的时候没有发现任何错误，数据同步也是正常的。大概过了 15 天进行数据比对的时候，发现了一个重要问题：新集群的主库可以正常同步旧集群主库的新增数据，但是新集群的从库无法同步新集群主库的新增数据，如图 19 所示。

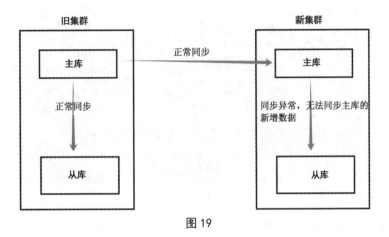

图 19

11.2.1 问题分析

(1) 由于从新集群的主库到新集群从库无法正常同步，所以我们先分析了新集群主库的 binlog 日志是否开启，还有 log_slave_update 是否也开启了，只有开启了，才能产生 binlog 做主从同步。发现都是开启的，所以只能从其他方面去找原因。

(2) 除此之外，我们还专门在旧集群主库上进行了创建库、插入数据操作，观察 position 位置点的变化信息。

创建库，插入数据之前，如图 20~22 所示。

图 20

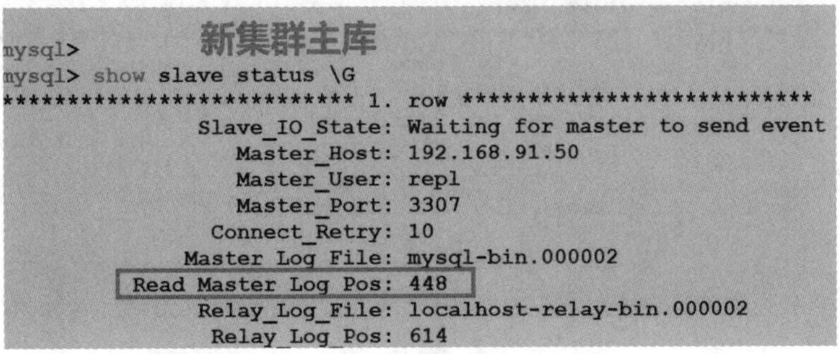

图 21

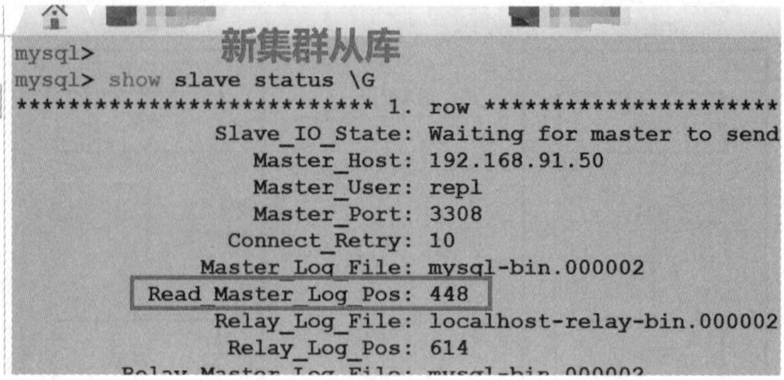

```
mysql>
mysql> show slave status \G
*********************** 1. row ***********************
               Slave_IO_State: Waiting for master to send
                  Master_Host: 192.168.91.50
                  Master_User: repl
                  Master_Port: 3308
                Connect_Retry: 10
              Master_Log_File: mysql-bin.000002
          Read_Master_Log_Pos: 448
               Relay_Log_File: localhost-relay-bin.000002
                Relay_Log_Pos: 614
        Relay_Master_Log_File: mysql-bin.000002
```

新集群从库

图 22

创建库，插入数据之后，如图 23~25 所示。

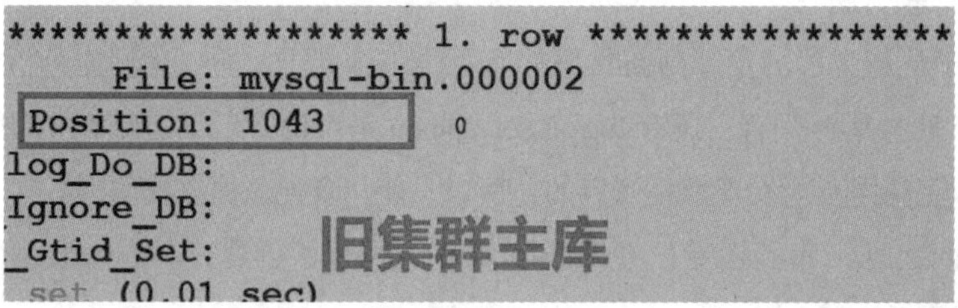

```
*********************** 1. row ***********************
             File: mysql-bin.000002
         Position: 1043                0
     log_Do_DB:
     Ignore_DB:
       _Gtid_Set:
set (0.01 sec)
```

旧集群主库

图 23

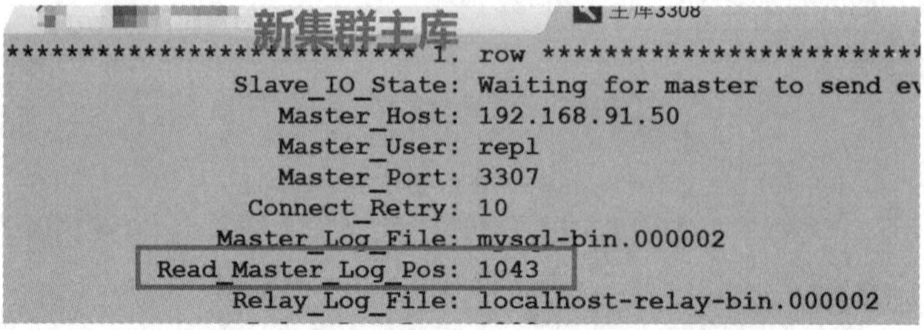

```
*********************** 1. row ***********************
               Slave_IO_State: Waiting for master to send ev
                  Master_Host: 192.168.91.50
                  Master_User: repl
                  Master_Port: 3307
                Connect_Retry: 10
              Master_Log_File: mysql-bin.000002
          Read_Master_Log_Pos: 1043
               Relay_Log_File: localhost-relay-bin.000002
```

新集群主库 主库3308

图 24

```
mysql>
mysql> show slave status \G
*********************** 1. row ***********************
               Slave_IO_State: Waiting for master to send
                  Master_Host: 192.168.91.50
                  Master_User: repl
                  Master_Port: 3308
                Connect_Retry: 10
              Master_Log_File: mysql-bin.000002
          Read_Master_Log_Pos: 448
               Relay_Log_File: localhost-relay-bin.000002
                Relay_Log_Pos: 614
```

新集群从库

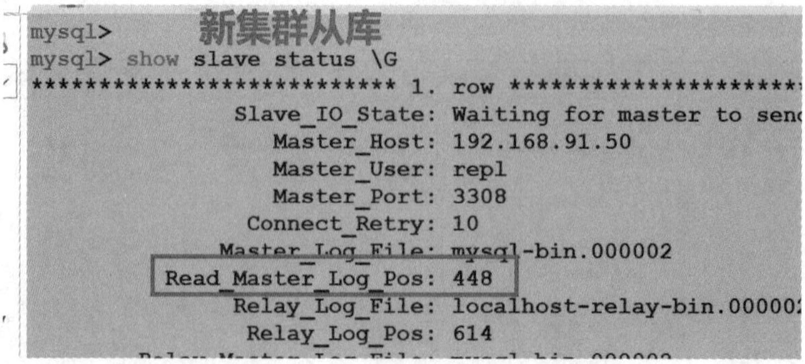

图 25

发现重要问题：插入数据的时候，旧集群主库和新集群主库的 binlog 位置点都发生了变化，只有新集群的从库的 binlog 位置点一直没变，这明显是不正常的。

(3) 前面也确认了 binlog 相关参数都是开启的。所以此时，我们只有把三台数据库的配置文件 my.cnf 拿出来对比一下，检查一下是不是配置文件的相关参数出了问题。

经过对比确认参数，发现了一个主要的问题：旧集群主库的 server_id 为 1，新集群主库的 server_id 为 2，新集群从库的 server_id 为 1（见图 26）。

这意味着什么？旧集群主库的 server_id 与新集群从库的 server_id 重复了。但是问题又来了，当时做主从的时候完全没有报错啊？那么，在级联复制中，是不是也要保证所有的 server_id 不同呢？

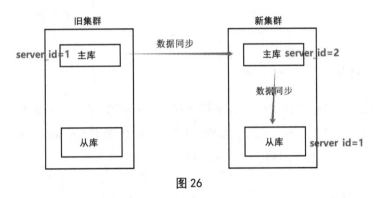

图 26

(4) 带着这个疑问，我们专门在本地环境搭建了一套类似于生产环境的级联复制，并且随意改动 server_id，然后插入数据，观察数据同步情况。我们验证了一条重要信息：在级联复制中，所有参与构建集群的 MySQL 数据库 server_id 不能相同，一旦相同，数据同步就会出现故障。

11.2.2 产生这一问题的根源

在项目中，数据集群众多，手动安装工作量较大，本次数据库是采用自动化安装的，在分配 server_id 的时候，也是随机分配 1 或者 2，所以才导致了本次新集群从库 server_id 跟旧集群冲突了。

11.3 整改步骤

数据已经同步 15 天了，但是我们的 binlog 只保存了 14 天，所以现在只有先修改 server_id，保证该级联复制中所有数据库的 server_id 都是不同的；然后再备份旧集群主库，恢复到新集群，重做级联复制。

11.4 带来的启示

(1) 使用级联复制时，一定要保证所有参与数据库的 server_id 不同。

(2) 要确保 binlog 日志以及相关参数是开启的。

(3) 由于级联复制存在各种小问题，所以日常生产中尽量少用级联复制。

12 TCP 缓存超负荷导致的 MySQL 连接中断

作者：龚唐杰

12.1 背景

在执行跑批任务的过程中，应用程序遇到了一个问题：部分任务的数据库连接会突然丢失，导致任务无法完成。从数据库的错误日志中，发现了 Aborted connection 的信息，这说明客户端和服务器之间的通信被异常中断了。

12.2 分析

为了找出问题的原因，我们首先根据经验，分析了可能导致连接被 Aborted 的几种常见情况：

(1) 客户端没有正确地关闭连接，没有调用 mysql_close() 函数。

(2) 客户端空闲时间超过了 wait_timeout 或 interactive_timeout 参数设置的秒数，服务器自动断开了连接。

(3) 客户端发送或接收的数据包大小超过了 max_allowed_packet 参数的值，导致连接中断。

(4) 客户端试图访问数据库，但没有权限，或者使用了错误的密码，或者连接包不包含正确的信息。

然而，经过排查，发现当前的问题都不属于以上情况。

(1) 因为任务之前都是正常运行的，而且程序也没有变动，所以可以排除第一种情况。

(2) 我们查看了 MySQL 的超时参数 wait_timeout 和 interactive_timeout，发现它们都是 28800，也就是 8 个小时，远远超过了任务执行时间，所以可以排除第二种情况。

(3) 我们检查了客户端和服务器的 max_allowed_packet 参数，发现它们都是 64M，也不太可能超过这个限制，所以可以排除第三种情况。

(4) 我们确认了客户端的数据库访问权限、密码、连接包等信息，都是正确的，所以可以排除第四种情况。

到此，我们初步判断 MySQL 层面应该没有问题，问题可能出在其他地方。

为了进一步定位问题，我们尝试了修改服务器的一些相关内核参数，如下：

```
net.ipv4.tcp_keepalive_intvl = 30
net.ipv4.tcp_keepalive_probes = 3
net.ipv4.tcp_keepalive_time = 120
net.core.rmem_default = 2097152
net.core.wmem_default = 2097152
net.ipv4.tcp_tw_reuse = 1
net.ipv4.tcp_max_syn_backlog = 16384
```

这些参数主要是为了优化网络连接的性能和稳定性，避免连接被意外关闭或超时。但是，修改后的结果并没有改善，连接还是会异常中断。

最后，我们尝试了进行抓包分析，通过 Wireshark 工具，发现了一个异常的现象：服务器会给客户端发送大量的 ACK 包，如图 27 所示。

```
TCP  8754 3306 → 42656 [ACK] Seq=2416874 Ack=1 Win=59 Len=8688 TSval=1169008248 TSecr=2576059429 [TCF
TCP  8754 3306 → 42656 [ACK] Seq=2425562 Ack=1 Win=59 Len=8688 TSval=1169008248 TSecr=2576059429 [TCF
TCP  8754 3306 → 42656 [ACK] Seq=2434250 Ack=1 Win=59 Len=8688 TSval=1169008248 TSecr=2576059429 [TCF
TCP  8754 3306 → 42656 [ACK] Seq=2442938 Ack=1 Win=59 Len=8688 TSval=1169008248 TSecr=2576059429 [TCF
TCP  8754 3306 → 42656 [ACK] Seq=2451626 Ack=1 Win=59 Len=8688 TSval=1169008248 TSecr=2576059429 [TCF
TCP  8754 3306 → 42656 [ACK] Seq=2460314 Ack=1 Win=59 Len=8688 TSval=1169008248 TSecr=2576059429 [TCF
TCP  8754 3306 → 42656 [ACK] Seq=2469002 Ack=1 Win=59 Len=8688 TSval=1169008248 TSecr=2576059429 [TCF
TCP  4410 3306 → 42656 [ACK] Seq=2477690 Ack=1 Win=59 Len=4344 TSval=1169008248 TSecr=2576059429 [TCF
TCP  8754 3306 → 42656 [ACK] Seq=2482034 Ack=1 Win=59 Len=8688 TSval=1169008248 TSecr=2576059429 [TCF
TCP  8754 3306 → 42656 [ACK] Seq=2490722 Ack=1 Win=59 Len=8688 TSval=1169008248 TSecr=2576059429 [TCF
TCP  8754 3306 → 42656 [ACK] Seq=2499410 Ack=1 Win=59 Len=8688 TSval=1169008248 TSecr=2576059429 [TCF
TCP  8754 3306 → 42656 [ACK] Seq=2508098 Ack=1 Win=59 Len=8688 TSval=1169008248 TSecr=2576059429 [TCF
TCP  8754 3306 → 42656 [ACK] Seq=2516786 Ack=1 Win=59 Len=8688 TSval=1169008248 TSecr=2576059429 [TCF
TCP  8754 3306 → 42656 [ACK] Seq=2525474 Ack=1 Win=59 Len=8688 TSval=1169008248 TSecr=2576059429 [TCF
TCP  8754 3306 → 42656 [ACK] Seq=2534162 Ack=1 Win=59 Len=8688 TSval=1169008248 TSecr=2576059429 [TCF
TCP  4410 3306 → 42656 [ACK] Seq=2542850 Ack=1 Win=59 Len=4344 TSval=1169008248 TSecr=2576059429 [TCF
TCP  8754 3306 → 42656 [ACK] Seq=2547194 Ack=1 Win=59 Len=8688 TSval=1169008248 TSecr=2576059429 [TCF
```

图 27

这些 ACK 包是 TCP 协议中的确认包，表示服务器已经收到了客户端的数据包，请求客户端继续发送数据。但是，为什么服务器会发送这么多的 ACK 包呢？我们猜测可能是网络有异常，导致客户端接收不到服务器返回的 ACK 包，所以服务器会反复发送，直到超时或收到客户端的响应。但是，经过网络人员的排查，未发现明显的问题。

继续分析抓包，我们又发现了另一个异常的现象：客户端会给服务器发送一些窗口警告，如图 28 所示。

Protocol	Length	Info
TCP	66	52164 → 3306 [ACK] Seq=439 Ack=53241881 Win=181760 Len=0 TSval=1744
MySQL	65226	ResponseResponse
MySQL	65226	ResponseResponse
TCP	66	52164 → 3306 [ACK] Seq=439 Ack=53372201 Win=51712 Len=0 TSval=17442
MySQL	51778	[TCP Window Full] ResponseResponse
TCP	66	[TCP ZeroWindow] 52164 → 3306 [ACK] Seq=439 Ack=53423913 Win=0 Len
TCP	66	[TCP Window Update] 52164 → 3306 [ACK] Seq=439 Ack=53423913 Win=400
MySQL	13514	ResponseResponse
MySQL	65226	ResponseResponse
MySQL	65226	ResponseResponse
MySQL	65226	ResponseResponse
TCP	66	52164 → 3306 [ACK] Seq=439 Ack=53632841 Win=191488 Len=0 TSval=1744
MySQL	65226	ResponseResponse
MySQL	65226	ResponseResponse
TCP	66	52164 → 3306 [ACK] Seq=439 Ack=53763161 Win=61440 Len=0 TSval=17442
MySQL	61506	[TCP Window Full] Response
TCP	66	[TCP ZeroWindow] 52164 → 3306 [ACK] Seq=439 Ack=53824601 Win=0 Len
TCP	66	[TCP Window Update] 52164 → 3306 [ACK] Seq=439 Ack=53824601 Win=399
MySQL	3786	ResponseResponse
MySQL	65226	ResponseResponse
MySQL	65226	ResponseResponse
MySQL	65226	ResponseResponse
TCP	66	52164 → 3306 [ACK] Seq=439 Ack=54023801 Win=200192 Len=0 TSval=1744
MySQL	65226	ResponseResponse
MySQL	65226	Response
MySQL	65226	ResponseResponse
TCP	66	52164 → 3306 [ACK] Seq=439 Ack=54219281 Win=5120 Len=0 TSval=174423
MySQL	5186	[TCP Window Full] ResponseResponse
TCP	66	[TCP ZeroWindow] 52164 → 3306 [ACK] Seq=439 Ack=54224401 Win=0 Len
TCP	66	[TCP Window Update] 52164 → 3306 [ACK] Seq=439 Ack=54224401 Win=393

图 28

这些窗口警告是 TCP 协议中的流量控制机制，表示服务器或客户端的接收窗口已经满了，不能再接收更多的数据。

[TCP Window Full] 是发送端向接收端发送的一种窗口警告，表示已经到数据接收端的极限了。

[TCP ZeroWindow] 是接收端向发送端发送的一种窗口警告，告诉发送者，接收端接收窗口已满，暂时停止发送。

根据以上信息，我们推测出问题的原因：由于 MySQL 需要发送的数据太大，客户端的 TCP 缓存已经满了，所以需要等待客户端把 TCP 缓存里面的数据消化掉，才能继续接收数据。但是，在这段时间内，MySQL 会一直向客户端请求继续发送数据，如果客户端在一定时间内（默认是 60 秒）没有响应，MySQL 就会认为发送数据超时，中断了连接。

为了验证推测，查看 MySQL 的慢日志，发现了很多 Last_errno: 1161 的记录。这些记录表示 MySQL 在发送数据时遇到了超时错误，而且出现的次数和应用程序失败的任务数很接近。根据 MySQL 官网的说明，这个错误的含义是：

> Error number: 1161; Symbol: ER_NET_WRITE_INTERRUPTED; SQLSTATE: 08S01
> Message: Got timeout writing communication packets

可知这个错误表示的意思是网络写入中断，而 MySQL 层面有个参数就是控制这个问题的，所以尝试更改 net_write_timeout 参数为 600，跑批任务正常运行。

所以，MySQL 连接被异常中断的原因在于客户端获取的数据太大，超过了客户端 TCP 缓存，客户端需要先处理缓存中的数据，在这段时间内，MySQL 会一直向客户端请求继续发送数据，但是客户端 60 秒内一直未能响应，导致 MySQL 发送数据超时，中断了连接。

12.3 结论

通过上述的分析和尝试，我们得出了以下的结论：

(1) 抓包信息中有很多 ACK 信息，是因为客户端的缓存满了而不能及时给服务端反馈，所以服务器会反复发送 ACK 信息，直到超过 60 秒（net_write_timeout 默认值是 60），导致 MySQL 把连接中断了。

(2) 慢日志中有很多 Last_errno: 1161 的记录，是因为该 SQL 实际已经在 MySQL 中执行完了，但是在发送数据到客户端时，由于数据量太大，超过了客户端的 TCP 缓存，然后客户端上的应用在 60 秒内未把缓存中的数据处理掉，导致 MySQL 往客户端发送数据超时。

(3) 在 MySQL 层面调整 net_write_timeout 参数只能缓解这个现象，其根本原因在于单个 SQL 获取的数据量太大，超过了客户端的缓存大小，应用程序不能在短时间内处理完缓存中的数据，进而导致后续的数据发送超时。

12.4 优化建议

(1) 业务层面分批处理数据时，避免单个 SQL 从服务器获取大量的数据，导致客户端的 TCP 缓存不足。

(2) 增加 MySQL 中的 net_write_timeout 参数值或者增加客户端的 TCP 缓存，可缓解此情况的发生，但不能彻底解决该问题，因为数据量太大仍然会影响性能和稳定性。

(3) 优化 SQL 语句，减少不必要的数据返回，比如使用 LIMIT、WHERE 等条件，或者使用聚合函数，分组函数等，以减少数据量和提高查询效率。

13 MySQL 通过 systemd 启动时无响应

作者：贲绍华

13.1 引言

正如题目所述，在自动化测试场景下，通过 systemd 无法启动 MySQL。连续 kill -9，结束实例进程，检测 mysqld 在退出后是否会被正确拉起。

具体信息如下：

(1) 主机信息：CentOS 8（Docker 容器）

(2) 使用 systemd 的方式管理 mysqld 进程

(3)systemd service 的运行模式为：forking

(4) 启动命令如下：

```
# systemd 启动命令
sudo -S systemctl start mysqld_11690.service

# systemd service 内的 ExecStart 启动命令
/opt/mysql/base/8.0.34/bin/mysqld --defaults-file=/opt/mysql/etc/11690/my.cnf
--daemonize --pid-file=/opt/mysql/data/11690/mysqld.pid --user=actiontech-mysql
--socket=/opt/mysql/data/11690/mysqld.sock --port=11690
```

13.2 现象描述

启动命令持续 hang 住，既不成功，也无任何返回信息，尝试几次后均无法手动复现该场景。

图 29 为复现场景，service 端口号不一致，请忽略。

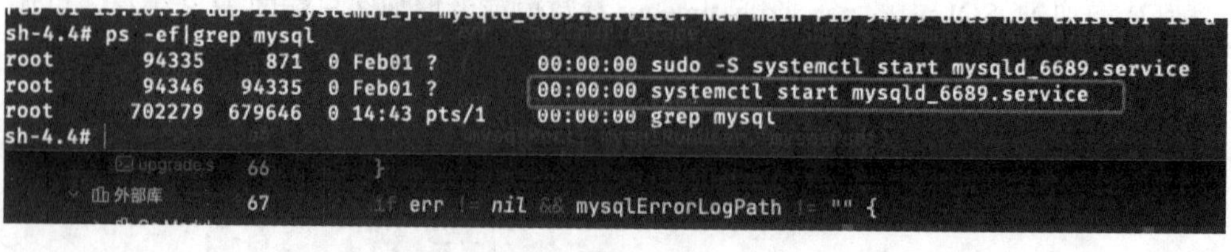
sh-4.4# ps -ef|grep mysql
root 94335 871 0 Feb01 ? 00:00:00 sudo -S systemctl start mysqld_6689.service
root 94346 94335 0 Feb01 ? 00:00:00 systemctl start mysqld_6689.service
root 702279 679646 0 14:43 pts/1 00:00:00 grep mysql
sh-4.4#
66 }
外部库 67 if err != nil && mysqlErrorLogPath != "" {

图 29

MySQL 错误日志无任何信息。查看 systemd service 状态，发现启动脚本中由于缺少参数 MAIN PID，执行失败。

systemd 最后输出的信息为：New main PID 31036 does not exist or is a zombie（见图 30、图 31）。

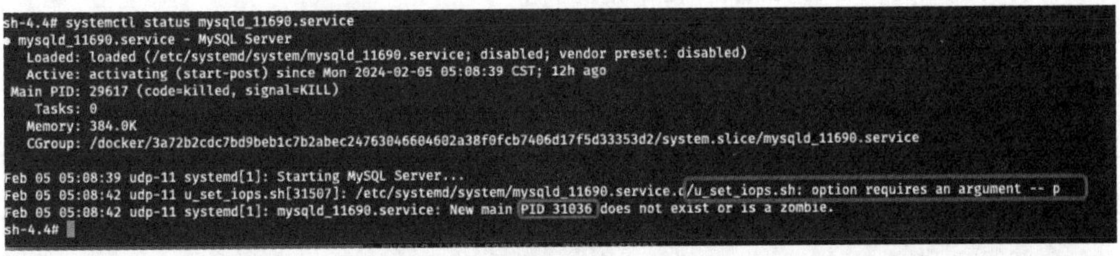

图 30

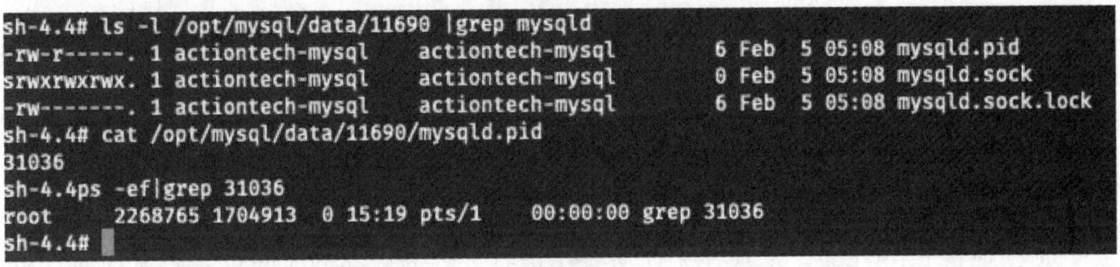

图 31

13.3 原因总结

systemd 启动 mysqld 的过程中，会先根据 service 模板中的配置，执行：

(1)ExecStart（启动 mysqld）。

(2)mysqld 启动创建 pid 文件。

(3)ExecStartPost（自定义的一些后置脚本，如调整权限、将 pid 写入 cgroup 等）。

在步骤 2~3 的中间态，也就是 pid 文件刚创建出来时，主机接收到自动化测试下发的命令：sudo -S kill -9 $(cat /opt/mysql/data/11690/mysqld.pid)。

由于这个 pid 文件和 pid 进程确实存在（如果不存在，kill 命令或 cat 会报错），自动化的 CASE 认为 kill 操作已成功结束。但由于 mysqld.pid 这个文件是由 MySQL 自身维护的，在 systemd 的视角中，还需要继续等待步骤 3 完成，才认为启动成功。

在 systemd 使用 forking 模式时，会根据子进程的 PID 值判断服务是否成功启动。

如果子进程成功启动，并且没有发生意外退出，则 systemd 会认为服务已启动，并将子进程的 PID 作为 MAIN PID。

而如果子进程启动失败或意外退出，则 systemd 会认为服务未能成功启动。

在执行 ExecStartPost 时，由于子进程 ID 31036 已经被 kill，后置 shell 缺少了启动参数，但 ExecStart 步骤已完成，导致 MAIN PID 31036 成了只存在于 systemd 里的僵尸进程。

13.4 排查过程

当遇到这个问题时，人是有点蒙的，我简单检查了一下内存、磁盘基本信息，发现这些都符合预期并没有出现资源不足的情况。

先从 MySQL 的 Error Log 入手，看看有什么发现。查看结果如下：

```
... 无关内容省略 ...
2024-02-05T05:08:42.538326+08:00 0 [Warning] [MY-010539] [Repl] Recovery
from source pos 3943309 and file mysql-bin.000001 for channel ''. Previous relay
log pos and relay log file had been set to 4, /opt/mysql/log/relaylog/11690/mysql-
relay.000004 respectively.
2024-02-05T05:08:42.548513+08:00 0 [System] [MY-010931] [Server] /opt/mysql/
base/8.0.34/bin/mysqld: ready for connections. Version: '8.0.34'  socket: '/opt/
mysql/data/11690/mysqld.sock'  port: 11690  MySQL Community Server - GPL.
2024-02-05T05:08:42.548633+08:00 0 [System] [MY-013292] [Server] Admin
interface ready for connections, address: '127.0.0.1' port: 6114
2024-02-05T05:08:42.548620+08:00 5 [Note] [MY-010051] [Server] Event
Scheduler: scheduler thread started with id 5
```

通过观察 Error Log，发现并无任何有用信息，因为启动的时间点之后无任何日志信息输出。

查看 systemctl status 确认服务当前状态（见图 32）：

图 32

图 33 为正常情况下的 status 信息：

图 33

通过对比，整理两条有用信息：

(1) 后置 shell 由于缺少 -p 参数导致执行失败（-p 参数为 MAIN PID，也就是 fork 子进程启动后的 PID）。

(2)systemd 无法获取 PID 31036，则此进程可能不存在或者为僵尸进程。

先来检查进程 ID 与 mysqld.pid 看看（见图 34、图 35）：

```
sh-4.4# ls -l /opt/mysql/data/11690 |grep mysqld
-rw-r-----. 1 actiontech-mysql    actiontech-mysql    6 Feb  5 05:08 mysqld.pid
srwxrwxrwx. 1 actiontech-mysql    actiontech-mysql    0 Feb  5 05:08 mysqld.sock
-rw-------. 1 actiontech-mysql    actiontech-mysql    6 Feb  5 05:08 mysqld.sock.lock
sh-4.4# cat /opt/mysql/data/11690/mysqld.pid
31036
sh-4.4ps -ef|grep 31036
root      2268765 1704913  0 15:19 pts/1    00:00:00 grep 31036
sh-4.4#
```

图 34

```
top - 17:18:56 up 5 days,  4:57,  0 users,  load average: 30.88, 23.59, 22.83
Tasks:  31 total,   2 running,  29 sleeping,   0 stopped,   0 zombie
%Cpu(s): 14.0 us, 14.6 sy,  0.0 ni, 68.3 id,  0.0 wa,  1.5 hi,  1.6 si,  0.0 st
MiB Mem :  257622.9 total,  132548.3 free,   89668.6 used,   35406.0 buff/cache
MiB Swap:       0.0 total,       0.0 free,       0.0 used.  166055.4 avail Mem

    PID USER      PR  NI    VIRT    RES    SHR S  %CPU  %MEM     TIME+ COMMAND
   1487 action+   20   0 4535880  40628  16256 S   4.7   0.0  29:28.18 ushard
    643 action+   20   0 5014388 100092  19444 S   3.7   0.0  36:51.61 ustats
    556 action+   20   0 4170744  62092  13556 S   1.3   0.0  12:37.68 uagent
   2588 action+   20   0   36.8g   3.8g  12456 S   1.3   1.5  16:40.18 java
   1075 action+   20   0 2909208  32456  14720 S   1.0   0.0   6:34.32 urman-agent
 506570 action+   20   0 1375048   5924   2852 S   1.0   0.0   7:34.05 redis-server
     51 dbus      20   0   54184   2952   2276 S   0.7   0.0   1:18.72 dbus-daemon
    852 action+   20   0 3290748  40656  17188 S   0.7   0.0   6:37.11 uguard-agent
    775 action+   20   0 2941904  42568  19008 S   0.3   0.0   4:40.35 udeploy
 612749 action+   20   0 1370440   5840   2844 S   0.3   0.0   2:56.86 redis-server
```

图 35

确认线索：

(1)PID 31036 不存在。

(2)mysqld.pid 文件存在，且文件内容为 31036。

(3) 用 top 命令查看，发现不存在僵尸进程。

还需要获取更多的线索来确认原因。检查 journalctl -u 内容，看看是否有帮助：

```
sh-4.4# journalctl -u mysqld_11690.service
-- Logs begin at Mon 2024-02-05 04:00:35 CST, end at Mon 2024-02-05 17:08:01 CST. --
Feb 05 05:07:54 udp-11 systemd[1]: Starting MySQL Server...
Feb 05 05:07:56 udp-11 systemd[1]: Started MySQL Server.
Feb 05 05:08:31 udp-11 systemd[1]: mysqld_11690.service: Main process exited, code=killed, status=9/KILL
Feb 05 05:08:31 udp-11 systemd[1]: mysqld_11690.service: Failed with result 'signal'.
Feb 05 05:08:32 udp-11 systemd[1]: Starting MySQL Server...
Feb 05 05:08:36 udp-11 systemd[1]: Started MySQL Server.
Feb 05 05:08:37 udp-11 systemd[1]: mysqld_11690.service: Main process exited, code=killed, status=9/KILL
```

```
    Feb 05 05:08:37 udp-11 systemd[1]: mysqld_11690.service: Failed with result
'signal'.
    Feb 05 05:08:39 udp-11 systemd[1]: Starting MySQL Server...
    Feb 05 05:08:42 udp-11 u_set_iops.sh[31507]: /etc/systemd/system/mysqld_11690.
service.d/u_set_iops.sh: option requires an argument -- p
    Feb 05 05:08:42 udp-11 systemd[1]: mysqld_11690.service: New main PID 31036
does not exist or is a zombie.
```

这里的 journalctl -u 内容也只描述了现象，无法分析具体原因，与 systemctl status 的内容相差不多，帮助不大。

查看 /var/log/messages 系统日志内容（见图 36）：

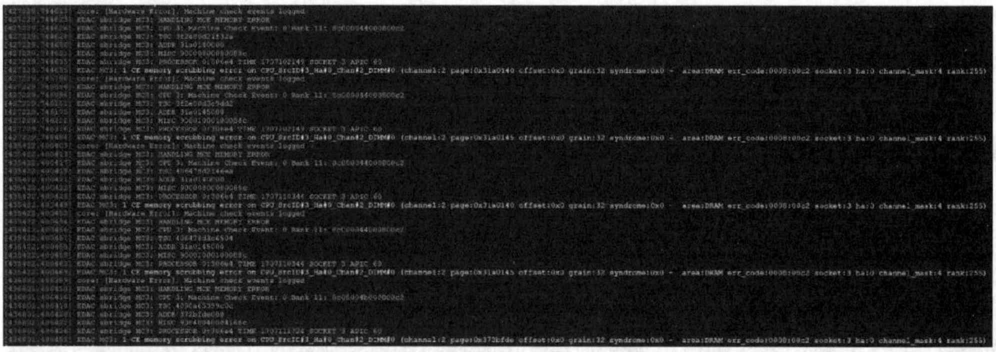

图 36

发现日志中循环报出了一些内存方面的错误信息，通过搜索后发现该错误可能为硬件问题。我询问自动化测试的同事后，得到结论：

(1) 场景为偶发问题，执行 4 次用例，2 次成功，2 次失败。

(2) 每次执行均为同一台宿主机，同一份容器镜像。

(3) 失败时 hang 住的容器为同一个。

既然有成功执行的结果，这里就先排除硬件问题。

既然提到了容器，想到 cgroup 会不会映射宿主机的时候出现了问题？观察前面排查的 systemctl status，可知 cgroup 映射的宿主机目录为：

```
    CGroup: /docker/3a72b2cdc7bd9beb1c7b2abec24763046604602a38f0fcb7406d17f5d3335
3d2/system.slice/mysqld_11690.service
```

检查父级文件夹 system.slice 的读写权限，并无异常。先暂时排除 cgroup 的映射问题（因为主机上还有其他 systemd 接管的 service 也在使用同一份 cgroup）。

尝试用 pstacksj 查看 systemd 具体 hang 在了哪个地方，其中 3048143 为 systemctl start 的 pid：

```
    sh-4.4# pstack 3048143
    #0  0x00007fdfaef33ade in ppoll () from /lib64/libc.so.6
```

```
    #1  0x00007fdfaf7768ee in bus_poll () from /usr/lib/systemd/libsystemd-
shared-239.so
    #2  0x00007fdfaf6a8f3d in bus_wait_for_jobs () from /usr/lib/systemd/
libsystemd-shared-239.so
    #3  0x000055b4c2d59b2e in start_unit ()
    #4  0x00007fdfaf7457e3 in dispatch_verb () from /usr/lib/systemd/libsystemd-
shared-239.so
    #5  0x000055b4c2d4c2b4 in main ()
```

发现 start_unit 比较可疑，此函数位于可执行文件中，它用于启动 systemd units，并没有什么帮助。

根据已有线索推测后可知：

(1)mysqld.pid 文件存在，表示之前确实有一个 mysqld 且进程号为 31036 的进程被启动了。

(2) 该进程启动后被自动化用例以 kill -9 结束。

(3)systemd 获取了一个已经被结束的 MAIN PID，导致后置 shell 执行失败，fork 流程失败。

通过梳理 systemd 启动流程的步骤，推测可能性。MySQL 实例只有在 mysqld 成功启动后才会生成 mysqld.pid 文件，所以可能是在后续步骤里被意外 kill -9 结束的。

13.5 复现方式

既然没什么其他头绪和线索，就根据推测结论尝试复现一下试试。

13.5.1 调整 systemd mysql serivce 模板

编辑模板文件 /etc/systemd/system/mysqld_11690.service。在 mysqld 启动后，sleep 10 秒，以便在这个时间窗口内模拟 kill 掉实例进程的场景（见图 37）。

```
Delegate=yes

# Set cgroups which systemd do not support
ExecReload=/etc/systemd/system/mysqld_11690.service.d/u_set_iops.sh -a add -p $MAINPID
ExecStartPost=/etc/systemd/system/mysqld_11690.service.d/u_set_iops.sh -a add -p $MAINPID
ExecStopPost=/etc/systemd/system/mysqld_11690.service.d/u_set_iops.sh -a delete

# Start main service
ExecStart=/bin/bash -c '/opt/mysql/base/8.0.34/bin/mysqld --defaults-file=/opt/mysql/etc/11690/my.cnf --dae
qld.pid --user=actiontech-mysql --socket=/opt/mysql/data/11690/mysqld.sock --port=11690 && sleep 10'

# Use this to switch malloc implementation
EnvironmentFile=-/etc/sysconfig/mysql

PrivateTmp=false
```

图 37

13.5.2 配置重载

执行 systemctl daemon-reload，使变更生效。

13.5.3 场景重现

(1)[ssh seesion A] 首先准备一个新的容器，做好相关配置后执行 sudo -S systemctl start mysqld_11690.service 启动一个 mysqld 进程，此时会因为 sleep 的原因 hang 住会话。

(2)[ssh seesion B] 在另一个会话窗口，当 start 命令 hang 住时，检查 mysqld.pid 文件，一旦文件被创建，立刻执行 sudo -S kill -9 $(cat /opt/mysql/data/11690/mysqld.pid)。

(3) 此时观察 systemctl status，其表现与预期一致（见图 38）。

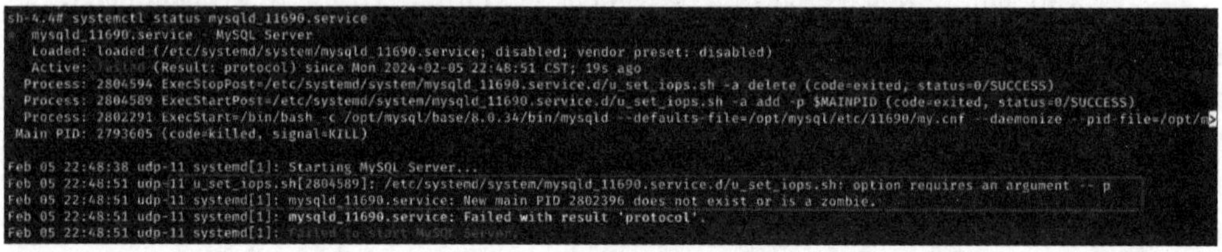

图 38

13.6 解决方式

先 kill 掉 hang 住的 systemctl start 命令，执行 systemctl stop mysqld_11690.service，这可以让 systemd 主动结束僵尸进程，虽然 stop 命令可能会报错，但这并不影响。

等待 stop 执行完成，再次使用 start 命令启动，恢复正常。

虽然文章跟 MySQL 没太大关系，但重要的是分析偶发故障的思考过程。

14 ERROR 1709: Index column size too large 引发的思考

作者：王田田

14.1 背景

某日，同事突然找到我，说测试环境中有张表无法访问，执行 SELECT、DML 和 DDL 均报错：

```
ERROR 1709 (HY000): Index column size too large. The maximum column size is 767 bytes.
```

其实看到 767 这个数字，大家可能会猜想估计和 compact/redundant 行格式有关系，后续也确实证实了和这个问题有点关系。

问题发生了就要想办法处理。当时第一反应是能不能做些"特殊操作"调整一下元数据，但能力有限无法实现。由于是测试环境，数据没那么重要，而且还是单节点，后续处理无非是利用备份重做这套库；若不想重做，而且该表不重要，也可以直接废弃该表，但是 xtrabackup 备份可能会报错。

既然问题一旦发生只能通过备份恢复来解决，那么我们应该探究一下如何提前避免该问题。

14.2 原因探究

以下为测试环境复现过程。

14.2.1 MySQL 5.6.21 原地升级至 MySQL5.7.20

先调整数据库配置文件，以下为简要升级步骤：

```
shell>/mysql/mysql-5.7.20/bin/mysqld_safe ... &
shell>/mysql/mysql-5.7.20/bin/mysql_upgrade ...
mysql>shutdown;
shell>/mysql/mysql-5.7.20/bin/mysqld_safe ... &
```

14.2.2 MySQL 5.7.20 原地升级至 MySQL8.0.21

先调整数据库配置文件，以下为简要升级步骤：

```
mysql>/mysql/mysql-8.0.21/bin/mysqld_safe ... &
mysql>shutdown;
shell>/mysql/mysql-8.0.21/bin/mysqld_safe ... &
```

14.2.3 MySQL 8.0.21 数据库添加字段并添加索引

表默认字符集为 utf8。

```
mysql> alter table sky.test add column test_col varchar(500);
Query OK, 0 rows affected (10.09 sec)
Records: 0  Duplicates: 0  Warnings: 0

mysql> alter table sky.test add index idx_test_col(test_col);
Query OK, 0 rows affected (0.02 sec)
Records: 0  Duplicates: 0  Warnings: 0
```

正常情况下，这个索引应无法创建成功，会立即抛出错误 ERROR 1071 (42000):Specified key was too long; max key length is 767 bytes。当然，一方面原因是 MySQL5.7 及 MySQL8.0 默认行格式为 dynamic，另一方面即使显式指定 row_format=compact，也会立即抛出错误。示例如下：

```
mysql>create table sky1 (id int);
```

```
Query OK, 0 rows affected (0.05 sec)

mysql>alter table sky1 add column test_col varchar(500);
Query OK, 0 rows affected (0.03 sec)
Records: 0  Duplicates: 0  Warnings: 0

mysql>alter table sky1  add index idx_test_col(test_col);
Query OK, 0 rows affected (0.03 sec)
Records: 0  Duplicates: 0  Warnings: 0

mysql>create table sky2(id int) row_format=compact;
Query OK, 0 rows affected (0.06 sec)

mysql>alter table sky2 add column test_col varchar(500);
Query OK, 0 rows affected (0.04 sec)
Records: 0  Duplicates: 0  Warnings: 0

mysql>alter table sky2 add index idx_test_col(test_col);
ERROR 1071 (42000): Specified key was too long; max key length is 767 bytes
```

数据库重启前，该表可正常访问。

14.2.3.1 重启数据库

```
systemctl stop mysqld_3306

systemctl start mysqld_3306
```

14.2.3.2 查看表情况

```
mysql> select *from sky.test limit 1;
ERROR 1709 (HY000): Index column size too large. The maximum column size is 767 bytes.

mysql> alter table sky.test row_format=dynamic;
ERROR 1709 (HY000): Index column size too large. The maximum column size is 767 bytes.

mysql> alter table sky.test engine=innodb;
ERROR 1709 (HY000): Index column size too large. The maximum column size is 767 bytes.

mysql> check table sky.test ;
+------------------+--------+----------+-------------------------------------------------------------+
| Table            | Op     | Msg_type | Msg_text                                                    |
+------------------+--------+----------+-------------------------------------------------------------+
| sky.test         | check  | Error    | Index column size too large. The
```

```
maximum column size is 767 bytes. |
    | sky.test          | check  | Error       | Table 'sky.test' doesn't exist
|
    | s k y . t e s t          | c h e c k    | e r r o r    | C o r r u p t
|
    +------------------+--------+-------------+--------------------------------
------------------------------+
    3 rows in set (0.01 sec)
```

14.2.3.3 查看相关信息

```
mysql>select TABLE_SCHEMA,TABLE_NAME,ROW_FORMAT,CREATE_OPTIONS from
information_schema.tables where table_schema='sky';
+--------------+------------+------------+---------------------+
| TABLE_SCHEMA | TABLE_NAME | ROW_FORMAT | CREATE_OPTIONS      |
+--------------+------------+------------+---------------------+
| sky          | test       | Compact    |                     |
| sky          | sky1       | Dynamic    |                     |
| sky          | sky2       | Compact    | row_format=COMPACT  |
+--------------+------------+------------+---------------------+
```

14.2.3.4 找不同的粗略猜想

sky2 表比 test 表多了一个 create_options 选项，所以不会触发 Bug。而且 create_options 是建表时显式指定的行格式 compact，而 test 表是在 MySQL5.6 版本隐式创建的行格式 compact；MySQL8.0 默认创建表的行格式为 Dynamic（由 innodb_default_row_format 参数控制），Dynamic 行格式不会存在 "767 bytes" 的限制。

碰到这样奇奇怪怪的问题，第一反应就是不走运，碰到了 Bug，因此先去 Bug 库中搜索一番，果不其然搜到了 Bug #99791，与我们测试环境的情况极为类似。

Bug #99791 表明官方在 MySQL 8.0.22 版本修复了非显式定义的 redundant 行格式表允许创建的索引列大小超 767 bytes 的 Bug。实际上，笔者在测试环境验证了一下，MySQL 8.0.22 确实已解决该问题，即隐式创建的 compact 行格式表在待创建的索引列超 767 bytes 时直接返回错误 ERROR 1071 (42000): Specified key was too long; max key length is 3072 bytes。因此，猜想虽然该 Bug 行格式与笔者本次环境对不上，但官方解决的应该是同一个问题，都是为了解决因隐式定义 compact/redundant 行格式而导致的问题。

14.3 解决方案

综上所述，我们可以得出以下解决方案：

(1)MySQL 5.6 升级至 MySQL 8.0.21 时，避免使用原地升级的方案，可新建一个 MySQL 8.0.21 的环境，将数据逻辑导入并搭建复制关系；若 MySQL 8.0.21 环境设置了 innodb_default_row_format=Dynamic 参数，在逻辑导入/复制时新环境会自动将行格式转为

Dynamic。

(2) 升级时选择高于 MySQL 8.0.21 版本的数据库，避免触发该 Bug。

(3) 若当前已经存在 MySQL 5.6 原地升级至 MySQL 8.0.21 的环境：

①可通过以下 SQL 语句排查是否存在超过 767 bytes 的问题表；若存在，可以趁现在数据库未重启，改造涉及的索引。

```
select s.table_schema,s.table_name,s.index_name,s.column_name from information_schema.statistics s,information_schema.columns c,information_schema.tables i where s.table_name=c.table_name and s.table_schema=c.table_schema and c.column_name=s.column_name and s.table_name=i.table_name and s.table_schema=i.table_schema and i.row_format in ('Redundant','Compact') and (s.sub_part is null or s.sub_part>255) and c.character_octet_length >767;
```

②筛选出隐式创建行格式为 compact/redundant 的表，并显式指定，如 alter table xx row_format=dynamic/compact。相关 SQL 如下：

```
select TABLE_SCHEMA,TABLE_NAME,ROW_FORMAT,CREATE_OPTIONS from information_schema.tables where ROW_FORMAT in ('Compact','Redundant') and CREATE_OPTIONS='';
```

由于笔者能力实在有限，如有错误还望大家能够批评指正。

15 MySQL 5.7 连续 Crash 引发 GTID 丢失

作者：许天云

15.1 问题现象

在生产环境中（MySQL 5.7.26 版本），当主库短时间内连续遇到 2 次 Crash 的特殊场景时，会导致备库重新建立复制时会抛出错误：Slave has more GTIDs than the master has，IO 线程复制报错。

```
2023-12-11T10:32:43.433707+08:00 1457551 [ERROR] Error reading packet from server for channel '': Slave has more GTIDs than the master has, using the master's SERVER_UUID. This may indicate that the end of the binary log was truncated or that the last binary log file was lost, e.g., after a power or disk failure when sync_binlog != 1. The master may or may not have rolled back transactions that were already replicated to the slave. Suggest to replicate any transactions that master has rolled back from slave to master, and/or commit empty transactions on master to account for transactions that have been (server_errno=1236)
```

如图 39、40 所示，备库比主库多了几十万个 GTID，而且解析对应的 Binlog，发现都是业务操作，更像是主库把这部分 GTID "丢掉了"。

- 主库：1-1080678246。
- 备库：1-1081067155。

图 39

图 40

15.2 问题分析

先来复习下 MySQL 5.7 GTID 持久化的原理。

(1)gtid_executed 变量：它是一个处于内存中的 GTID SET，表示数据库中执行了哪些 GTID，会实时更新，但是一旦重启就会丢失。show slave status 的 Executed_Gtid_Set 和 show master status 中的 Executed_Gtid_Set 都来自这个变量。

(2)mysql.gtid_executed 表：GTID 持久化的介质，只有在 binlog 切换时才会触发更新。将该 binlog 中的 GTID SET 记录到表中，所以该表中会记录所有历史 binlog 中的 GTID SET。

当 MySQL 启动时，会初始化 gtid_executed 变量。通过读取 mysql.gtid_executed 表的持久化记录（已持久化的 GTID），再加上扫描最后一个 binlog 的 GTID（未持久化的 GTID）合并后完成初始化。新的 GTID 会基于 gtid_executed 变量递增产生（见图 41）。

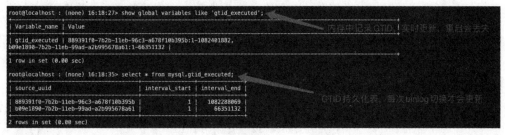

图 41

基于上述 GTID 持久化的原理，我们就有理由怀疑是主库没有持久化最后 1 个 binlog 中的所有 GTID，导致备库比主库多了很多的 GTID。

通过对比 Crash 前后 binlog 中的 GTID，发现主库确实并没有持久化到 mysql-bin.001499 中的 binlog，导致后续新产生的 GTID 反而比之前的还小（正常情况下 GTID 肯定是随着 binlog 中事务的记录不断增大的）（见图 42）。

图 42

15.3 GDB 调试复现

基于 MySQL 5.7.26 版本，通过 GDB 调试复现了上述问题现象，即主库连续崩溃恢复后会丢失最后 1 个 binlog 中的 GTID，引发备库 GTID 大于主库。

先讲下原因：

(1) 在 MySQL 第一次崩溃恢复过程中，会先创建新的 binlog，再将崩溃前最后 1 个 binlog 中的 GTID 持久化到表中。

(2) 如果在这个间隙，再次发生崩溃，就可能会导致 MySQL 产生新的 binlog，但是还未将第一次崩溃前最后 1 个 binlog 持久化到表中。

(3) MySQL 再次启动时，就不会再读取第一次崩溃前最后 1 个 binlog 做持久化了，而是读取新产生的 binlog 做持久化，那么就会丢失第一次崩溃前最后 1 个 binlog 中的 GTID。

```
#mysql5.7.26 crash 启动流程
|main
 |mysqld_main
  |ha_recover            #mysqld.cc:4256 恢复数据流程
  |open_binlog           #mysqld.cc:4282 生成新的 binlog
  |Gtid_state::save      #mysqld.cc:4870 读取最后 1 个 binlog 写入 mysql.gtid_executed
   |Gtid_table_persistor::save
```

```
|Gtid_table_persistor::write_row
```

15.4 本地 GDB 模拟复现过程

(1) 使用 MySQL 5.7.26 mysqld-debug 版本启动 MySQL（见图 43）。

```
[root@xutianyun ~]# /opt/mysql_base_5_7_26/bin/mysqld --version
/opt/mysql_base_5_7_26/bin/mysqld  Ver 5.7.26-debug for Linux on x86_64 (Source distribution)
[root@xutianyun ~]# /opt/mysql_base_5_7_26/bin/mysqld --defaults-file=/etc/my.cnf
```

图 43

(2) 连入 MySQL，创建 test 库，生成 1 个 GTID。当前 binlog 为 mysql-bin.000001，mysql.gtid_executed 表为空，GTID 还未持久化（见图 44）。

```
root@localhost : (none) 13:45:01> reset master ;
Query OK, 0 rows affected (0.03 sec)

root@localhost : (none) 13:45:06> create database test;
Query OK, 1 row affected (0.01 sec)

root@localhost : (none) 13:45:12>
root@localhost : (none) 13:45:12> show master status ;
+------------------+----------+--------------+------------------+-------------------------------------------+
| File             | Position | Binlog_Do_DB | Binlog_Ignore_DB | Executed_Gtid_Set                         |
+------------------+----------+--------------+------------------+-------------------------------------------+
| mysql-bin.000001 |      318 |              |                  | a001535e-f06d-11ed-aeb3-5254007f48c1:1    |
+------------------+----------+--------------+------------------+-------------------------------------------+
1 row in set (0.01 sec)

root@localhost : (none) 13:45:15> show binary logs;
+------------------+-----------+
| Log_name         | File_size |
+------------------+-----------+
| mysql-bin.000001 |       318 |
+------------------+-----------+
1 row in set (0.00 sec)

root@localhost : (none) 13:45:28> select * from mysql.gtid_executed;
Empty set (0.00 sec)
```

图 44

(3) kill -9 mysql 进程，模拟 OOM（见图 45）。

```
[root@xutianyun ~]# ps -ef | grep mysqld
mysql    14993 13985  2 13:42 pts/2    00:00:05 /opt/mysql_base_5_7_26/bin/mysqld --defaults-file=/etc/my.cnf
root     16556 13626  0 13:47 pts/1    00:00:00 grep --color=auto mysqld
[root@xutianyun ~]# kill -9 14993
```

图 45

(4) 使用 GDB 启动 MySQL，并设置 2 个断点，分别是 MYSQL_BIN_LOG::open_binlog（创建新的 binlog）和 Gtid_state::save（崩溃恢复过程中持久化最后 1 个 binlog 的 GTID）（见图 46）。

```
(gdb) info b
Num    Type           Disp Enb Address    What
6      breakpoint     keep y   <MULTIPLE>
6.1                       y    0x00000000017907c in Gtid_state::save(THD*) at /opt/source/mysql-5.7.26/sql/rpl_gtid_state.cc:710
6.2                       y    0x000000000017910aa in Gtid_state::save(Gtid_set const*)
                                   at /opt/source/mysql-5.7.26/sql/rpl_gtid_state.cc:735
7      breakpoint     keep y   <MULTIPLE>
7.1                       y    0x00000000017db7e5 in MYSQL_BIN_LOG::open_binlog(char const*, char const*, unsigned long, bool, bool, bool, Format_description_log_event*)
                                   at /opt/source/mysql-5.7.26/sql/binlog.cc:4977
7.2                       y    0x000000000017e3d7e in MYSQL_BIN_LOG::open_binlog(char const*)
                                   at /opt/source/mysql-5.7.26/sql/binlog.cc:8388
```

图 46

(5) 执行 RUN 后，会停在第一个断点 MYSQL_BIN_LOG::open_binlog 处，此时可以看到还未产生新的 binlog（见图 47、图 48）。

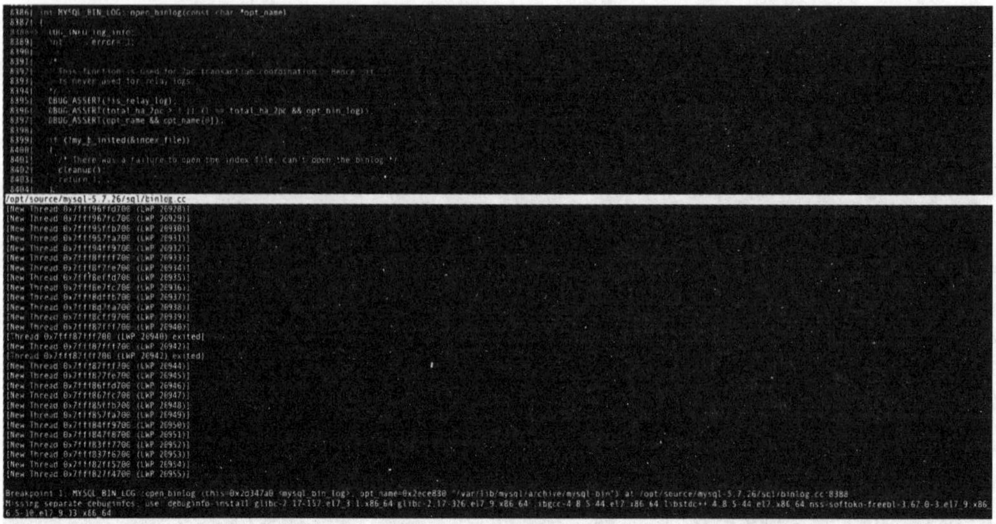

图 47

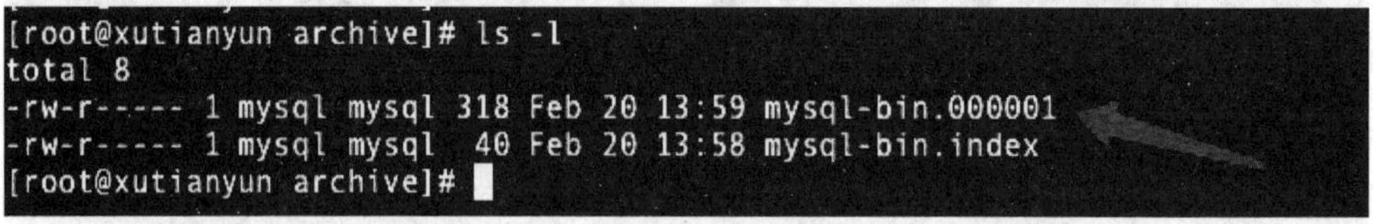

图 48

（6）执行 continue 继续运行，会卡在第二个断点 Gtid_state::save。此时已经生成了新的 binlog，但是还未将 mysql-bin.000001 中的 GTID 持久化（见图 49、图 50）。

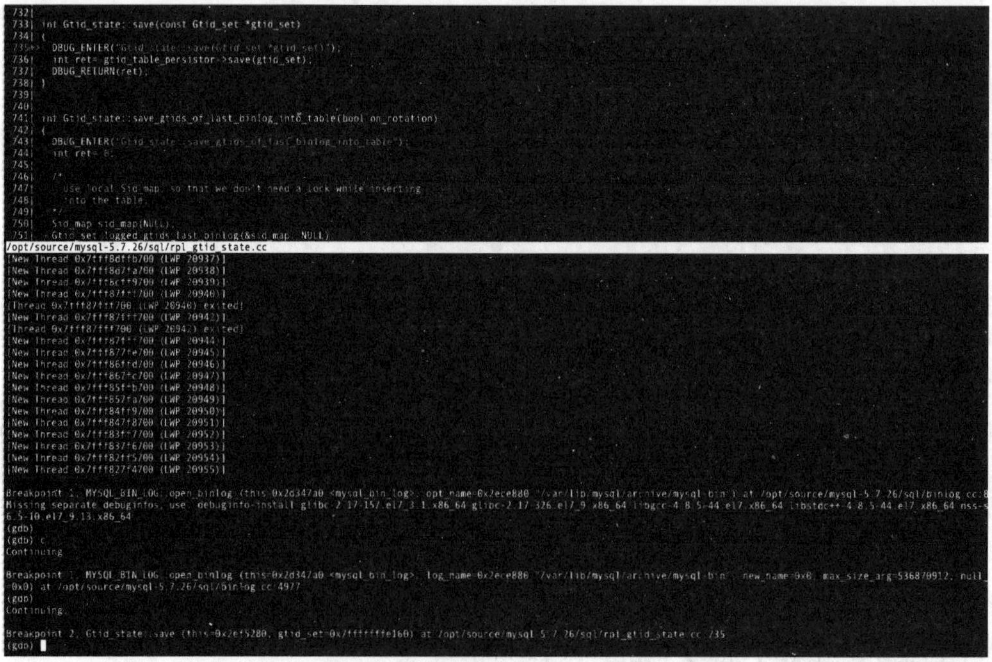

图 49

```
[root@xutianyun archive]# ls -l
total 12
-rw-r----- 1 mysql mysql 318 Feb 20 14:04 mysql-bin.000001
-rw-r----- 1 mysql mysql 123 Feb 20 14:04 mysql-bin.000002
-rw-r----- 1 mysql mysql  80 Feb 20 14:04 mysql-bin.index
[root@xutianyun archive]#
```

图 50

(7) 正常情况下，Gtid_state::save 执行完以后，就会把 mysql-bin.000001 中的 GTID 持久化到表中。这里模拟把 mysql 给 kill 了，此时产生了新的 binlog，但是还未做完 GTID 持久化（见图 51）。

```
[root@xutianyun archive]# ps -ef |grep mysqld
root      20656 12566  0 13:59 pts/0    00:00:00 cgdb --args /opt/mysql_base_5_7_26/bin/mysqld
root      20657 20656  0 13:59 pts/4    00:00:04 gdb --nw --annotate=2 -x /root/.tgdb/a2_gdb_init --args /opt/mysql_base_5_7_26/bin/mysqld
mysql     20906 20657  0 14:00 pts/5    00:00:00 /opt/mysql_base_5_7_26/bin/mysqld
root      23318 13626  0 14:07 pts/1    00:00:00 grep --color=auto mysqld
[root@xutianyun archive]# kill -9 20906
```

图 51

(8) 再次正常启动 MySQL，就会发现“丢掉了”mysql-bin.000001 中的 GTID。此次启动并不会再次读取 mysql-bin.000001 中的 GTID 做持久化，所以备库会比主库多整整 1 个 binlog 的 GTID（见图 52）。

```
root@localhost : (none) 14:10:05>
root@localhost : (none) 14:10:06> show binary logs;
+------------------+-----------+
| Log_name         | File_size |
+------------------+-----------+
| mysql-bin.000001 |       318 |
| mysql-bin.000002 |       123 |
| mysql-bin.000003 |       154 |
+------------------+-----------+
3 rows in set (0.00 sec)

root@localhost : (none) 14:10:09> show master status;
+------------------+----------+--------------+------------------+-------------------+
| File             | Position | Binlog_Do_DB | Binlog_Ignore_DB | Executed_Gtid_Set |
+------------------+----------+--------------+------------------+-------------------+
| mysql-bin.000003 |      154 |              |                  |                   |
+------------------+----------+--------------+------------------+-------------------+
1 row in set (0.00 sec)

root@localhost : (none) 14:10:14> select * from mysql.gtid_executed;
Empty set (0.00 sec)

root@localhost : (none) 14:10:18>
root@localhost : (none) 14:10:19> show binlog events in 'mysql-bin.000001';
+------------------+-----+----------------+-----------+-------------+--------------------------------------------------------------------+
| Log_name         | Pos | Event_type     | Server_id | End_log_pos | Info                                                               |
+------------------+-----+----------------+-----------+-------------+--------------------------------------------------------------------+
| mysql-bin.000001 |   4 | Format_desc    |      1001 |         123 | Server ver: 5.7.26-debug-log, Binlog ver: 4                        |
| mysql-bin.000001 | 123 | Previous_gtids |      1001 |         154 |                                                                    |
| mysql-bin.000001 | 154 | Gtid           |      1001 |         219 | SET @@SESSION.GTID_NEXT= 'a001535e-f06d-11ed-aeb3-5254007f48c1:1'  |
| mysql-bin.000001 | 219 | Query          |      1001 |         318 | create database test                                               |
+------------------+-----+----------------+-----------+-------------+--------------------------------------------------------------------+
4 rows in set (0.00 sec)
```

图 52

同样的测试在 MySQL v5.7.26、v5.7.36、v5.7.44 中均可以复现，说明 MySQL 5.7 中都存在该现象。而 MySQL 8.0 因为 GTID 持久化做了优化，所以不会有此类问题。

15.5 总结

在 MySQL 5.7 版本下，因为 GTID 持久化机制的原因，当 MySQL 处于崩溃恢复阶段时，如果再次遇到 Crash，就可能会丢失最后 1 个 binlog 中的 GTID。

16 为什么你的 show slave status 会卡住？

作者：徐文梁

16.1 问题背景

在数据库 MySQL 的日常运维中，生产环境一般都是一主多从的高可用架构，涉及主从同步的问题。通常执行 show slave status 命令就可以了解主从实例之间的同步状态，但是凡事总会有意外。

最近，在生产环境中遇到 show slave status 命令执行卡住了的情况。以前在测试环境中也遇到过，当时并没有深究，但这次是被客户问到了，发现自己说不清楚。如果此时此刻正在阅读本文的读者朋友也不说清楚，那请随我去源码中一探究竟吧！

16.2 问题分析

为了更全面地了解 MySQL 的状态，通过 pstack 拿到了相应的线程信息。这里保留了关键信息，如下：

```
Thread 6 (Thread 0xxxx (LWP xxx)):
#0  0x..... in __lll_lock_wait () from /lib64/libpthread.so.0
#1  0x..... in _L_lock_975 () from /lib64/libpthread.so.0
#2  0x..... in pthread_mutex_lock () from /lib64/libpthread.so.0
#3  0x..... in inline_mysql_mutex_lock (that=0x1358520 <LOCK_active_mi>, src_file=<optimized out>, src_line=<optimized out>) at .../mysql-5.6.40/include/mysql/psi/mysql_thread.h:688
#4  0x..... in mysql_execute_command (thd=thd@entry=0x5d0abb0) at .../mysql-5.6.40/sql/sql_parse.cc:2853
#5  0x..... in mysql_parse (thd=thd@entry=0x5d0abb0, rawbuf=<optimized out>, length=<optimized out>, parser_state=parser_state@entry=0x7fe44c644690) at .../mysql-5.6.40/sql/sql_parse.cc:6453
#6  0x..... in dispatch_command (command=<optimized out>, thd=0x5d0abb0, packet=0x5d0def1 "show slave status", packet_length=17) at .../mysql-5.6.40/sql/sql_parse.cc:1374
```

通过上面的信息可知，卡住是因为 show slave status 在获取某种 mutex 锁的时候被阻塞所导致的。我们根据这个结论在测试环境中模拟和分析。

16.2.1 测试环境

客户的 MySQL 版本比较旧,所以我自己准备了 MySQL 5.7.41 的 debug 环境。后续的测试分析都基于该版本,其他版本可能存在差异,但分析过程类似。

16.2.2 源码分析

执行 show slave status 时需要哪些 mutex 锁?我们去源码中看看 show slave status 的执行逻辑。

当我们在客户端执行 show slave status 命令后,经过一系列兜兜转转,会来到 mysql_execute_command 函数,进入 SQLCOM_SHOW_SLAVE_STAT 分支,对应文件为 mysql-5.7.41\sql\sql_parse.cc。

```
case SQLCOM_SHOW_SLAVE_STAT:
{
/* Accept one of two privileges */
if (check_global_access(thd, SUPER_ACL | REPL_CLIENT_ACL))
    goto error;
res= show_slave_status_cmd(thd);
break;
}
```

show_slave_status_cmd 函数是 show slave status 命令的入口,可以看出会涉及 channel_map 锁,对应文件为 mysql-5.7.41\sql\rpl_slave.cc。

```
bool show_slave_status_cmd(THD *thd)
{
Master_info *mi= 0;
LEX *lex= thd->lex;
bool res;
......
channel_map.rdlock();
if (!lex->mi.for_channel)
    res= show_slave_status(thd);
else
{
......
res= show_slave_status(thd, mi);
}
channel_map.unlock();
......
}
```

show_slave_status 函数负责执行 show slave status 语句,可以看出涉及 global_sid_lock 锁,对应文件为 mysql-5.7.41\sql\rpl_slave.cc。

```
bool show_slave_status(THD* thd, Master_info* mi)
```

```
{
......
if (mi != NULL)
{
global_sid_lock->wrlock();
......
global_sid_lock->unlock();
}
......
show_slave_status_metadata(field_list, io_gtid_set_size, sql_gtid_set_size);
......
if (mi != NULL && mi->host[0])
{
    if (show_slave_status_send_data(thd, mi,io_gtid_set_buffer, sql_gtid_set_
buffer))
    ......
    }
```

show_slave_status_send_data 函数将数据发送到 Master_info 的客户端，这部分涉及的锁相对较多，对应文件为 mysql-5.7.41\sql\rpl_slave.cc。

```
bool show_slave_status_send_data(THD *thd, Master_info *mi,char* io_gtid_set_
buffer,char* sql_gtid_set_buffer)
{
......
mysql_mutex_lock(&mi->info_thd_lock);
......
mysql_mutex_unlock(&mi->info_thd_lock);

mysql_mutex_lock(&mi->rli->info_thd_lock);
......
mysql_mutex_unlock(&mi->rli->info_thd_lock);

mysql_mutex_lock(&mi->data_lock);
mysql_mutex_lock(&mi->rli->data_lock);
mysql_mutex_lock(&mi->err_lock);
mysql_mutex_lock(&mi->rli->err_lock);
......
mysql_mutex_unlock(&mi->rli->err_lock);
mysql_mutex_unlock(&mi->err_lock);
mysql_mutex_unlock(&mi->rli->data_lock);
mysql_mutex_unlock(&mi->data_lock);
}
```

show_master_status 函数负责执行 show master status 语句，可以看出涉及 global_sid_lock 锁，对应文件为 mysql-5.7.41\sql\rpl_slave.cc。

```
bool show_master_status(THD* thd)
```

```
{
......
global_sid_lock->wrlock();
......
global_sid_lock->unlock();
......
}
```

关于主从复制中的 mutex 锁的功能及涉及该锁的大部分操作，官方文档还是很贴心地给出了详细的解释，就不一一赘述了，对应文件为 mysql-5.7.41\sql\rpl_slave.h。

16.3 模拟场景

直接模拟 show slave status 卡住的场景不太容易，但是通过 debug+断点调试可以轻松实现。

从前文源码分析可知，show slave status 命令和 show master status 命令执行过程中都涉及获取 global_sid_lock 锁，因此可以在 show master status 命令对应的函数中获取 global_sid_lock 锁（此时还没释放的中间位置设置断点），然后执行 show slave status 命令时获取 global_sid_lock 锁便会被阻塞。另外，可以同步在释放 global_sid_lock 锁函数结束之前设置断点，形成对比。详细流程及步骤如表 6。

表 6

GDB 操作	会话 A	A 状态	会话 B	B 状态
	login in		login in	
set breakpoint1				
set breakpoint2				
			show slave status	正常，立即返回
	show master status	hit breakpoint1 阻塞		
			show slave status	阻塞
continue				
		hit breakpoint2 继续阻塞	show slave status 返回结果	
			show slave status	正常，立即返回
continue		show master status 返回结果		

（1）通过源码编译的 debug 模式启动 MySQL 服务，并启动两个会话。

```
# 查看MySQL进程
[root@localhost ~]# ps -ef|grep mysql
root      14949 14919  0 23:13 pts/3    00:00:00 grep --color=auto mysql
mysql     31658     1  0 06:52 ?        00:01:14 /root/mysql-5.7.41/build/sql/mysqld --defaults-file=/etc/my.cnf --user=mysql
```

```
# 会话 A
mysql> show processlist;
+----+------+-----------+------+---------+------+----------+------------------+
| Id | User | Host      | db   | Command | Time | State    | Info             |
+----+------+-----------+------+---------+------+----------+------------------+
|  4 | root | localhost | NULL | Query   |    0 | starting | show processlist |
|  5 | root | localhost | NULL | Sleep   |    9 |          | NULL             |
+----+------+-----------+------+---------+------+----------+------------------+
2 rows in set (0.00 sec)

# 会话 B
mysql> show processlist;
+----+------+-----------+------+---------+------+----------+------------------+
| Id | User | Host      | db   | Command | Time | State    | Info             |
+----+------+-----------+------+---------+------+----------+------------------+
|  4 | root | localhost | NULL | Sleep   |  154 |          | NULL             |
|  5 | root | localhost | NULL | Query   |    0 | starting | show processlist |
+----+------+-----------+------+---------+------+----------+------------------+
2 rows in set (0.00 sec)

# 查看 thread 表
mysql> select thread_id,processlist_id ,thread_os_id, name from performance_schema.threads where  name="thread/sql/one_connection";
+-----------+----------------+--------------+---------------------------+
| thread_id | processlist_id | thread_os_id | name                      |
+-----------+----------------+--------------+---------------------------+
|        30 |              4 |        16055 | thread/sql/one_connection |
|        31 |              5 |        16090 | thread/sql/one_connection |
+-----------+----------------+--------------+---------------------------+
2 rows in set (0.00 sec)
```

(2) 通过 gdb 关联 MySQL 进程并设置断点。

```
# 关联 MySQL 进程
Type "apropos word" to search for commands related to "word".
(gdb) show non-stop
Controlling the inferior in non-stop mode is off.
(gdb) set non-stop on
(gdb) show non-stop
Controlling the inferior in non-stop mode is on.
(gdb) attach 31658
Attaching to process 31658

# 查看相关线程
(gdb) info threads
  Id   Target Id         Frame
* 1    Thread 0x2b0ca2618480 (LWP 31658) "mysqld" 0x00002b0ca4113ddd in poll () from /lib64/libc.so.6
  ......
```

```
31    Thread 0x2b0caff6a700 (LWP 16055) "mysqld" (running)
32    Thread 0x2b0caff28700 (LWP 16090) "mysqld" (running)


# 切换到指定线程，对应会话 A
(gdb) t 31
[Switching to thread 31 (Thread 0x2b0caff6a700 (LWP 16055))](running)

# 设置断点，持有锁而不释放
(gdb) b rpl_master.cc:647
Breakpoint 1 at 0x181a565: file /root/mysql-5.7.41/sql/rpl_master.cc, line 647.
# 设置断点，释放锁
(gdb) b rpl_master.cc:649
Breakpoint 2 at 0x181a577: file /root/mysql-5.7.41/sql/rpl_master.cc, line 649.
# 查看断点
(gdb) info breakpoints
Num     Type           Disp Enb Address            What
1       breakpoint     keep y   0x000000000181a565 in show_master_status(THD*)
at /root/mysql-5.7.41/sql/rpl_master.cc:647
2       breakpoint     keep y   0x000000000181a577 in show_master_status(THD*)
at /root/mysql-5.7.41/sql/rpl_master.cc:649
```

(3) 会话 B 执行 show slave status 命令。

```
# 结果很快返回，不会被阻塞
mysql> show slave status;
Empty set (0.00 sec)
```

(4) 先在会话 A 执行 show master status 命令，然后在会话 B 执行 show slave status 命令。

```
# 会话 A 执行 "show master status" 命令卡住
mysql> select now();
+---------------------+
| now()               |
+---------------------+
| 2024-08-02 23:55:02 |
+---------------------+
1 row in set (0.00 sec)

mysql> show master status;


# 会话 B 执行 "show slave status" 命令卡住
mysql> select now();
+---------------------+
| now()               |
+---------------------+
| 2024-08-02 23:55:18 |
+---------------------+
```

```
    1 row in set (0.00 sec)

    mysql> show slave status;


    # 会话 A 触发断点一
    Thread 31 "mysqld" hit Breakpoint 1, show_master_status (thd=0x2b0e040009a0)
at /root/mysql-5.7.41/sql/rpl_master.cc:648
    648         global_sid_lock->unlock();

    Num     Type           Disp Enb Address            What
    1       breakpoint     keep y   0x000000000181a565 in show_master_status(THD*)
at /root/mysql-5.7.41/sql/rpl_master.cc:647
        breakpoint already hit 1 time
    2       breakpoint     keep y   0x000000000181a577 in show_master_status(THD*)
at /root/mysql-5.7.41/sql/rpl_master.cc:649
```

(5)gdb 调试继续运行会话 A，观察会话 A 和会话 B 状态。

```
    # 继续运行，会话 A 触发断点二
    (gdb) c
    Continuing.

    Thread 31 "mysqld" hit Breakpoint 2, show_master_status (thd=0x2b0e040009a0)
at /root/mysql-5.7.41/sql/rpl_master.cc:650
    650         field_list.push_back(new Item_empty_string("File", FN_REFLEN));


    # 观察会话 A，"show master status" 命令继续卡住
    mysql> select now();
    +---------------------+
    | now()               |
    +---------------------+
    | 2024-08-02 23:55:02 |
    +---------------------+
    1 row in set (0.00 sec)

    mysql> show master status;

    # 观察会话 B，"show show status" 命令返回
    mysql> select now();
    +---------------------+
    | now()               |
    +---------------------+
    | 2024-08-02 23:55:18 |
    +---------------------+
    1 row in set (0.00 sec)

    mysql> show slave status;
```

```
Empty set (4 min 13.76 sec)

# 会话 B 继续执行 "show show status" 命令均立即返回
mysql> select now();
+---------------------+
| now()               |
+---------------------+
| 2024-08-03 00:03:53 |
+---------------------+
1 row in set (0.00 sec)

mysql> show slave status;
Empty set (0.00 sec)

mysql> show slave status;
Empty set (0.00 sec)

# 继续运行，观察会话 A，因为后面没有其他断点，所以成功返回
mysql> show master status;
+------------------+----------+--------------+------------------+-------------------+
| File             | Position | Binlog_Do_DB | Binlog_Ignore_DB | Executed_Gtid_Set |
+------------------+----------+--------------+------------------+-------------------+
| mysql-bin.000044 |      154 |              |                  |                   |
+------------------+----------+--------------+------------------+-------------------+
1 row in set (9 min 56.99 sec)

mysql> select now();
+---------------------+
| now()               |
+---------------------+
| 2024-08-03 00:05:34 |
+---------------------+
1 row in set (0.00 sec)
```

注意：如果断点设置在 show master status 命令获取 global_sid_lock 锁之前，或者释放 global_sid_lock 锁之后，则无法阻塞 show slave status 命令。

16.4 总结

通过代码和实验，对 show slave status 命令有了更清晰的了解。可能导致 show slave status 卡住的场景很多，例如前面的模拟测试，无法穷举。类似问题可以从以下方向进行排查：

(1) 执行 show slave status 过程中需要获取 channel_map.rdlock()、global_sid_lock → wrlock、mi → data_lock 等相关 mutex 锁。如果此时这些锁中的一个或多个被其

他线程持有，show slave status 就会阻塞。

（2）当 show slave status 命令执行时出现阻塞，可以借助 pstack 工具和 performance_schema.threads 表（MySQL 5.7 之前的版本不支持 thread_os_id 字段）保留当时的信息，便于后期进行排查，分析 show slave status 命令阻塞的位置。

17 MySQL 含有下画线的数据库名在特殊情况下导致权限丢失

作者：芬达

在 MySQL 的授权操作中，通配符 "_" 和 "%" 用于匹配单个或多个字符的数据库对象名。然而，许多 DBA 在进行授权时可能忽视了这些通配符的特殊作用，导致数据库权限错配。这篇文章将讨论通配符误用所带来的潜在风险，并提供避免此类问题的解决方案。

17.1 误用通配符导致权限授予错误

在授权数据库权限时，如果数据库名中含有下画线 "_"，可能会引发意想不到的结果。我们来看一个常见的授权语句：

```
GRANT ALL ON `db_1`.* test_user;
```

表面上看，这个语句似乎是授予用户 test_user 对数据库 db_1 的全部权限。然而，通配符 "_" 在 MySQL 中具有特殊含义，它用于匹配任意单个字符。因此，这条授权语句实际上可能会匹配多个数据库，而不仅仅是 db_1。例如，以下数据库名都可能被匹配：

（1）数据库名匹配数字：db01、db11、db21……db91。

（2）数据库名匹配英文字符：dba1、dbb1。

（3）数据库名匹配特殊字符：db-1、db+1、db?1，等等。

这种误操作可能导致某些用户意外获得不该有的权限，从而带来严重的安全隐患。实际上，按照常见的数据库命名规范，数据库名中的字符通常是 26 个英文小写字母或 10 个数字，也包括 2 种特殊字符（中划线或下画线）。因此，这个授权错误可能将权限的应用范围扩大到 38 倍之多。这是基于对命名模式的分析做出的估算，具体情况可能因实际使用的命名规则而有所不同。

17.2 授权带来的隐患

当库名中有多个"_"时，情况更为复杂。假设数据库名称是 db_1_1，那么授权就不仅是扩大到 38 倍，而是 38×38=1444 倍，权限扩大的规模超出想象。如果这些库中有不应该公开的敏感数据，安全性风险将非常严重。

17.3 如何避免这个问题？

17.3.1 正确的做法：转义通配符

为了避免这种授权滥用的风险，应该将通配符作为普通字符来处理。MySQL 支持使用反斜杠（\）对通配符进行转义，例如：

```
GRANT ALL ON `db\_1`.* TO 'test_user';
```

通过这种方式，"_"将被解释为字面量，而不是通配符，从而确保授权的仅是特定的 db_1 数据库。

> 接下来，本文会多次提到"通配符"（_）和"转义通配符"（_）这两个术语，理解它们的区别有助于避免常见授权错误。

17.3.2 阿里云 DMS 等连接工具的优势

值得注意的是，在使用阿里云 DMS 授权时，系统底层会自动将通配符进行转义，这也就是为什么很多 DBA 并没有意识到自己授权时遇到的潜在风险。阿里云的这种机制为用户省去了手动转义的烦恼，保证了授权的准确性。

> 然而，阿里云允许绕过 DMS，在底层手动授权，所以本文内容依然适用于使用阿里云的 DBA。

17.4 整改过程中的风险

在意识到这个问题后，你可能会急于对现有授权进行整改，但需要注意两种场景：

(1) 遗漏整改：部分库可能没有彻底整改，仍然使用了通配符授权。

(2) 保留通配符功能：有些场景下，你希望保留部分通配符授权。

在这两种场景下，会碰到本文要讨论的主要问题——含有下画线的数据库名在特殊情况下会有权限丢失的风险。

17.5 模拟场景：遗漏整改导致权限丢失

现在我们来模拟一个场景，展示如何由于遗漏整改而导致权限问题的发生。

假设在权限整改过程中，不需要保留通配符的授权，于是你对几百个数据库的授权进行了整改，但你还是遗漏了其中一个数据库。我认为这类情况很有可能发生。该数据库名为 app_db，其授权如下：

```
GRANT SELECT, INSERT, UPDATE, DELETE ON `app_db`.* TO `app_user`@`%`;
```

然后，随着业务的扩展，你意识到应用程序需要自动维护分区表的能力，因此你希望新增 CREATE、DROP、ALTER 权限。因为你的授权平台已经经过改造，可以正确地对通配符进行转义，所以新的授权语句如下：

```
GRANT CREATE, DROP, ALTER ON `app\_db`.* TO `app_user`@`%`;
```

之后，app_user 的授权状态如下：

```
mysql> show grants for app_user;
+-----------------------------------------------------------------------+
| Grants for app_user@%                                                 |
+-----------------------------------------------------------------------+
| GRANT USAGE ON *.* TO `app_user`@`%`                                  |
| GRANT CREATE, DROP, ALTER ON `app\_db`.* TO `app_user`@`%`            |
| GRANT SELECT, INSERT, UPDATE, DELETE ON `app_db`.* TO `app_user`@`%`  |
+-----------------------------------------------------------------------+
3 rows in set (0.01 sec)
```

于是，产生了一种，通配符（_）和转义通配符（_）混合使用的场景。

表面上看，两个授权并没有合并到一条语句中，但根据我们前面学到的知识，不难理解这两个授权是希望表达：

(1) app_user 拥有对 app_db 的 CREATE、DROP、ALTER 权限。

(2) app_user 也拥有对 app_db 本身及其他符合通配符匹配的数据库的 SELECT、INSERT、UPDATE、DELETE 权限。

表面看似一切正常，但实际上在操作中却发现了问题。

17.5.1 权限测试

我们来实际测试一下授权效果：

```
    ERROR 1142 (42000): SELECT command denied to user 'app_user'@'127.0.0.1' for table 't'
    mysql> insert into `app_db`.t values (1);
    ERROR 1142 (42000): INSERT command denied to user 'app_user'@'127.0.0.1' for table 't'
    mysql> update `app_db`.t set a=1;
    ERROR 1142 (42000): UPDATE command denied to user 'app_user'@'127.0.0.1' for table 't'
    mysql> delete from `app_db`.t;
    ERROR 1142 (42000): DELETE command denied to user 'app_user'@'127.0.0.1' for
```

```
table 't'

mysql> create table `app_db`.t2(a int);
Query OK, 0 rows affected (0.01 sec)
mysql> alter table `app_db`.t2 engine=innodb;
Query OK, 0 rows affected (0.02 sec)
Records: 0  Duplicates: 0  Warnings: 0
mysql> drop table `app_db`.t2;
Query OK, 0 rows affected (0.01 sec)
```

尽管新增的 CREATE、DROP、ALTER 权限生效了，但原来的 SELECT、INSERT、UPDATE 和 DELETE 权限却全部丢失了！

17.5.2 解释与分析

这显然会在生产环境中引发严重问题。那么这是一个 MySQL 的 Bug 吗？

最初，我也认为这可能是个 S2 级别的 Bug，并向官方提交了报告。但深入调查后发现，这实际上是 MySQL 授权机制的一个已知行为，而不是 Bug。根据官方文档：

The use of the wildcard characters%and_as described in the next few paragraphs is deprecated, and thus subject to removal in a future version of MySQL.

其大意为，接下来几段中描述的使用通配符 "%" 和 "_" 的方式已被弃用，因此在未来的 MySQL 版本中可能会被移除。

这意味着 MySQL 未来会彻底废弃通配符在授权中的使用。更进一步，官方文档提到：

Issuing multiple GRANT statements containing wildcards may not have the expected effect on DML statements; when resolving grants involving wildcards, MySQL takes only the first matching grant into consideration. In other words, if a user has two database-level grants using wildcards that match the same database, the grant which was created first is applied. Consider the databasedband tabletcreated using the statements shown here:

简单来说，即当多个授权中涉及通配符时，MySQL 只会考虑第一个匹配的授权。

17.5.3 我的案例

我遇到的情况与官方文档中描述的多个通配符授权略有不同。

官方文档提到，当涉及多个通配符授权时，MySQL 只会应用第一个匹配的授权，后续的通配符授权将不会生效。然而，在我的案例中，情况有所不同：我只使用了一个通配符授权，之后又添加了一个经过正确转义的授权。结果是，MySQL 仅识别并应用了转义后的授权，而原本的通配符授权则被忽略。

这表明，MySQL 在处理通配符和转义字符时存在文档不完善的情况。尽管官方文档中提到通配符授权的局限性，但并未具体说明在混合使用通配符和转义后的授权时，通配符授权可能会被转义后的授权所取代。这种情况下，开发者容易误认为这是一种 Bug，而实际上是 MySQL 授权机制的已知行为。

17.5.4 更进一步测试

以下是基于 MySQL 5.7 测试的结论：

(1) 单一授权生效：测试和官方文档一致，MySQL 只会匹配并使其中一条授权生效，不会同时应用两条授权——这是我的案例里踩到的"坑"。

(2) 优先级问题：当通配符授权和转义通配符授权混合使用时，MySQL 优先应用不含通配符的授权。

然而，在 MySQL 8.0 的测试中，结果又有所不同：哪个授权生效取决于 mysql.db 的加载顺序，先进行的授权将优先生效。

为更清楚说明这一问题，我在多个 MySQL 版本中进行了进一步测试，结论如下表所示。

表 7

MySQL 版本	混合一个通配符和一个转义通配符授权的情况，哪个授权最终生效？
5.5	MySQL 会优先使用不含通配符的那个授权
5.7	MySQL 会优先使用不含通配符的那个授权
8.0	先进行的授权将优先生效
8.4	MySQL 会优先使用不含通配符的那个授权
9.0	MySQL 会优先使用不含通配符的那个授权

关于通配符还有一些奇怪的"例外"设置。

In privilege assignments, MySQL interprets occurrences of unescaped_and%SQL wildcard characters in database names as literal characters under these circumstances:

- When a database name is not used to grant privileges at the database level, but as a qualifier for granting privileges to some other object such as a table or routine (for example,GRANT ... ON db_name.tbl_name).

- Enabling partial_revokes causes MySQL to interpret unescaped_and%wildcard characters in database names as literal characters, just as if they had been escaped as_and\%. Because this changes how MySQL interprets privileges, it may be advisable to avoid unescaped wildcard characters in privilege assignments for installations where partial_revokes may be enabled. For more information, see Section 8.2.12, "Privilege Restriction Using Partial Revokes".

其大意为，如果库名中的"_"未转义，它会被解释为通配符；但是库名只是用作表名或存储过程的限定符时，库名里的"_"就不再是通配符，而是字面量。

如果库名中的"_"未转义，它会被解释为通配符；但是库名只是用作表名（表级授权场景）、函数、存储过程的限定符时，库名里的"_"就不再是通配符，而是字面量。

此外，如果你启用了 MySQL 的部分撤销授权参数 partial_revokes，数据库名中的"_"不需要转义，它会被直接解释为字面量。

一会儿是字面量，一会儿又是通配符，难怪官方打算放弃这个功能，他们自己可能都被搞晕了。通配符的设定确实让人难以理解。

17.6 隐患排查

我们应该和官方一样，放弃使用通配符授权，使用正确的转义授权。排查所有使用"_"或"%"通配符的情况，统一整改为"_"或"\\%"。

以下 SQL 脚本由 AI 生成，请测试和谨慎使用（见图 53）。

```sql
SELECT
-- 库名是否含有 _ 或 % 通配符
CASE
WHEN EXISTS (
SELECT 1
FROM information_schema.schemata
WHERE INSTR(schema_name, '_') > 0 OR INSTR(schema_name, '%') > 0
) THEN '是'
ELSE '否'
END AS '库名是否含有_或%通配符',

-- 授权里库名是否使用了 "_" 或 "%" 通配符
CASE
WHEN EXISTS (
SELECT 1
FROM mysql.db
WHERE (INSTR(Db, '_') > 0 OR INSTR(Db, '%') > 0)
AND (INSTR(Db, '\\_') = 0 AND INSTR(Db, '\\%') = 0)
) THEN '是'
ELSE '否'
END AS '授权里库名是否使用了"_"或"%"通配符',

-- 授权里库名是否使用了 "\_" 或 "\%" 转义通配符
CASE
WHEN EXISTS (
SELECT 1
FROM mysql.db
WHERE INSTR(Db, '\\_') > 0 OR INSTR(Db, '\\%') > 0
```

```
    ) THEN '是'
    ELSE '否'
    END AS '授权里库名是否使用了"\\_"或"\\%"转义通配符',

    -- 授权里是否存在使用了表级授权的情况（排除指定的两条记录）
    CASE
    WHEN EXISTS (
    SELECT 1
    FROM mysql.tables_priv
    WHERE NOT (
    (Host = 'localhost' AND Db = 'mysql' AND User = 'mysql.session' AND Table_name
= 'user')
    OR
    (Host = 'localhost' AND Db = 'sys' AND User = 'mysql.sys' AND Table_name =
'sys_config')
    )
    ) THEN '是'
    ELSE '否'
    END AS '授权里是否存在使用了表级授权的情况';
```

库名是否含有_或%通配符	授权里库名是否使用了"_"或"%"通配符	授权里库名是否使用了"_"或"\%"转义通配符	授权里是否存在使用了表级授权的情况
是	是	是	否

1 row in set (0.00 sec)

图 53

17.7 如何规避

(1) 不要使用通配符授权，多数人不知道“_”“%”这个是通配符，用错了，要转义！

(2) 不要使用通配符授权，这是官方打算放弃的功能。

(3) 如果仍然需要使用通配符授权，不要混合使用。既不要混合使用转义通配符授权，也不要混合使用多个通配符授权。

(4) 如果仍然需要使用通配符授权，且打算混合使用，要考虑本文中的测试结论，并测试，例如我的案例里，我可以在保留通配符授权情况下，这样授权：

```
mysql> SHOW GRANTS FOR app_user;
+----------------------------------------------------------------------------------+
| Grants for app_user@%                                                            |
+----------------------------------------------------------------------------------+
| GRANT USAGE ON *.* TO `app_user`@`%`                                             |
| GRANT SELECT, INSERT, UPDATE, DELETE, CREATE, DROP, ALTER ON `app\_db`.* TO `app_user`@`%` |
| GRANT SELECT, INSERT, UPDATE, DELETE ON `app_db`.* TO `app_user`@`%`             |
+----------------------------------------------------------------------------------+
3 rows in set (0.01 sec)
```

18 企业如何做好 SQL 质量管理？

作者：SQLE 项目组

用 SQL 操作数据库对软件研发人员而言是一项基础且常见的工作内容。如何避免"问题" SQL 流转到生产环境，保证数据质量？这个问题值得被研发、DBA、运维所重视。

18.1 什么是 SQL 问题？

对于研发人员来说，日常工作的大部分都需要使用数据库。项目中的很多业务都需要进行增删改查等常见数据库操作，这些数据库操作对应的 SQL 语句主要都是由研发人员编写的。作为数据库管理和运维人员，DBA 负责数据库的日常运维工作。当出现问题 SQL 时，DBA 通常首当其冲，负责问题诊断。

那么，什么是 SQL 问题或者说问题 SQL 呢？从一个更广泛的定义来看，SQL 问题指影响业务正常运行的各种 SQL 相关问题。比如图 15 举的案例，它们都可以被归类为 SQL 问题。爱可生作为一家数据库公司，从客户那里获得了大量的反馈，我们发现每年都会发生因为 SQL 问题导致的高级别的生产事故。类似的 SQL 问题导致业务中断的新闻也时不时会出现（见图 54）。

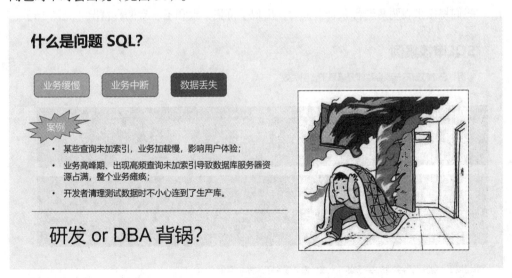

图 54

对研发人员来说，这些 SQL 问题在开发阶段很难被发现，因为他们的首要任务是保证需求实现。此外，按我们对研发人群和客户的调研结果显示，研发人员在新功能开发阶段通常没有时间对 SQL 进行优化，只要能完成需求交付就已经很不错了。一方面是项目进度压力大，另一方面研发人员自身水平和经验也有高有低。所以，研发人员很难对所有的 SQL 全面进行优化。下面我们举一个真实且典型的案例。

18.2 一个典型的 SQL 问题案例

图 55 中有三张表，请注意表的字符集不同，分别是 utf8 和 utf8mb4。

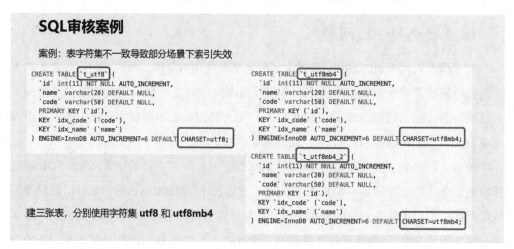

图 55

我们将三张表两两进行联合查询时，出现字符集一致和不一致两种情况（见图 56）。

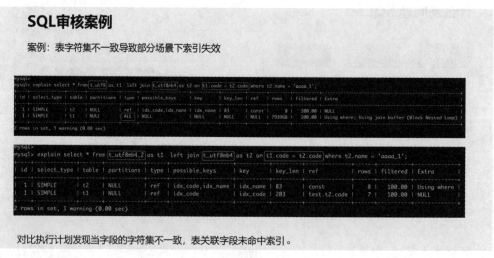

图 56

当字符集不一致时，从执行计划中可见，进行了全表扫描，表关联字段未命中索引

（见图 57）。

SQL审核案例

案例：表字符集不一致导致部分场景下索引失效

```
mysql> select count(*) from t_utf8 as t1 left join t_utf8mb4 as t2 on t1.code = t2.code where t2.name = 'aaaa_1';
| count(*) |
|    64    |
1 row in set (0.93 sec)

mysql> select count(*) from t_utf8mb4_2 as t1 left join t_utf8mb4 as t2 on t1.code = t2.code where t2.name = 'aaaa_1';
| count(*) |
|    64    |
1 row in set (0.00 sec)
```

每张表插入80万数据，执行时间差异大

图 57

当每张表有 80 万条测试数据时，执行时间差异明显，字符集不一致的 SQL 执行时间达到了 0.9 秒。随着数据量的增加，该 SQL 的问题会愈加明显，两表联查时表的字符集不匹配会导致查询效率大幅下降。这就是一个典型的 SQL 问题。

可能你会觉得本案例的 SQL 问题看似不该发生，但我们的团队曾多次在客户的生产环境中遇到类似的情况。根据大家所掌握的慢 SQL 优化习惯来看，引发问题的有些因素是反直觉的。就算是本案例，研发人员也不一定会立刻将问题定位和排查出来。所以，我们需要一种更高效的方式来帮助研发人员解决这类问题。

18.3 全方位提高 SQL 质量

SQLE 是一款全方位的 SQL 质量管理平台，覆盖开发至生产环境的 SQL 审核和管理，支持主流的开源、商业、国产数据库，为开发和运维提供流程自动化能力，提升 SQL 上线效率，提高数据质量（见图 58）。

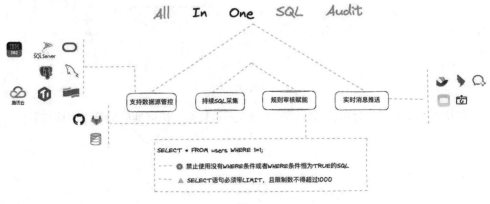

图 58

SQLE 于 2021 年 10 月 24 日这个属于开发者的日子正式开源，至今已经两年多。我们保持每个月发布一个新版本的频率，不断更新和迭代产品功能（见图 59）。

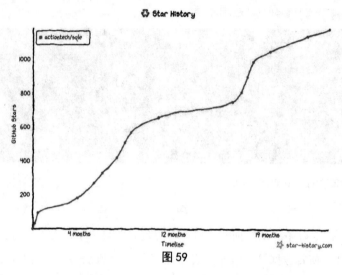

图 59

前面的典型 SQL 问题案例，在 SQLE 中如何解决呢（见图 60）？

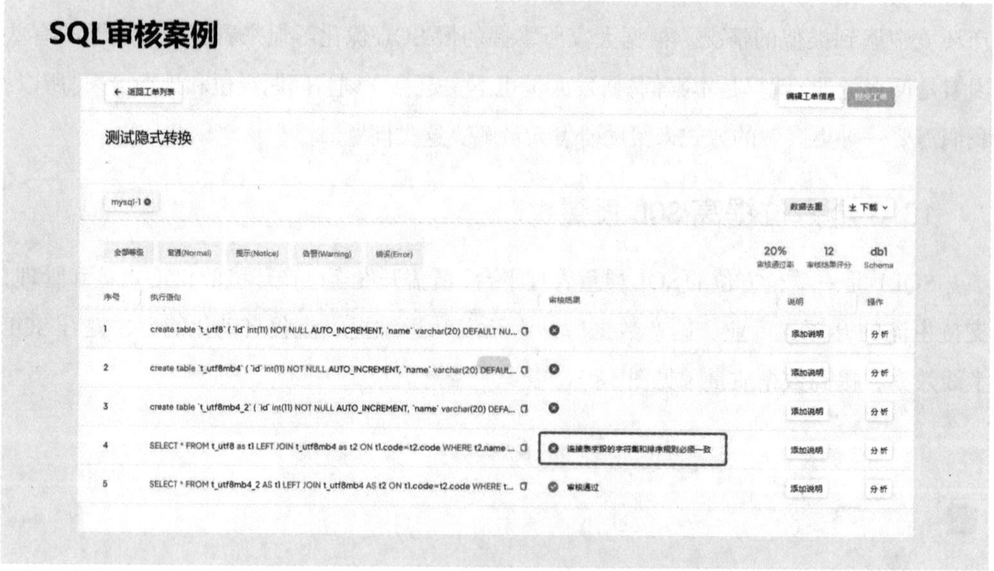

图 60

根据图 61，SQLE 会在相同的情况下触发审核规则，快速准确地给出审核结果。

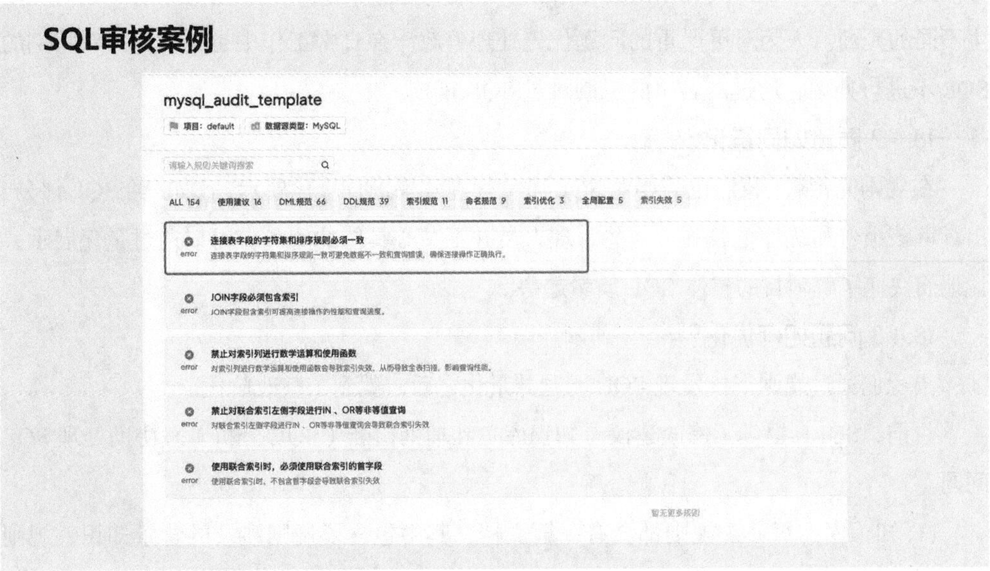

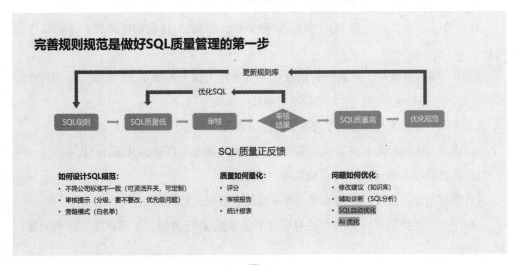

图 61

18.4 做好 SQL 质量管理的第一步

在 SQLE 中有非常丰富的 SQL 规则，上面的案例触发了索引失效类规则中的一条。制定一套完善的 SQL 规则规范，是做好 SQL 质量管理的第一步（见图 62）。

图 62

18.4.1 如何设计 SQL 规范？

不同的公司和业务场景对 SQL 规范都会有不同的要求。想要设计一款通用的 SQL 质量管理平台，对于 SQL 规范的设计要做到按需配置，支持通过规则模板给不同的业务配置不同的规则集。而且不同的规则集应该有分级匹配机制，以避免触发多条规则产生不同的判断。人为对更严重的问题优先处理整改。在日常工作中也同样允许对特例的 SQL 不进行处理，通过白名单的机制跳过 SQL 审核。

18.4.2 质量如何量化？

在规则完善后，我们也需要对 SQL 质量处理效果进行量化展示，比如：给 SQL 评分、出具审核报告和统计报表等。一些管理人员并不关心具体的业务，可以通过量化展示，让他们快速了解项目的整体 SQL 质量趋势。

18.4.3 问题如何优化？

当我们通过规则审核发现 SQL 问题并量化之后，就到了整改阶段。

目前，SQLE 提供了修改建议（知识库）和辅助诊断（SQL 分析）来协助处理 SQL 问题。

(1) 知识库：每一条规则都会有一篇文档，其中包含了规则涉及的背景知识、规则设置的原理和常见解决方案。

(2) SQL 分析：使用者在进行 SQL 优化前，会将 SQL 问题涉及的数据（表结构、索引使用情况、SQL 执行计划）进行整理，实现辅助诊断。

未来，SQLE 会增加主动优化的功能（SQL 改写、引入专用大模型能力），敬请期待。

18.5 SQL 质量管理具体怎么做？

在日常工作中，如何将 SQL 质量管理的理念落地呢？让我们先回顾一下软件生命周期。

抛开一些具体差异，每家公司的软件开发流程大体上都如图 63 所示，分为设计与实现、测试、部署与发布、生产与运维等阶段（见图 63）。

站在 SQL 质量管理的角度，不同阶段的工作如下：

(1) 设计与实现阶段：开发人员需要完成表结构和业务逻辑 SQL 的设计。

(2) 测试阶段：测试人员验证 SQL 的正确性。

(3) 部署与发布阶段：运维人员对库表结构和数据进行初始化。

(4) 生产与运维阶段：运维人员对环境中的 SQL 进行监控，发现问题，诊断问题，解决问题。

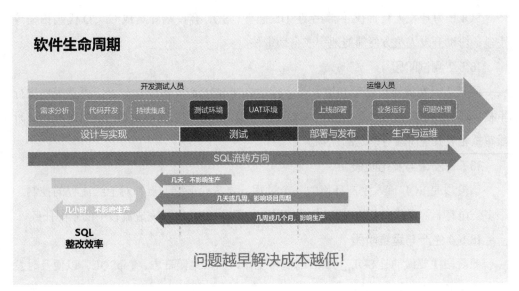

图 63

通过对各阶段 SQL 流转中各岗位工作内容的分析可知，SQL 问题越早解决，成本越低！

我们都希望将问题消灭在萌芽中，但无法保证在不同阶段都不会发生问题。所以，需要在不同阶段准备对应的审核手段（见图 64）。

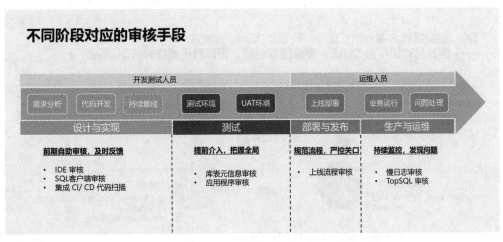

图 64

18.5.1 设计与实现阶段

在设计与实现阶段完成自助审核，尽早发现问题。前面说过，这个阶段主要任务是完成业务功能的开发，能进行 SQL 审核的都是非常优秀的开发人员。在尽量保持低成本且不改变开发习惯的情况下完成自助审核为主要需求。

SQLE 为开发人员提供了常用的 IDE 插件、SQL 客户端和集成 CI/CD 代码扫描等手段，协助开发人员方便简捷地完成自助审核。

18.5.2 测试阶段

测试人员在该阶段已经知道业务运行的具体 SQL 和库表结构，可以更直观地进行审核，还可以审核通过网络层抓包或者云平台提供的审计功能抓取到的具体数据。此阶段审核相较其他阶段有一定的优势。

18.5.3 部署与发布阶段

该阶段是 SQL 流入生产环境的一个过程，要实现审核卡点和对上线流程的控制。很多公司有非常规范专业的 SQL 上线流程，由开发和 DBA 来完成流程中的不同任务。

18.5.4 生产与运维阶段

此阶段主要做的是 SQL 上线后的监督工作，如采集慢日志、TopSQL，以便及时发现生产环境中的问题。

18.6 总结

最后，我们以 SQLE 为例总结如下：在软件生命周期中以 SQL 流转的角度，在四个不同的阶段通过建立规范、上线前控制、标准发布、上线后监督，完成闭环渐进式的 SQL 质量提升（见图 65）。

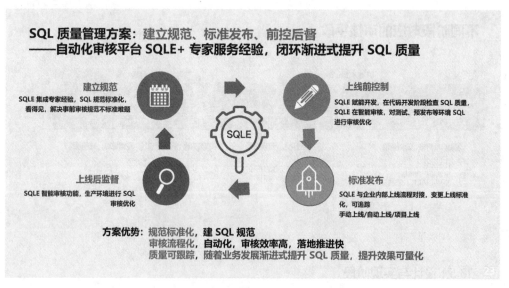

图 65

欢迎大家来体验 SQLE 社区版。

03 OceanBase 篇

在软件国产化的浪潮中，数据库技术作为信息化建设的基石，其重要性日益凸显。过去的一年，我们目睹了 OceanBase 社区的蓬勃发展，DBA 们对这一国产分布式关系型数据库的热情持续高涨。为此，社区提高了 OceanBase 内容的发布频率，使得今年收录的文章数量总计达到了 16 篇。

在内容上，我们延续了上一版的风格，专注于分享故障案例分析和性能优化的实践经验。这些内容为读者们生动地展现了故障现场，为 OceanBase 的运维和优化提供了实用的指导。此外，我们特别推荐安全审计系列文章，这些内容是上一版"OceanBase 篇"中安全审计系列的后续，建议读者结合阅读，以获得更全面的理解。此外，本篇还特别收录了与 ActionDB 和 ActionOMS 相关的精彩内容。

ActionDB 是上海爱可生信息技术股份有限公司发布的基于 OceanBase 开源内核的企业级原生分布式数据库，其底层基于开源内核 OceanBase 4.x，具有原厂授权和内核支持。ActionDB 不仅继承了 OceanBase 高性能、海量数据分析处理、平滑伸缩等优点，还增强了 MySQL 的兼容性和安全性，提供企业级运维管理工具，可以解决数据库性能瓶颈、降本增效等难题，广泛应用于金融、政企、电商、能源电力等行业。

1 OceanBase 安全审计之传输加密

作者：金长龙、陈慧明

本文主要实践如何配置传输加密并验证是否真的加密。

1.1 环境准备

(1) 企业版 OceanBase 4.1 集群（3 节点）+OBProxy。

(2) 配置 CA、服务端及客户端证书。

> OceanBase 社区版也可以实现。

1.2 OBServer 传输加密

1.2.1 开启加密

OceanBase 传输加密通过多个配置项组合开启。

(1) 通过 root 用户登录 sys 租户。

(2) 指定私钥、证书、CA 证书的获取方式。

```
altersystemsetssl_external_kms_info='
{
"ssl_mode":"file"
}';
```

(3) 配置 MySQL 端口的 SSL 通信。

```
altersystemsetssl_client_authentication='TRUE';
# 配置为 TRUE 后，MySQL 通信 SSL 即时开启
```

(4) 配置 RPC 通信的 SSL 白名单。

由于 OBServer 之间的 TCP 连接都是长连接，因此需要重启 OBServer 才能开启 RPC 的 SSL 加密通信。

```
# 配置 RPC 通信 SSL 白名单
# 整个集群都开启
altersystemset_ob_ssl_invited_nodes='ALL';
```

```
# 指定 IP 的 OBServer 开启 SSL
altersystemset_ob_ssl_invited_nodes='135.xxx.xx.xx, 128.xxx.xx.xx';
```

1.2.2 验证加密

1.2.2.1 MySQL 端口（2881）

(1) 通过 \s 查看（见图 1）。

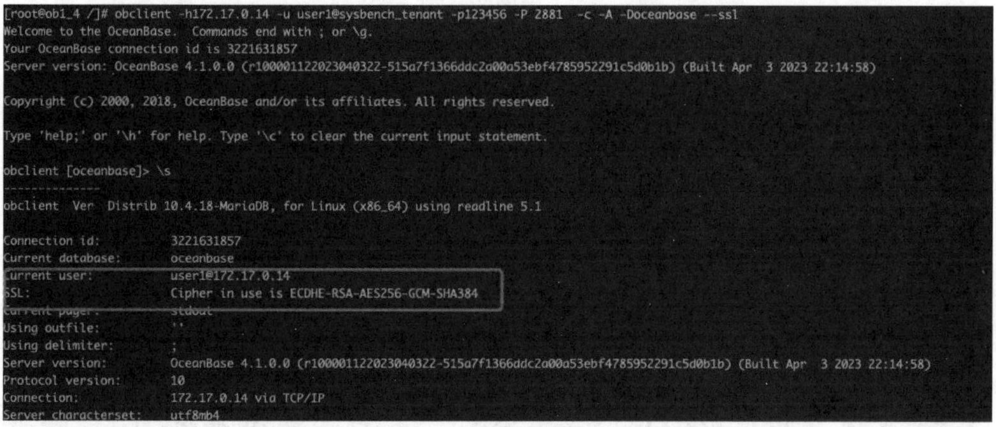

图 1

(2) 抓包（见图 2）。

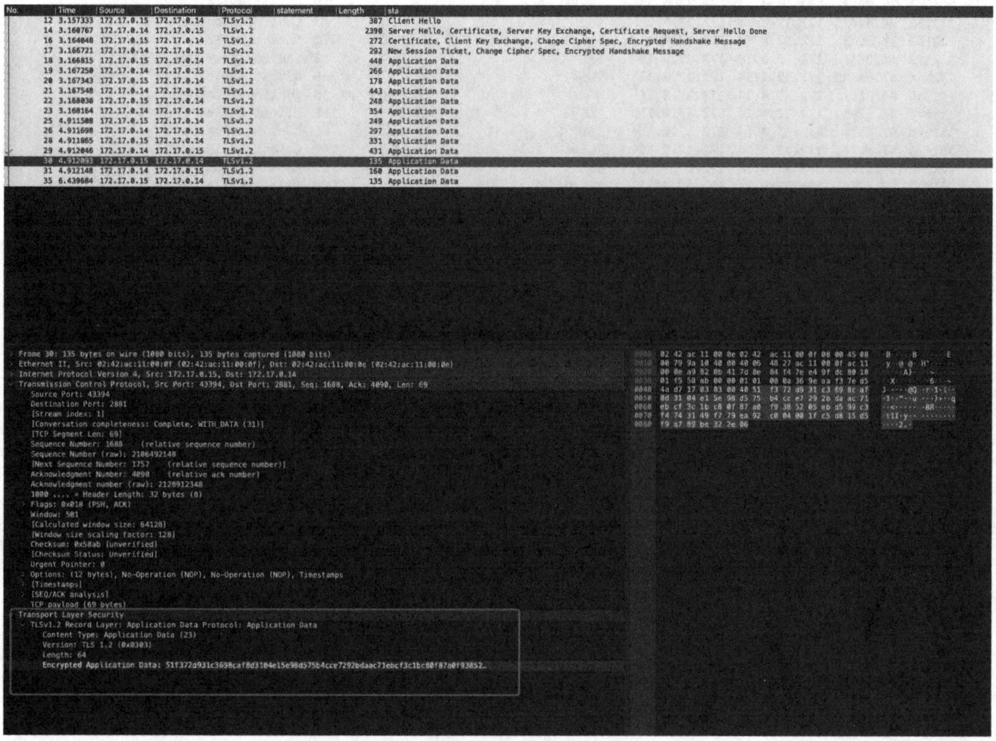

图 2

1.2.2.2 RPC 端口（2882）

(1) 通过日志检索 rpc connection accept，查看 use_ssl 的值是 True 还是 False（见图 3）。

图 3

(2) 抓包（见图 4）。

No.	Time	Source	Destination	Protocol	statement	Length	sta
8406	2.739470	172.17.0.14	172.17.0.15	SSL		85	Continuation Data
8408	2.739548	172.17.0.14	172.17.0.15	TLSv1		268	Client Hello
18349	5.745907	172.17.0.14	172.17.0.15	SSL		85	Continuation Data
18351	5.745996	172.17.0.14	172.17.0.15	TLSv1		268	Client Hello
28016	8.752380	172.17.0.14	172.17.0.15	SSL		85	Continuation Data
28018	8.752472	172.17.0.14	172.17.0.15	TLSv1		268	Client Hello
37220	11.7586...	172.17.0.14	172.17.0.15	SSL		85	Continuation Data
37222	11.7586...	172.17.0.14	172.17.0.15	TLSv1		268	Client Hello
46621	14.7604...	172.17.0.14	172.17.0.15	SSL		85	Continuation Data
46631	14.7605...	172.17.0.14	172.17.0.15	TLSv1		268	Client Hello
56666	17.7672...	172.17.0.14	172.17.0.15	SSL		85	Continuation Data
56668	17.7673...	172.17.0.14	172.17.0.15	TLSv1		268	Client Hello
65947	20.7734...	172.17.0.14	172.17.0.15	SSL		85	Continuation Data
65949	20.7735...	172.17.0.14	172.17.0.15	TLSv1		268	Client Hello
75200	23.7767...	172.17.0.14	172.17.0.15	SSL		85	Continuation Data
75202	23.7768...	172.17.0.14	172.17.0.15	TLSv1		268	Client Hello
85085	26.7828...	172.17.0.14	172.17.0.15	SSL		85	Continuation Data
85087	26.7829...	172.17.0.14	172.17.0.15	TLSv1		268	Client Hello
94595	29.7842...	172.17.0.14	172.17.0.15	SSL		85	Continuation Data
94597	29.7844...	172.17.0.14	172.17.0.15	TLSv1		268	Client Hello
103878	32.7927...	172.17.0.14	172.17.0.15	SSL		85	Continuation Data
103880	32.7928...	172.17.0.14	172.17.0.15	TLSv1		268	Client Hello
113536	35.7972...	172.17.0.14	172.17.0.15	SSL		85	Continuation Data
113543	35.7973...	172.17.0.14	172.17.0.15	TLSv1		268	Client Hello
123181	38.8033...	172.17.0.14	172.17.0.15	SSL		85	Continuation Data
123183	38.8033...	172.17.0.14	172.17.0.15	TLSv1		268	Client Hello
132352	41.8047...	172.17.0.14	172.17.0.15	SSL		85	Continuation Data
132354	41.8048...	172.17.0.14	172.17.0.15	TLSv1		268	Client Hello
141869	44.8107...	172.17.0.14	172.17.0.15	SSL		85	Continuation Data
141871	44.8108...	172.17.0.14	172.17.0.15	TLSv1		268	Client Hello
151974	47.8176...	172.17.0.14	172.17.0.15	SSL		85	Continuation Data
151976	47.8177...	172.17.0.14	172.17.0.15	TLSv1		268	Client Hello
161398	50.8209...	172.17.0.14	172.17.0.15	SSL		85	Continuation Data
161400	50.8210...	172.17.0.14	172.17.0.15	TLSv1		268	Client Hello
170659	53.8257...	172.17.0.14	172.17.0.15	SSL		85	Continuation Data
170661	53.8258...	172.17.0.14	172.17.0.15	TLSv1		268	Client Hello
180369	56.8319...	172.17.0.14	172.17.0.15	SSL		85	Continuation Data
180371	56.8320...	172.17.0.14	172.17.0.15	TLSv1		268	Client Hello
189827	59.8359...	172.17.0.14	172.17.0.15	SSL		85	Continuation Data
189829	59.8360...	172.17.0.14	172.17.0.15	TLSv1		268	Client Hello

图 4

1.3 ODP 传输加密

使用了 OBProxy 之后，客户端跟 OceanBase 建立加密连接，实际是跟 OBProxy 建立加密连接，然后 OBProxy 跟 OBServer 再建立加密连接。按照这个逻辑，前面服务端 OceanBase 集群开启 SSL 客户端认证也成为一个必要的前提了。

1.3.1 开启加密

(1) 使用 OBProxy 的 root@proxysys 账号登录。

(2) 设置证书、公钥、私钥。

```
UPDATEproxyconfig.security_configSETCONFIG_VAL='{"sourceType" : "FILE", "CA" : "certs/ca.pem", "publicKey" : "certs/client-cert.pem", "privateKey" : "certs/client-key.pem"}'WHEREAPP_NAME='obproxy'andVERSION='1';
```

注意：这里配置的公钥和私钥，是前面生成的客户端证书，而不是服务端的。因为 OBProxy 作为客户端和服务端链路中间重要的一环，是客户端的"服务端"，同时也是 OceanBase 服务端的"客户端"。

(3) 检查是否设置成功（见图 5）。

图 5

(4) 配置客户端和 OBProxy，开启 SSL 连接。

```
alterproxyconfigsetenable_client_ssl=true;
```

(5) 配置 OBProxy 和 OBServer，开启 SSL 连接。

```
alterproxyconfigsetenable_server_ssl=true;
```

(6) 用业务租户的管理员账户登录，设置 SSL 白名单。

```
altersystemsetob_ssl_invited_common_names="obclient";
# 这个参数是租户级别的，需要在要连接的租户里设置，立即生效，不需要重启实例或者集群。
```

注意：ob_ssl_invited_common_names 的值要设置成和客户端证书 subject 中的 cn(common name) 字段一致。

具体代码如图 6 所示。

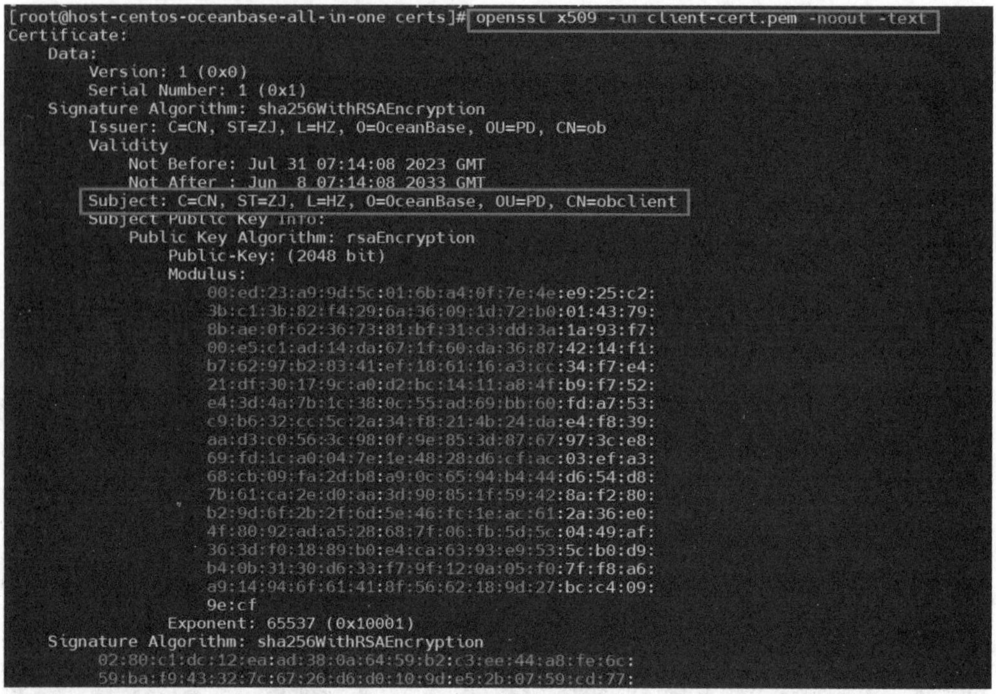

```
[root@host-centos-oceanbase-all-in-one certs]# openssl x509 -in client-cert.pem -noout -text
Certificate:
    Data:
        Version: 1 (0x0)
        Serial Number: 1 (0x1)
    Signature Algorithm: sha256WithRSAEncryption
        Issuer: C=CN, ST=ZJ, L=HZ, O=OceanBase, OU=PD, CN=ob
        Validity
            Not Before: Jul 31 07:14:08 2023 GMT
            Not After : Jun  8 07:14:08 2033 GMT
        Subject: C=CN, ST=ZJ, L=HZ, O=OceanBase, OU=PD, CN=obclient
        Subject Public Key Info:
            Public Key Algorithm: rsaEncryption
                Public-Key: (2048 bit)
                Modulus:
                    00:ed:23:a9:9d:5c:01:6b:a4:0f:7e:4e:e9:25:c2:
                    3b:c1:3b:82:f4:29:6a:36:09:1d:72:b0:01:43:79:
                    8b:ae:0f:62:36:73:81:bf:31:c3:dd:3a:1a:93:f7:
                    00:e5:c1:ad:14:da:67:1f:60:da:36:87:42:14:f1:
                    b7:62:97:b2:83:41:ef:18:61:16:a3:cc:34:f7:e4:
                    21:df:30:17:9c:a0:d2:bc:14:11:a8:4f:b9:f7:52:
                    e4:3d:4a:7b:1c:38:0c:55:ad:69:bb:60:fd:a7:53:
                    c9:b6:32:cc:5c:2a:34:f8:21:4b:24:da:e4:f8:39:
                    aa:d3:c0:56:3c:98:0f:9e:85:3d:87:67:97:3c:e8:
                    69:fd:1c:a0:04:7e:1e:48:28:d6:cf:ac:03:ef:a3:
                    68:cb:09:fa:2d:b8:a9:0c:65:94:b4:44:d6:54:d8:
                    7b:61:ca:2e:d0:aa:3d:90:85:1f:59:42:8a:f2:80:
                    b2:9d:6f:2b:2f:6d:5e:46:fc:1e:ac:61:2a:36:e0:
                    4f:80:92:ad:a5:28:68:7f:06:fb:5d:5c:04:49:af:
                    36:3d:f0:18:89:b0:e4:ca:63:93:e9:53:5c:b0:d9:
                    b4:0b:31:30:d6:33:f7:9f:12:0a:05:f0:7f:f8:a6:
                    a9:14:94:6f:61:41:8f:56:62:18:9d:27:bc:c4:09:
                    9e:cf
                Exponent: 65537 (0x10001)
    Signature Algorithm: sha256WithRSAEncryption
         02:80:c1:dc:12:ea:ad:38:0a:64:59:b2:c3:ee:44:a8:fe:6c:
         59:ba:f9:43:32:7c:67:26:d6:d0:10:9d:e5:2b:07:59:cd:77:
```

图 6

1.3.2 验证加密

(1) 客户端和 OBProxy 的连接如图 7 所示。

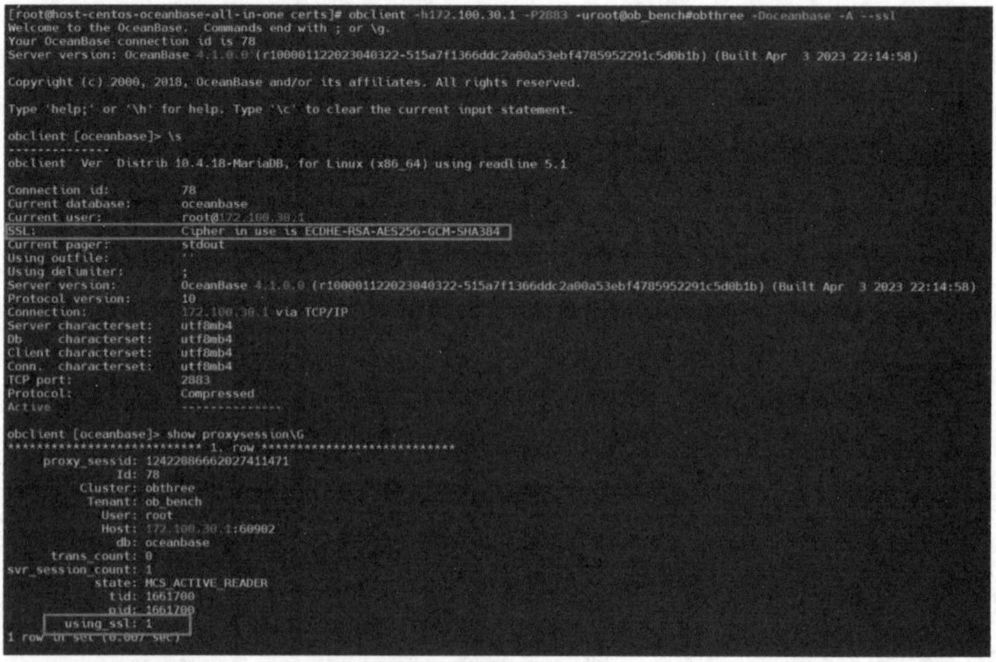

```
[root@host-centos-oceanbase-all-in-one certs]# obclient -h172.100.30.1 -P2883 -uroot@ob_bench#obthree -Doceanbase -A --ssl
Welcome to the OceanBase.  Commands end with ; or \g.
Your OceanBase connection id is 78
Server version: OceanBase 4.1.0.0 (r100001122023040322-515a7f1366ddc2a00a53ebf4785952291c5d0b1b) (Built Apr  3 2023 22:14:58)

Copyright (c) 2000, 2018, OceanBase and/or its affiliates. All rights reserved.

Type 'help;' or '\h' for help. Type '\c' to clear the current input statement.

obclient [oceanbase]> \s
--------------
obclient  Ver  Distrib 10.4.18-MariaDB, for Linux (x86_64) using readline 5.1

Connection id:          78
Current database:       oceanbase
Current user:           root@172.100.30.1
SSL:                    Cipher in use is ECDHE-RSA-AES256-GCM-SHA384
Current pager:          stdout
Using outfile:          ''
Using delimiter:        ;
Server version:         OceanBase 4.1.0.0 (r100001122023040322-515a7f1366ddc2a00a53ebf4785952291c5d0b1b) (Built Apr  3 2023 22:14:58)
Protocol version:       10
Connection:             172.100.30.1 via TCP/IP
Server characterset:    utf8mb4
Db     characterset:    utf8mb4
Client characterset:    utf8mb4
Conn.  characterset:    utf8mb4
TCP port:               2883
Protocol:               Compressed
Active                  --------------

obclient [oceanbase]> show proxysession\G
*************************** 1. row ***************************
     proxy_sessid: 12422086662027411471
               Id: 78
          Cluster: obthree
           Tenant: ob_bench
             User: root
             Host: 172.100.30.1:60902
               db: oceanbase
      trans_count: 0
svr_session_count: 1
            state: MCS_ACTIVE_READER
              tid: 1661700
              pid: 1661700
        using_ssl: 1
1 row in set (0.007 sec)
```

图 7

(2)OBProxy 和 OBServer 的连接如图 8 所示。

图 8

1.4 总结

在实际配置和验证的过程中也踩了几个"坑",还是要对文档多加理解和消化。

2 OceanBase 安全审计之透明加密

作者：张乾

本文主要实践数据透明加密，并验证加密是否有效。

环境版本：OceanBase 4.1.0.0 企业版。

2.1 加密配置

详细的加密步骤在此略过，本次使用 MySQL 租户。

2.1.1 开启透明加密并创建表空间

以管理员用户身份登录到集群的 MySQL 租户。

```
# 开启 internal 方式的透明加密
# tde_method 默认值为 none，表示关闭透明表空间加密
obclient [oceanbase]>ALTERSYSTEMSETtde_method='internal';
Query OK,0rowsaffected (0.022sec)


obclient [oceanbase]>SHOWPARAMETERSLIKE'tde_method';
+--------+---------+-------------+----------+------------+---------+--------
--+-----------------------------------------------------------------------
```

```
+------+----------+------------+----------+------------+-----------+-------+------+---------+-------+--------+------------+
|zone  |svr_type  |svr_ip      |svr_port  |name        |data_type  |value  |info  |section  |scope  |source  |edit_level  |
+------+----------+------------+----------+------------+-----------+-------+------+---------+-------+--------+------------+
|zone1 |observer  |172.17.0.13 |2882      |tde_method  |NULL       |internal|none: transparent encryption is none, none means cannot use tde, internal : transparent encryption is in the form of internal tables, bkmi : transparent encryption is in the form of external bkmi|OBSERVER|TENANT|DEFAULT|DYNAMIC_EFFECTIVE|
+------+----------+------------+----------+------------+-----------+-------+------+---------+-------+--------+------------+
1 row in set (0.017 sec)

# 执行该语句，生成主密钥
obclient [oceanbase]> ALTER INSTANCE ROTATE INNODB MASTER KEY;
Query OK, 0 rows affected (0.028 sec)

# 创建表空间并指定加密算法，其中 'y' 表示默认使用 aes-256 算法
obclient [oceanbase]> CREATE TABLESPACE sectest_ts1 encryption='y';
Query OK, 0 rows affected (0.021 sec)
```

2.1.2 在加密表空间内创建新表

以普通用户身份登录到数据库的 MySQL 租户，创建新表 t1。

```
# 创建表并指定表空间
obclient [sysbenchdb]> CREATE TABLE t1 (id1 int, id2 int) TABLESPACE sectest_ts1;
Query OK, 0 rows affected (0.076 sec)

# 确认表空间内的表是否标记为加密
# encryptionalg 为 aes-256, 且 encrypted 为 YES 则表示表加密配置成功
obclient [oceanbase]> SELECT table_name, encryptionalg, encrypted FROM oceanbase.V$OB_ENCRYPTED_TABLES;
+------------+---------------+-----------+
|table_name  |encryptionalg  |encrypted  |
+------------+---------------+-----------+
|t1          |aes-256        |YES        |
+------------+---------------+-----------+
1 row in set (0.048 sec)
```

往表内插入一条值，并执行大合并，使值落盘 SSTable。

```
# 插入值
obclient [sysbenchdb]>insertintot1values(147852369,999999991);
Query OK,1rowaffected (0.005sec)

# 执行大合并
ALTERSYSTEMMAJOR FREEZE TENANT=ALL;

# 查看合并进度
SELECT*FROMoceanbase.CDB_OB_ZONE_MAJOR_COMPACTION\G
```

2.1.3 创建一个不加密的表用以对比

以普通用户身份登录到数据库的 MySQL 租户，创建不指定加密空间的新表 tttttt2。

同样插入一条数据，并执行大合并。

```
obclient [sysbenchdb]>CREATETABLEtttttttt2 (id1int, id2int);
Query OK,0rowsaffected (0.076sec)
obclient [sysbenchdb]>insertintotttttttt2values(147852369,999999991);
Query OK,1rowaffected (0.005sec)

# 执行大合并
ALTERSYSTEMMAJOR FREEZE TENANT=ALL;

# 查看合并进度
SELECT*FROMoceanbase.CDB_OB_ZONE_MAJOR_COMPACTION\G
```

2.2 加密验证

方式是借助工具 ob_admin 验证，其 dumpsst 功能可以显示 block_file 文件中的内容。

使用 dumpsst 来查看加密表的内容，验证是否加密。使用前，需要知道目标数据的 macro block id。接下来，先找到上面数据对应的 macro block id。

2.2.1 查找 macro block id

先根据 oceanbase.DBA_OB_TABLE_LOCATIONS 找到两张表的 TABLET_ID，其中加密表 t1 的 TABLET_ID 为 200001，未加密表 tttttt2 的 TABLET_ID 为 200002。

```
    obclient [oceanbase]>select*fromoceanbase.DBA_OB_TABLE_LOCATIONSwhereTABLE_
NAME='t1';
    +---------------+------------+----------+------------+---------------+-------
-----------+------------+--------------+------------+--------+--------+----------
--+---------+--------+--------------+
    |DATABASE_NAME|TABLE_NAME|TABLE_ID|TABLE_TYPE|PARTITION_NAME|SUBPARTITION_
NAME|INDEX_NAME|DATA_TABLE_ID|TABLET_ID|LS_ID|ZONE|SVR_IP|SVR_PORT|ROLE|REPLICA_
TYPE|
```

```
        +----------------+------------+----------+------------+----------------+------
------------+------------+--------------+------------+--------+---------+-------
-+---------+-------+--------------+
        |sysbenchdb|t1|500006|USERTABLE|NULL|NULL|NULL|NULL|200001|1001|zone1|172.17.0
.13|2882|LEADER|FULL|
        +----------------+------------+----------+------------+----------------+------
------------+------------+--------------+------------+--------+---------+-------
-+---------+-------+--------------+
    1rowinset(0.005sec)

    obclient [oceanbase]>select*fromoceanbase.DBA_OB_TABLE_LOCATIONSwhereTABLE_
NAME='ttttttt2';
        +----------------+------------+----------+------------+----------------+------
------------+------------+--------------+------------+--------+---------+-------
-+---------+-------+--------------+
        |DATABASE_NAME|TABLE_NAME|TABLE_ID|TABLE_TYPE|PARTITION_NAME|SUBPARTITION_
NAME|INDEX_NAME|DATA_TABLE_ID|TABLET_ID|LS_ID|ZONE|SVR_IP|SVR_PORT|ROLE|REPLICA_
TYPE|
        +----------------+------------+----------+------------+----------------+------
------------+------------+--------------+------------+--------+---------+-------
-+---------+-------+--------------+
        |sysbenchdb|ttttttt2|500007|USERTABLE|NULL|NULL|NULL|NULL|200002|1001|zone1|17
2.17.0.13|2882|LEADER|FULL|
        +----------------+------------+----------+------------+----------------+------
------------+------------+--------------+------------+--------+---------+-------
-+---------+-------+--------------+
    1rowinset(0.005sec)
```

根据 TABLET_ID 与合并时间，在 GV$OB_TABLET_COMPACTION_HISTORY 中找到 MACRO_ID_LIST，其中记录的 ID 即我们需要的 macro block id。

从输出中，我们可以看到加密表 t1 对应的 macro block id 为 387，未加密表 ttttttt2 对应的 macro block id 为 718。

```
    obclient [oceanbase]>select*fromGV$OB_TABLET_COMPACTION_HISTORYwhereTABLET_
ID=200001andTYPE='MAJOR_MERGE'orderbySTART_TIME \G
    ***************************1.row***************************
    SVR_IP:172.17.0.13
    SVR_PORT:2882
    TENANT_ID:1004
    LS_ID:1001
    TABLET_ID:200001
    TYPE: MAJOR_MERGE
    COMPACTION_SCN:1685093467526445446
    START_TIME:2023-05-2617:31:22.478149
    FINISH_TIME:2023-05-2617:31:22.482045
    TASK_ID: YB42AC11000D-0005FC95091493EB-0-0
    OCCUPY_SIZE:432
```

```
    MACRO_BLOCK_COUNT:1
    MULTIPLEXED_MACRO_BLOCK_COUNT:0
    NEW_MICRO_COUNT_IN_NEW_MACRO:1
    MULTIPLEXED_MICRO_COUNT_IN_NEW_MACRO:0
    TOTAL_ROW_COUNT:1
    INCREMENTAL_ROW_COUNT:1
    COMPRESSION_RATIO:0.67
    NEW_FLUSH_DATA_RATE:100
    PROGRESSIVE_COMPACTION_ROUND:1
    PROGRESSIVE_COMPACTION_NUM:0
    PARALLEL_DEGREE:1
    PARALLEL_INFO:-
    PARTICIPANT_TABLE: table_cnt=4,[MAJOR]scn=1;[MINI]start_scn=1,end_
scn=1685093478867382402;
    MACRO_ID_LIST:387
    COMMENTS: serialize_medium_list:{cnt=1;1685093467526445446}|time_guard=EXECUTE
=4.20ms|(0.79)|CREATE_SSTABLE=648us|(0.12)|total=5.32ms;
    ***************************2.row***************************
    SVR_IP:172.17.0.13
    SVR_PORT:2882
    TENANT_ID:1004
    LS_ID:1001
    TABLET_ID:200001
    TYPE: MAJOR_MERGE
    COMPACTION_SCN:1685094492266634220
    START_TIME:2023-05-2617:48:27.276906
    FINISH_TIME:2023-05-2617:48:27.282468
    TASK_ID: YB42AC11000D-0005FC9509149878-0-0
    OCCUPY_SIZE:432
    MACRO_BLOCK_COUNT:1
    MULTIPLEXED_MACRO_BLOCK_COUNT:0
    NEW_MICRO_COUNT_IN_NEW_MACRO:1
    MULTIPLEXED_MICRO_COUNT_IN_NEW_MACRO:0
    TOTAL_ROW_COUNT:1
    INCREMENTAL_ROW_COUNT:1
    COMPRESSION_RATIO:0.67
    NEW_FLUSH_DATA_RATE:71
    PROGRESSIVE_COMPACTION_ROUND:1
    PROGRESSIVE_COMPACTION_NUM:0
    PARALLEL_DEGREE:1
    PARALLEL_INFO:-
    PARTICIPANT_TABLE: table_cnt=3,[MAJOR]scn=1685093467526445446;[MINI]start_
scn=1685093467530410154,end_scn=1685094504683817069;
    MACRO_ID_LIST:718
    COMMENTS: serialize_medium_list:{cnt=1;1685094492266634220}|time_guard=EXECUTE
=5.92ms|(0.45)|CREATE_SSTABLE=5.94ms|(0.45)|total=13.10ms;

    obclient [oceanbase]>select*fromGV$OB_TABLET_COMPACTION_HISTORYwhereTABLET_
```

```
ID=200002andTYPE='MAJOR_MERGE'orderbySTART_TIME \G
*************************1.row***************************
    SVR_IP:172.17.0.13
    SVR_PORT:2882
    TENANT_ID:1004
    LS_ID:1001
    TABLET_ID:200002
    TYPE: MAJOR_MERGE
    COMPACTION_SCN:1685094492266634220
    START_TIME:2023-05-2617:48:27.277801
    FINISH_TIME:2023-05-2617:48:27.284542
    TASK_ID: YB42AC11000D-0005FC9509149879-0-0
    OCCUPY_SIZE:424
    MACRO_BLOCK_COUNT:1
    MULTIPLEXED_MACRO_BLOCK_COUNT:0
    NEW_MICRO_COUNT_IN_NEW_MACRO:1
    MULTIPLEXED_MICRO_COUNT_IN_NEW_MACRO:0
    TOTAL_ROW_COUNT:1
    INCREMENTAL_ROW_COUNT:1
    COMPRESSION_RATIO:0.61
    NEW_FLUSH_DATA_RATE:40
    PROGRESSIVE_COMPACTION_ROUND:1
    PROGRESSIVE_COMPACTION_NUM:0
    PARALLEL_DEGREE:1
    PARALLEL_INFO:-
    PARTICIPANT_TABLE: table_cnt=4,[MAJOR]scn=1685093467526445446;[MINI]start_
scn=1,end_scn=1685094504683817070;
    MACRO_ID_LIST:718
    COMMENTS: serialize_medium_list:{cnt=1;1685094492266634220}|time_guard=EXECUTE
=10.20ms|(0.86)|total=11.87ms;
```

2.2.2 解析 block_file 文件

安装完 ob_admin，使用 dumpsst 解析 2.2.1 节中获取的 macro block id。

注意：

（1）ob_admin dumpsst 必须在 ${path_to_oceanbase}/oceanbase 层级运行，原因是读取 etc/observer.config.bin 时使用的是相对路径。

（2）目前测试显示，必须指定 --macro-id，否则会报错（报错内容需在 ob_admin.log 中查看）。

本次使用的几个参数如下：

（1）-f：指定 data 目录。

（2）-d：宏块类型，目前仅支持 macro_block。

（3）-a：即 macro-id，填写在 2.2.1 节中获取的值。

（4）-t：指定 tablet_id，进一步缩小解析范围。

(5)-i：即 micro block id，-1 表示所有 micro blocks。

2.2.3 解析 t1 表，即加密表

可以看到输出中 tablet_id 为 200001，row_count 为 1，对应我们插入的那一条数据。其中并未展示这行数据内容，验证了数据成功加密。

```
[admin@ob_4 oceanbase]$ ob_admin dumpsst-f/home/admin/oceanbase/store/obdemo/-d macro_block-a387-t200001-i-1
succtoopen, filename=ob_admin.log, fd=3, wf_fd=2
oldlog_file needclose,old=ob_admin.lognew=ob_admin.log
succtoopen, filename=ob_admin.log, fd=3, wf_fd=2
succtoopen, filename=ob_admin_rs.log, fd=4, wf_fd=2
-----------------------------{Common Header}-----------------------------
|header_size|24
|version|1
|magic|1001
|attr|1
|payload_size|952
|payload_checksum|-1027413104
--------------------------------------------------------------------------
-----------------------------{SSTable Macro Block Header}-----------------------------
|header_size|208
|version|1
|magic|1007
|tablet_id|200001
|logical_version|1685093467526445446
|data_seq|0
|column_count|5
|rowkey_column_count|3
|row_store_type|1
|row_count|1
|occupy_size|432
|micro_block_count|1
|micro_block_data_offset|232
|data_checksum|2617981320
|compressor_type|6
|master_key_id|500004
--------------------------------------------------------------------------
--------{column_index         column_type       column_order column_checksum collation_type}----------
|[0ObUInt64TypeASC334486997463]
|[1ObIntTypeASC31365443363]
|[2ObIntTypeASC238884235363]
|[3ObInt32TypeASC277679507263]
|[4ObInt32TypeASC8253742263]
--------------------------------------------------------------------------
```

2.2.4 解析 tttttt2 表，即未加密的表

将命令中 tablet_id 和 macro block id 替换为 tttttt2 表的 id，进行解析。对比加密表 t1，未加密表输出的信息更丰富，并且可以看到具体的数据内容。

此处精简展示，可以看到 Total Rows 中显示了前面插入的那条数据 [{"INT":147852369}] [{"INT":999999991}]。

```
[admin@ob_4 oceanbase]$ ob_admin dumpsst-f/home/admin/oceanbase/store/
obdemo/-d macro_block-a718-t200002-i-1
    succtoopen, filename=ob_admin.log, fd=3, wf_fd=2
    oldlog_file needclose,old=ob_admin.lognew=ob_admin.log
    succtoopen, filename=ob_admin.log, fd=3, wf_fd=2
    succtoopen, filename=ob_admin_rs.log, fd=4, wf_fd=2
    -----------------------------{Common Header}-----------------------------
    |header_size|24
    |version|1
    |magic|1001
    |attr|1
    |payload_size|892
    |payload_checksum|-1696352947
    --------------------------------------------------------------------------
    -----------------------------{SSTable Macro Block Header}-----------------------------
    |header_size|208
    |version|1
    |magic|1007
    |tablet_id|200002
    |logical_version|1685094492266634220
    |data_seq|0
    |column_count|5
    |rowkey_column_count|3
    |row_store_type|1
    |row_count|1
    |occupy_size|424
    |micro_block_count|1
    |micro_block_data_offset|232
    |data_checksum|725485397
    |compressor_type|6
    |master_key_id|0
    --------------------------------------------------------------------------
    ……
    -----------------------------{Total Rows[1]}-----------------------------
    |ROW[0]:trans_id=[{txid:0}],dml_flag=[N|INSERT],mvcc_flag=[]|[{"BIGINT
UNSIGNED":1}][{"BIGINT":-1685094482154160502}][{"BIGINT":0}][{"INT":147852369}]
[{"INT":999999991}]
    ……
    -----------------------------{Encoding Column Header[4]}-----------------------------
    |type|0
```

```
            |attribute|0
            |isfix length|0
            |has extendvalue|0
            |isbit packing|0
            |islastvar field|0
            |extendvalueindex|65542
            |store object type|0
            |offset|0
            |length|0
            --------------------------------------------------------------------------------
            ----------------------------{Index Micro Block[0]}-----------------------------
            ------------------------------{Total Rows[1]}----------------------------
            |ROW[0]:trans_id=[{txid:0}],dml_flag=[N|INSERT],mvcc_flag=[]|[{"BIGINT
 UNSIGNED":1}][{"BIGINT":-1685094482154160502}][{"BIGINT":0}][{"VARCHAR":"
            ",collation:"binary", coercibility:"NUMERIC"}]
            |Index BlockRowHeader|[{version:1, row_store_type:1, compressor_type:6,
 is_data_index:1, is_data_block:1, is_leaf_block:0, is_major_node:1, is_pre_
 aggregated:0, is_deleted:0, contain_uncommitted_row:0, is_macro_node:0, has_string_
 out_row:0, all_lob_in_row:1, macro_id:[-1](ver=0,mode=0,seq=0), block_offset:232,
 block_size:192, master_key_id:0, encrypt_id:0, encrypt_key:"data_size:16, data:0
 0000000000000000000000000000000", row_count:1, schema_version:1685094464567160,
 macro_block_count:0, micro_block_count:1}]
            ----------------------------{Macro Meta Micro Block}----------------------------
            ----------------------------{Encoding Micro Header}-----------------------------
            |header_size|96
            |version|2
            |magic|1005
            |column_count|4
            |rowkey_column_count|3
            |row_count|1
            |row_store_type|2
            |row_index_byte|0
            |var_column_count|0
            |row_data_offset|357
            |column_chksum[0]|3344869974
            |column_chksum[1]|1868627082
            |column_chksum[2]|2388842353
            |column_chksum[3]|1583982749
            --------------------------------------------------------------------------------
......
```

2.3 总结

本文主要介绍了如何使用 ob_admin 工具的 dumpsst 功能解析 block_file，以验证 OceanBase 的数据透明加密功能。

如果在使用 dumpsst 过程中遇到问题，建议多查看 ob_admin.log，对于排查问题比较有帮助。

3 OceanBase 建表分区数超限报错

作者：何文超

3.1 背景

ERROR 1499 (HY000): Too many partitions (including subpartitions) were defined

创建表时报错，虽然是内部错误，但是错误信息显示：创建了太多的分区。

```
    [root@observer04 ~]# mysql -h10.186.64.125 -P2883 -uroot@wenchao_mysql#hwc_cluster:1682755171 -p"xxxx"
    MySQL [lss]> CREATE TABLE `wms_order` (
        `A1` varchar(100) CHARACTER SET utf8mb4 COLLATE utf8mb4_bin DEFAULT NULL COMMENT 'A1',
        `A2` varchar(100) CHARACTER SET utf8mb4 COLLATE utf8mb4_bin DEFAULT NULL COMMENT 'A2',
        `A3` varchar(100) CHARACTER SET utf8mb4 COLLATE utf8mb4_bin DEFAULT NULL COMMENT 'A3',
        `A4` varchar(100) CHARACTER SET utf8mb4 COLLATE utf8mb4_bin DEFAULT NULL COMMENT 'A4',
        `A5` varchar(100) CHARACTER SET utf8mb4 COLLATE utf8mb4_bin DEFAULT NULL COMMENT 'A5',
        `A6` varchar(100) CHARACTER SET utf8mb4 COLLATE utf8mb4_bin DEFAULT NULL COMMENT 'A6',
        `A7` varchar(100) CHARACTER SET utf8mb4 COLLATE utf8mb4_bin DEFAULT NULL COMMENT 'A7',
        `A8` varchar(100) CHARACTER SET utf8mb4 COLLATE utf8mb4_bin DEFAULT NULL COMMENT 'A8',
        `A9` varchar(100) CHARACTER SET utf8mb4 COLLATE utf8mb4_bin DEFAULT NULL COMMENT 'A9',
        `A10` varchar(100) CHARACTER SET utf8mb4 COLLATE utf8mb4_bin DEFAULT NULL COMMENT 'A10'
    ) DEFAULT CHARSET = utf8mb4 ROW_FORMAT = DYNAMIC COMPRESSION = 'zstd_1.0' REPLICA_NUM = 3 BLOCK_SIZE = 16384 USE_BLOOM_FILTER = FALSE TABLET_SIZE = 134217728 PCTFREE = 0 COMMENT = '物流订单表'
    MySQL [lss]> ERROR 1499 (HY000): Too many partitions (including subpartitions) were defined
```

接下来我们分析一下问题的原因。

3.2 排查

3.2.1 检查参数

(1) 检查每个 OBServer 上可以创建最大的分区数量，当前是 500000。

```
[root@observer04 ~]# mysql -h10.186.64.125 -P2883 -uroot@sys#hwc_cluster:16827
55171 -p"xxxx" -A oceanBase
    MySQL [oceanBase]> select * from __all_virtual_sys_parameter_
stat where name like '%_max_partition_%';
+-------+----------+---------------+----------+--------------------------------+
-----------+--------+---------------+-----------+------------------------------+
-------------+--------+---------------+----------+---------+--------+-----------+
    | zone  | svr_type | svr_ip        | svr_port | name
| data_type | value  | value_strict  | info
| need_reboot | section | visible_level | scope    | source  | edit_level |
+-------+----------+---------------+----------+--------------------------------+
-----------+--------+---------------+-----------+------------------------------+
-------------+--------+---------------+----------+---------+--------+-----------+
    | zone1 | observer | 10.186.64.122 |     2882 | _max_partition_cnt_per_server
| NULL      | 500000 | NULL          | specify max partition count on one observer
|        NULL | OBSERVER | NULL        | CLUSTER  | DEFAULT | DYNAMIC_EFFECTIVE |
+-------+----------+---------------+----------+--------------------------------+
-----------+--------+---------------+-----------+------------------------------+
-------------+--------+---------------+----------+---------+--------+-----------+
```

(2) 检查当前分区数量的和，目前并没有超过这个限制（500000）。

```
MySQL [oceanBase]> select count(*) from v$partition;
+----------+
| count(*) |
+----------+
|   421485 |
+----------+
```

3.2.2 检查回收站

(1) 检查回收站是否开启。

```
[root@observer04 ~]# mysql -h10.186.64.125 -P2883 -uroot@wenchao_mysql#hwc_
cluster:1682755171 -p"xxxx"

MySQL [lss]> show variables like '%recy%';
+---------------+-------+
| Variable_name | Value |
+---------------+-------+
| recyclebin    | ON    |
+---------------+-------+
1 row in set (0.01 sec)
```

(2) 检查回收站中是否存在未删除的分区表。

```
MySQL [lss]> show recyclebin;
+------------------------------------------+---------------+-------+------------------------------+
| OBJECT_NAME                              | ORIGINAL_NAME | TYPE  | CREATETIME                   |
+------------------------------------------+---------------+-------+------------------------------+
| __recycle_$_1682755171_1689139725669688  | mytable_1     | TABLE | 2023-07-12 13:28:45.687379   |
| __recycle_$_1682755171_1689139737584112  | mytable_1     | TABLE | 2023-07-12 13:28:57.584660   |
| __recycle_$_1682755171_1689139750594392  | t1            | TABLE | 2023-07-12 13:29:10.594118   |
+------------------------------------------+---------------+-------+------------------------------+
3 rows in set (0.01 sec)
```

如果存在，需要和业务侧沟通是否可以清理。清理回收站的表后，发现分区表数量减少，但是创建表时依旧报错。

(3) 查看回收站中对象的保留天数。

```
MySQL [lss]> SHOW PARAMETERS LIKE 'recyclebin_object_expire_time'\G;
*************************** 1. row ***************************
      zone: zone1
  svr_type: observer
    svr_ip: 10.186.64.122
  svr_port: 2882
      name: recyclebin_object_expire_time
 data_type: NULL
     value: 0s
      info: recyclebin object expire time, default 0 that means auto purge recyclebin off. Range: [0s, +∞)
   section: ROOT_SERVICE
     scope: CLUSTER
    source: DEFAULT
edit_level: DYNAMIC_EFFECTIVE
1 row in set (0.02 sec)
```

配置项 recyclebin_object_expire_time 的取值说明如下：

① 当其值为 0s 时，表示关闭自动 Purge 回收站功能。

② 当其值不为 0s 时，表示回收对应时间内进入回收站的数据库对象。

3.2.3 检查租户内存

(1) 找到分区数最多的 10 个租户。

```
[root@observer04 ~]# mysql -h10.186.64.125 -P2883 -uroot@sys#hwc_
cluster:1682755171 -p"xxxx"
    SELECT t2.tenant_name,t2.tenant_id, t1.replica_count
    FROM
     (SELECT tenant_id, COUNT(*) AS replica_count
      FROM __all_virtual_partition_info
      GROUP BY tenant_id
      ORDER BY replica_count DESC
      LIMIT 10) t1
    JOIN
     (SELECT tenant_id, tenant_name
      FROM __all_tenant) t2
    ON t1.tenant_id=t2.tenant_id
    ORDER BY replica_count DESC;
+-------------------+-----------+---------------+
| tenant_name       | tenant_id | replica_count |
+-------------------+-----------+---------------+
| wenchao_mysql     |      1100 |        107853 |
| wenchao_01        |      1088 |         99846 |
| wenchao_02        |      1104 |         15873 |
| wenchao_03        |         1 |          3867 |
| wenchao_04        |      1044 |          3270 |
| wenchao_05        |      1066 |          2811 |
| wenchao_06        |      1079 |          2658 |
| wenchao_07        |      1103 |          2103 |
| wenchao_08        |      1057 |          2040 |
| wenchao_09        |      1016 |          1950 |
+-------------------+-----------+---------------+
10 rows in set (0.13 sec)
```

(2) 查找租户有多少表。

```
    select count(*),svr_Ip from __all_virtual_meta_table where tenant_
id=1100 and role=1 group by svr_ip;
+----------+--------------+
| count(*) | svr_Ip       |
+----------+--------------+
|    11921 |10.186.64.103 |
|    11868 |10.186.64.104 |
|    12013 |10.186.64.105 |
+----------+--------------+
3 rows in set (0.35 sec)
```

(3) 计算租户需要扩容的内存大小。

①租户当前分区总数：num=107853/ 副本数。

②租户可用内存上限：(1-memstore_limit_percentage) × 租户 unit 的内存大小 =(1-0.8) × 24GB=4.8GB。

③单个副本分区所需总内存：

partition_mem=128kB×(107853/3)+max(1000,(107853/3)/10)×400kB=5.75GB。

注意：单个副本分区所需总内存＞租户可用内存上限，若租户所需内存超限，需要对租户内存进行扩容。

(4) 根据租户内存计算最大分区数量。

①单机租户允许创建的最大分区数量：(max_memory-memstore_limit)/partition_mem_n（partition_mem_n 指的是单个分区所需总内存）。

②单机租户允许创建的最大分区数量：(24-24*0.8)/(5.75/(107853/3))=4.8/(5.75/(107853/3))=30011。

③临时处理方案：扩容租户内存。

④根源治理：不可能无限扩内存。为业务方设定合理的分区数量限制，建议业务侧合理使用分区表，制定合理的定期清理策略。

3.3 总结

根据上述计算得出：单机租户允许创建的最大分区数量为 30011，建议业务侧注意控制分区数量，以免超限对业务造成影响。

4 MySQL 迁移至 OceanBase 场景中的自增主键实践

作者：赵黎明

4.1 背景

在 MySQL 迁移到 OceanBase（Oracle 模式）的场景中，通常需要考虑 OceanBase（Oracle 模式）中自增主键如何实现的问题。本文将从解决实际问题的角度出发，验证并总结一个比较可行的实施方案。

4.2 方案一

我们将通过创建自定义序列的方式来实现自增主键。

4.2.1 MySQL 端创建测试表

```
zlm@10.186.60.68 [zlm]> desc t;
+-------+-------------+------+-----+---------+----------------+
| Field | Type        | Null | Key | Default | Extra          |
```

```
+-------+-------------+------+-----+---------+----------------+
| id    | bigint(20)  | NO   | PRI | NULL    | auto_increment |
| name  | varchar(10) | YES  |     | NULL    |                |
+-------+-------------+------+-----+---------+----------------+
2 rows in set (0.00 sec)

zlm@10.186.60.68 [zlm]> select * from t;
+----+------+
| id | name |
+----+------+
|  1 | a    |
|  2 | b    |
|  3 | c    |
+----+------+
3 rows in set (0.00 sec)
```

4.2.2 使用 DBCAT 导出表结构

DBCAT 是 OceanBase 提供的命令行工具，主要用于异构数据库迁移场景中非表对象的 DDL 导出和转换，如 Oracle 中的序列、函数、存储过程、包、触发器、视图等对象。

```
cd /opt/oceanbase_package/tools/dbcat-1.9.1-RELEASE/bin
./dbcat convert -H 10.186.60.68 -P 3332 --user=zlm --password=zlm --database=
zlm --no-schema --no-quote --from mysql57 --to oboracle32x --table t --file=/tmp
Parsed args:
[--no-quote] true
[--no-schema] true
[--table] [t]
[--host] 10.186.60.68
[--port] 3332
[--user] zlm
[--password] ******
[--database] zlm
[--file] /tmp
[--from] mysql57
[--to] oboracle32x
2023-08-16 14:41:58 INFO Init convert config finished.
2023-08-16 14:41:58 INFO {dataSource-1} inited
2023-08-16 14:41:58 INFO Init source druid connection pool finished.
2023-08-16 14:41:58 INFO Register c.o.o.d.m.c.m.MySql56ObOracle22xColumnConverter
2023-08-16 14:41:58 INFO Register c.o.o.d.m.c.m.MySql56ObOracle22xIndexConverter
2023-08-16 14:41:58 INFO Register c.o.o.d.m.c.m.MySql56ObOracle22xPrimaryKeyConverter
2023-08-16 14:41:58 INFO Register c.o.o.d.m.c.m.MySql56ObOracle22xUniqueKeyConverter
2023-08-16 14:41:58 INFO Register c.o.o.d.m.c.m.MySql56ObOracle22xPartitionConverter
2023-08-16 14:41:59 INFO Load meta/mysql/mysql56.xml, meta/mysql/mysql57.xml successed
2023-08-16 14:42:09 INFO Query 0 dependencies elapsed 17.35 ms
2023-08-16 14:42:09 INFO Query table: "t" attr finished. Remain: 0
2023-08-16 14:42:09 INFO Query 1 tables elapsed 69.71 ms
```

```
2023-08-16 14:42:09 WARN Include types is empty. Ignore schema: ZLM
2023-08-16 14:42:09 WARN Skip to compare/convert sequences as SEQUENCE is unsu
pported
2023-08-16 14:42:09 INFO Starting to convert schema to path: "/tmp/
dbcat-2023-08-16-144209/ZLM"
2023-08-16 14:42:09 INFO Successed to generate report in the path: "/tmp/
dbcat-2023-08-16-144209/ZLM-conversion.html"
2023-08-16 14:42:09 INFO {dataSource-1} closing ...
2023-08-16 14:42:09 INFO {dataSource-1} closed
cd /tmp/dbcat-2023-08-16-144209/ZLM
cat TABLE-schema.sql
CREATE TABLE t (
    id NUMBER(19,0),
    name VARCHAR2(30 BYTE),
    CONSTRAINT PRIMARY PRIMARY KEY (id)
);

-- CREATE SEQUENCE xxx START WITH 1 INCREMENT BY 1 ... for t
```

DBCAT 会对目标表的表结构做转换，使其符合 Oracle 的语法，并在导出的 DDL 语句中写入一行创建序列的伪 SQL。可见，此工具也是建议创建序列来处理 MySQL 自增列的。

4.2.3 在 OceanBase 目标端创建序列

(1) 对于含有自增列的每个表，都需要创建一个序列与之对应。

(2) 创建序列时，建议以 SEQ_<表名>_<字段名> 的方式命名。

(3) 当不指定序列的 CYCLE 和 ORDER 属性时，其默认值都是 N，即不循环，不排序。

(4) 当不指定序列的 CACHE 属性时，默认缓存 20 个序列。

(5) 字段 MIN_VALUE 对应创建序列时 MIN_VALUE 属性的值。

(6) 字段 LAST_NUMBER 对应创建序列时 START WITH 属性的值。

```
ZLM[ZLM]> CREATE SEQUENCE SEQ_T_ID MINVALUE 1 MAXVALUE 999999 INCREMENT BY 1 S
TART WITH 1;
Query OK, 0 rows affected (0.03 sec)

ZLM[ZLM]> SELECT SEQUENCE_NAME,MIN_VALUE,LAST_NUMBER,CYCLE_FLAG,ORDER_
FLAG,CACHE_SIZE FROM DBA_SEQUENCES WHERE SEQUENCE_OWNER='ZLM';
+---------------+-----------+-------------+------------+------------+------------+
| SEQUENCE_NAME | MIN_VALUE | LAST_NUMBER | CYCLE_FLAG | ORDER_FLAG | CACHE_SIZE |
+---------------+-----------+-------------+------------+------------+------------+
| SEQ_T_ID      |         1 |           1 | N          | N          |         20 |
+---------------+-----------+-------------+------------+------------+------------+
1 row in set (0.01 sec)

ZLM[ZLM]> drop sequence SEQ_T_ID;
```

```
    Query OK, 0 rows affected (0.03 sec)

    ZLM[ZLM]> CREATE SEQUENCE SEQ_T_ID MINVALUE 1 MAXVALUE 999999 INCREMENT BY 1 S
TART WITH 10;
    Query OK, 0 rows affected (0.03 sec)

    ZLM[ZLM]> SELECT SEQUENCE_NAME,MIN_VALUE,LAST_NUMBER,CYCLE_FLAG,ORDER_
FLAG,CACHE_SIZE FROM DBA_SEQUENCES WHERE SEQUENCE_OWNER='ZLM';
    +---------------+-----------+-------------+------------+------------+------------+
    | SEQUENCE_NAME | MIN_VALUE | LAST_NUMBER | CYCLE_FLAG | ORDER_FLAG | CACHE_SIZE |
    +---------------+-----------+-------------+------------+------------+------------+
    | SEQ_T_ID      |         1 |          10 | N          | N          |         20 |
    +---------------+-----------+-------------+------------+------------+------------+
    1 row in set (0.03 sec)
```

4.2.4 在 OceanBase 目标端建表

基于 4.2.3 节中第 1 步获取的 DDL 和第 2 步创建的序列，执行以下 SQL 语句：

```
ZLM[ZLM]> CREATE TABLE "ZLM"."T" (
    ->     "ID" NUMBER(19,0) DEFAULT SEQ_T_ID.NEXTVAL,
    ->     "NAME" VARCHAR2(30 BYTE),
    ->     CONSTRAINT "PRIMARY" PRIMARY KEY ("ID"));
Query OK, 0 rows affected (0.15 sec)
```

通常表结构及数据都是通过 OMS 来完成迁移的，很少会直接用 DBCAT 生成的 DDL 建表语句在目标端手动建表，除非一些较特殊的场景，如以上这种给字段增加缺省属性为序列值情况。

> 建表时注意：
>
> （1）应将表名和字段名都应为大写，因为 Oracle 严格区分数据库对象的大小写。
>
> （2）ID 列的 DEFAULT 值，指定为 4.2.3 节中第 2 步所创建序列的下一个值，即 SEQ_T_ID.NEXTVAL。

4.2.5 使用 DataX 迁移数据

DataX 是阿里开源的离线数据同步工具，支持多种异构数据源，可以通过 OceanBase 的 Reader 和 Writer 插件实现 OceanBase 与异构数据库之间的数据迁移。

```
-- 创建 DataX 配置文件 (存放在 dataX 的 ./job 目录下)
cat t.json
{
  "job": {
    "setting": {
      "speed": {
        "channel": 4
```

```
      },
      "errorLimit": {
        "record": 0,
        "percentage": 0.1
      }
    },
    "content": [
      {
        "reader": {
          "name": "mysqlreader",
          "parameter": {
            "username": "zlm",
            "password": "zlm",
            "column": [
              "*"
            ],
            "connection": [
              {
                "table": [
                  "t"
                ],
                "jdbcUrl": ["jdbc:mysql://10.186.60.68:3332/zlm?useUnicode=true&characterEncoding=utf8"]
              }
            ]
          }
        },
        "writer": {
          "name": "oceanbasev10writer",
          "parameter": {
            "obWriteMode": "insert",
            "column": [
              "*"
            ],
            "preSql": [
              "truncate table T"
            ],
            "connection": [
              {
                "jdbcUrl": "||_dsc_ob10_dsc_||jingbo_ob:ob_oracle||_dsc_ob10_dsc_||jdbc:oceanbase://10.186.65.22:2883/ZLM?useLocalSessionState=true&allowBatch=true&allowMultiQueries=true&rewriteBatchedStatements=true",
                "table": [
                  "T"
                ]
              }
            ],
            "username": "ZLM",
```

```
                "password":"zlm",
                "writerThreadCount":10,
                "batchSize": 1000,
                "memstoreThreshold": "0.9"
            }
          }
        }
      ]
    }
}
```

-- 执行数据迁移
./bin/datax.py job/t.json
DataX (20220610-external), From Alibaba !
Copyright (C) 2010-2017, Alibaba Group. All Rights Reserved.
full db is not specified.
schema sync is not specified.
java -server -Xms4g -Xmx16g -XX:+HeapDumpOnOutOfMemoryError -XX:HeapDumpPath=/home/admin/datax3/log -DENGINE_VERSION=20220610-external -Xms4g -Xmx16g -XX:+HeapDumpOnOutOfMemoryError -XX:HeapDumpPath=/home/admin/datax3/log -Dloglevel=info -Dproject.name=di-service -Dfile.encoding=UTF-8 -Dlogback.statusListenerClass=ch.qos.logback.core.status.NopStatusListener -Djava.security.egd=file:///dev/urandom -Ddatax.home=/home/admin/datax3 -Dlogback.configurationFile=/home/admin/datax3/conf/logback.xml -classpath /home/admin/datax3/lib/*:. -Dlog.file.name=in_datax3_job_t_json com.alibaba.datax.core.Engine -mode standalone -jobid -1 -job /home/admin/datax3/job/t.json -fulldb false -schema false
    2023-08-16 14:58:41.088 [main] INFO Engine - running job from /home/admin/datax3/job/t.json
    2023-08-16 14:58:41.374 [main] INFO VMInfo - VMInfo# operatingSystem class => sun.management.OperatingSystemImpl
    2023-08-16 14:58:41.382 [main] INFO Engine - the machine info =>
    （略）
    2. record average count and max count task info :
    PHASE | AVERAGE RECORDS | AVERAGE BYTES | MAX RECORDS | MAX RECORD`S BYTES | MAX TASK ID | MAX TASK INFO
    READ_TASK_DATA | 3 | 6B | 3 | 6B | 0-0-0 | t,jdbcUrl:[jdbc:mysql://10.186.60.68:3332/zlm]
    2023-08-16 14:58:45.189 [job-0] INFO MetricReportUtil - reportJobMetric is turn off
    2023-08-16 14:58:45.189 [job-0] INFO StandAloneJobContainerCommunicator - Total 3 records, 6 bytes | Speed 3B/s, 1 records/s | Error 0 records, 0 bytes | All Task WaitWriterTime 0.000s | All Task WaitReaderTime 0.000s | Percentage 100.00%
    2023-08-16 14:58:45.190 [job-0] INFO LogReportUtil - report datax log is turn off

```
2023-08-16 14:58:45.190 [job-0] INFO JobContainer -
任务启动时刻              : 2023-08-16 14:58:41
任务结束时刻              : 2023-08-16 14:58:45
任务总计耗时              :                  3s
任务平均流量              :                3B/s
记录写入速度              :              1rec/s
读出记录总数              :                   3
读写失败总数              :                   0
2023-08-16 14:58:45.190 [job-0] INFO PerfTrace - reset PerfTrace.
```

4.2.6 验证效果

验证主键列能否实现自增。

```
SYS[ZLM]> select * from t;
+----+------+
| ID | NAME |
+----+------+
|  1 | a    |
|  2 | b    |
|  3 | c    |
+----+------+
3 rows in set (0.01 sec)

SYS[ZLM]> insert into t(name) values('d');
Query OK, 1 row affected (0.02 sec)

SYS[ZLM]> select * from t;
+----+------+
| ID | NAME |
+----+------+
|  1 | a    |
|  2 | b    |
|  3 | c    |
|  4 | d    |
+----+------+
4 rows in set (0.00 sec)
```

新插入的数据每次都会先获取 ID 列上序列的 NEXTVAL 值，于是就实现了主键自增的需求。

使用自定义序列的 NEXTVAL 作为主键列的 DEFAULT 值后，在迁移过程中不必关心源端表中记录的自增列最大值，将表迁移到目标端后，直接插入新数据时，不会与原来的数据发生冲突。

4.3 方案二

利用 GENERATED BY DEFAULT AS IDENTITY 属性生成序列的方案是否好用？先

来看一个测试吧！

```
-- 删除并重建测试表
ZLM[ZLM]> DROP TABLE T;
Query OK, 0 rows affected (0.10 sec)
ZLM[ZLM]> CREATE TABLE "ZLM"."T" (
    ->      "ID" NUMBER(19,0) GENERATED BY DEFAULT AS IDENTITY MINVALUE 1 MAXVALUE 999999,
    ->      "NAME" VARCHAR2(30 BYTE),
    ->      CONSTRAINT "PRIMARY" PRIMARY KEY ("ID"));
Query OK, 0 rows affected (0.15 sec)

-- 查看序列
ZLM[ZLM]> SELECT SEQUENCE_NAME,MIN_VALUE,LAST_NUMBER,CYCLE_FLAG,ORDER_FLAG,CACHE_SIZE FROM DBA_SEQUENCES WHERE SEQUENCE_OWNER='ZLM';
+-----------------+-----------+-------------+------------+------------+------------+
| SEQUENCE_NAME   | MIN_VALUE | LAST_NUMBER | CYCLE_FLAG | ORDER_FLAG | CACHE_SIZE |
+-----------------+-----------+-------------+------------+------------+------------+
| SEQ_T_ID        |         1 |          21 | N          | N          |         20 |
| ISEQ$$_50034_16 |         1 |           1 | N          | N          |         20 |
+-----------------+-----------+-------------+------------+------------+------------+
2 rows in set (0.00 sec)

# 此时，系统自动创建了名为ISEQ$$_50034_16的序列，其他默认值与自定义创建的序列一致

-- 查看表结构
ZLM[ZLM]> desc t;
+-------+--------------+------+-----+-----------------+-------+
| FIELD | TYPE         | NULL | KEY | DEFAULT         | EXTRA |
+-------+--------------+------+-----+-----------------+-------+
| ID    | NUMBER(19)   | NO   | PRI | SEQUENCE.NEXTVAL| NULL  |
| NAME  | VARCHAR2(30) | YES  |     | NULL            | NULL  |
+-------+--------------+------+-----+-----------------+-------+
2 rows in set (0.02 sec)

# 注意，ID列的DEFAULT值为SEQUENCE.NEXTVAL，而不是ISEQ$$_50034_16.NEXTVAL

-- 重新导入数据
./bin/datax.py job/t.json
（略）

-- 插入数据
ZLM[ZLM]> insert into t(name) values('d');
ORA-00001: unique constraint '1' for key 'PRIMARY' violated

ZLM[ZLM]> insert into t(name) values('d');
ORA-00001: unique constraint '2' for key 'PRIMARY' violated
```

```
ZLM[ZLM]> insert into t(name) values('d');
ORA-00001: unique constraint '3' for key 'PRIMARY' violated

ZLM[ZLM]> insert into t(name) values('d');
Query OK, 1 row affected (0.01 sec)

ZLM[ZLM]> select "ISEQ$$_50034_16".CURRVAL from dual;
+---------+
| CURRVAL |
+---------+
|    4    |
+---------+

1 row in set (0.00 sec)
```

表中有3条数据，当执行插入时，START WITH 实际上还是从默认值1开始的。

每次执行插入，ID 都会获取序列的 NEXTVAL 值，直到执行至第4次，未与表中已有记录冲突，才能插入成功。

对于这种场景，解决方案有两种，不过都比较烦琐：

(1) 插入记录前先获取当前序列的 NEXTVAL 值（需多次执行，执行次数＝源端表记录数）。

(2) 创建序列时，根据源端表上自增列最大值来指定 START WITH 属性。

注意: 使用 GENERATED BY DEFAULT AS IDENTITY 属性生成的序列无法直接删除，会报错。

```
ORA-32794: cannot drop a system-generated sequence
```

4.3.1 获取表中自增列最大值

4.3.1.1 方法1：MAX 函数

```
zlm@10.186.60.68 [zlm]> SELECT MAX(id)+1 as AUTO_INCREMENT FROM t;
+----------------+
| AUTO_INCREMENT |
+----------------+
|       4        |
+----------------+

1 row in set (0.00 sec)
```

4.3.1.2 方法2：系统视图

```
zlm@10.186.60.68 [zlm]> select AUTO_INCREMENT from information_schema.tables where table_name='t';
+----------------+
```

```
| AUTO_INCREMENT |
+----------------+
|              4 |
+----------------+

1 row in set (0.00 sec)
```

4.3.1.3 方法 3：show create table 命令

```
zlm@10.186.60.68 [zlm]> show create table t\G
*************************** 1. row ***************************
    Table: t
Create Table: CREATE TABLE `t` (
 `id` bigint(20) NOT NULL AUTO_INCREMENT,
 `name` varchar(10) DEFAULT NULL,
 PRIMARY KEY (`id`)
) ENGINE=InnoDB AUTO_INCREMENT=4 DEFAULT CHARSET=utf8
1 row in set (0.00 sec)
```

4.3.2 脚本初始化序列的一个示例

```
-- 删除并重建表
ZLM[ZLM]> drop table t;
Query OK, 0 rows affected (0.02 sec)

ZLM[ZLM]> CREATE TABLE "ZLM"."T" (
    ->          "ID" NUMBER(19,0) GENERATED BY DEFAULT AS IDENTITY MINVALUE 1 MAXVALUE 999999,
    ->          "NAME" VARCHAR2(30 BYTE),
    ->          CONSTRAINT "PRIMARY" PRIMARY KEY ("ID"));
Query OK, 0 rows affected (0.04 sec)

-- 导入数据
./bin/datax.py job/t.json
（略）

-- 执行脚本并确认返回结果正常
[root@10-186-65-73 ~]# cat init_sequence.sh
#!/bin/bash

## 获取当前表的自增列最大值
i=$(obclient -h10.186.60.68 -P3332 -uzlm -pzlm -Nse "SELECT MAX(id)+1 FROM zlm.t;" 2>/dev/null | head -1)

## 循环执行SQL初始化序列值
for ((j=1; j<=$i; j++))
do
  obclient -h10.186.65.43 -P2883 -uZLM@ob_oracle#bobo_ob:1675327512 -pzlm -Ac -DZLM -Nse "select ISEQ\$\$_50037_16.nextval from dual;" 1>/dev/null 2>&1
```

```
done
[root@10-186-65-73 ~]# sh init_sequence.sh
[root@10-186-65-73 ~]# echo $?
0

-- 执行插入
ZLM[ZLM]> insert into t(name) values('d');
Query OK, 1 row affected (0.01 sec)
```

序列经过初始化处理后，在完成数据导入并直接插入新增记录时，就不会再产生唯一性冲突的报错了。同样地，先用脚本获取自增列的最大值，在创建序列时指定 START WITH 与自增列最大值一致，也可以解决以上问题，这里不展开阐述。

总体而言，通过 GENERATED BY DEFAULT AS IDENTITY 属性创建的序列（方案二）不如自定义序列（方案一）好用。

4.4 总结

本文验证并阐述了在 OceanBase Oracle 中实现自增主键的两种方法：创建自定义序列和利用 GENERATED BY DEFAULT AS IDENTITY 属性生成序列。

4.4.1 方案一

创建自定义序列的时候，需要为每张有自增列的表创建一个单独的序列，建议将序列名与表名关联，但无需关注 START WITH 的取值。当插入新记录时，系统会自动获取下一个可用的序列值。

4.4.2 方案二

利用 GENERATED BY DEFAULT AS IDENTITY 属性生成序列时，存在一些限制：

(1) 因序列由系统自动创建并管理，需要查询系统视图才能获取序列名，无法与业务表名直接对应。

(2) 创建序列时需要根据自增列最大值来指定 START WITH 的取值，当有大量表需要处理时，较烦琐。

利用 GENERATED BY DEFAULT AS IDENTITY 属性生成的序列名，在内部有一个计数器，会累计增加，即使删除了原来的序列，原有的名字也不会被重用。删除表时，会自动清理由 GENERATED BY DEFAULT AS IDENTITY 属性生成的序列（直接删除该序列会报错），但不会影响之前创建的其他自定义序列。

采用 GENERATED BY DEFAULT AS IDENTITY 属性生成序列的方案时，还要额外考虑源端待迁移表当前自增列最大值的问题，这无疑增加了迁移的复杂性。

综上所述，更推荐使用自定义序列实现自增主键的方案。

5 如何通过日志观测冻结转储流程？

作者：陈慧明

本文旨在通过日志解析 OceanBase 的冻结转储流程，以其冻结检查线程为切入点，以租户（1002）的线程名为例。

> 以下内容基于版本：5.7.25 OceanBase_CE 4.2.0.0 （r100000152023080109-8024d8ff45c45cf7c62a548752b985648a5795c3）。

基本流程如图 9 所示。

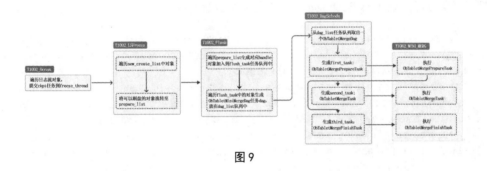

图 9

5.1 T1002_0ccam

5.1.1 线程介绍

这是冻结检查线程，每 2 秒执行一次检查，一旦需要进行冻结操作，就会生成一个检查点任务，并由冻结线程负责处理。可以通过在日志中检索 "tenant freeze timer task" 来验证该线程是否正常运行。

5.1.2 日志流程

当需要进行冻结操作时，系统会记录日志输出：[TenantFreezer] A minor freeze is needed。触发条件为租户的 active_memstore_used_ 超过了 memstore_freeze_trigger 阈值。在触发后，系统会遍历租户日志流，生成并提交相应的冻结任务到冻结线程中。

```
succeed to start ls_freeze_task(ret=0, ls_id={id:xxx})
```

5.2 T1002_LSFreeze

5.2.1 线程介绍

该线程的主要职责是将满足刷盘条件的冻结检查点从 new_create_list 流转至 prepare_list。在这一执行过程中，它会依据 road_to_flush 方法和 ready_for_flush_ 方法所定义的判断条件进行操作。这些条件包括检查 memtable 的 rec_scn 是否处于冻结状态以及是否存在回放引用等因素。

> 注意：每当初始化一个 memtable 后，会将与之关联的冻结检查点注册到一个名为 new_create_list 的双向链表中。这一过程的具体实现可以在 ObTabletMemtableMgr::create_memtable() 方法中找到。

5.2.2 日志流程

日志记录的信息并不会详细展示流程的所有细节，但可以通过以下信息来判断流程是否正常执行，road_to_flush end 也标志着冻结流程完成。

```
    [2023-08-18 06:44:51.285827] INFO  [STORAGE] road_to_flush (ob_data_checkpoint.
cpp:333) [1553][T1002_LSFreeze1][T1002][Y0-0000000000000000-0-0] [lt=7] [Freezer]
road_to_flush begin(ls_->get_ls_id()={id:1001})
    [2023-08-18 06:44:51.285846] INFO  [STORAGE] road_to_flush (ob_data_checkpoint.
cpp:341) [1553][T1002_LSFreeze1][T1002][Y0-0000000000000000-0-0] [lt=16] [Freezer]
new_create_list to ls_frozen_list success(ls_->get_ls_id()={id:1001})
    [2023-08-18 06:44:51.285861] INFO  [STORAGE] road_to_flush (ob_data_checkpoint.
cpp:345) [1553][T1002_LSFreeze1][T1002][Y0-0000000000000000-0-0] [lt=3] [Freezer]
ls_frozen_list to active_list success(ls_->get_ls_id()={id:1001})
    [2023-08-18 06:44:51.285867] INFO  [STORAGE] road_to_flush (ob_data_checkpoint.
cpp:355) [1553][T1002_LSFreeze1][T1002][Y0-0000000000000000-0-0] [lt=6] [Freezer]
active_list to ls_frozen_list success(ls_->get_ls_id()={id:1001})
    [2023-08-18 06:44:51.337395] INFO  [STORAGE] road_to_flush (ob_data_checkpoint.
cpp:358) [1553][T1002_LSFreeze1][T1002][Y0-0000000000000000-0-0] [lt=16] [Freezer]
road_to_flush end(ls_->get_ls_id()={id:1001})
```

5.3 T1002_Flush

5.3.1 线程介绍

这是 Flush 线程，每 5 秒运行一次，其运行状态可以通过日志信息"traversal_flush timer task"来标识。该线程的主要任务是遍历 prepare_list 中的检查点对象，并生成相应的 ObTabletMiniMergeDag 对象作为 DAG 任务执行。

5.3.2 日志流程

转储的执行对象为数据分片（Tablet），每次转储操作可能涉及多个数据分片。以下以数据分片 ID 为 200001 的数据分片为例来描述流程。

首先，针对 ID 为 200001 的数据分片，创建并添加相应的 DAG（有向无环图）至任务队列中。

```
    [2023-08-18 06:44:51.335124] INFO    [COMMON] inner_add_dag (ob_dag_scheduler.
cpp:3377)  [1655][T1002_Flush][T1002][Y0-0000000000000000-0-0] [lt=29] add dag
success(dag=0x7fa95f358b20, start_time=0, id=Y0-0000000000000000-0-0, dag-
>hash()=7887337314793470841, dag_cnt=23, dag_type_cnts=22)
    [2023-08-18 06:44:51.335132] INFO    [COMMON] create_and_add_dag (ob_dag_
scheduler.h:1119) [1655][T1002_Flush][T1002][Y0-0000000000000000-0-0] [lt=3]
success to create and add dag(ret=0, dag=0x7fa95f358b20)
```

如果 DAG 创建成功，会记录相应的成功标志，即日志中会出现"schedule tablet merge dag successfully"。同时，该 DAG 的任务类型会被标记为"MINI_MERGE"。

```
    [2023-08-18 06:44:51.335134] INFO    [STORAGE.TRANS] flush (ob_memtable.cpp:2095)
[1655][T1002_Flush][T1002][Y0-0000000000000000-0-0] [lt=2] schedule tablet merge
dag successfully(ret=0, param={merge_type:"MINI_MERGE", merge_version:0, ls_
id:{id:1001}, tablet_id:{id:200001}, report_:null, for_diagnose:false,
...
recommend_snapshot_version:{val:18446744073709551615, v:3}})
```

5.4 T1002_DagScheduler

5.4.1 线程介绍

根据 DAG 队列中的任务类型，系统会创建对应的线程来执行任务。在这个过程中，会创建一个名为 T1002_MINI_MERGE 的线程来执行转储任务；同时，会创建第一个任务，即 ObTabletMergePrepareTask，这个任务的执行最终会触发生成另外两个任务：ObTabletMergeTask 和 ObTabletMergeFinishTask。

5.4.2 日志流程

在 T1002_DagScheduler 线程中，通过 tablet_id 可以筛选出对应的日志。可以找到类型为"DAG_MINI_MERGE"的记录，并记录下对应的 task_id（YB427F000001-0006032C0D448715-0-0）。

```
    [2023-08-18 06:44:51.420180] INFO    [SERVER] add_task (ob_sys_task_stat.
cpp:142) [1597][T1002_DagSchedu][T1002][Y0-0000000000000000-0-0] [lt=9] succeed
to add sys task(task={start_time:1692341091420175, task_id:YB427F000001-
0006032C0D448715-0-0, task_type:3, svr_ip:"127.0.0.1:2882", tenant_
id:1002, is_cancel:false, comment:"info="DAG_MINI_MERGE";ls_id=1001;tablet_
id=200001;compaction_scn=0;extra_info=merge_type="MINI_MERGE"";"})
```

在线程 T1002_DagScheduler 中，通过筛选任务标识 task_id，可以明确看到整个

DAG 任务的调度过程，总计调度了 3 个任务。

```
    [2023-08-18 06:44:51.420180] INFO  [SERVER] add_task (ob_sys_task_stat.
cpp:142) [1597][T1002_DagSchedu][T1002][Y0-0000000000000000-0-0] [lt=9] succeed
to add sys task(task={start_time:1692341091420175, task_id:YB427F000001-
0006032C0D448715-0-0, task_type:3, svr_ip:"127.0.0.1:2882", tenant_
id:1002, is_cancel:false, comment:"info="DAG_MINI_MERGE";ls_id=1001;tablet_
id=200001;compaction_scn=0;extra_info="merge_type="MINI_MERGE"";"})
    [2023-08-18 06:44:51.420192] INFO  [COMMON] schedule_one (ob_dag_scheduler.
cpp:2997) [1597][T1002_DagSchedu][T1002][YB427F000001-0006032C0D448715-0-0] [lt=12]
schedule one task(task=0x7fa9264c8080, priority="PRIO_COMPACTION_HIGH", group id=0,
total_running_task_cnt=6, running_task_cnts_[priority]=6, low_limits_[priority]=6,
up_limits_[priority]=6, task->get_dag()->get_dag_net()=NULL)
    [2023-08-18 06:44:51.421879] INFO  [COMMON] schedule_one (ob_dag_scheduler.
cpp:2997) [1597][T1002_DagSchedu][T1002][YB427F000001-0006032C0D448715-0-0] [lt=8]
schedule one task(task=0x7fa9264c81b0, priority="PRIO_COMPACTION_HIGH", group id=0,
total_running_task_cnt=6, running_task_cnts_[priority]=6, low_limits_[priority]=6,
up_limits_[priority]=6, task->get_dag()->get_dag_net()=NULL)
    [2023-08-18 06:44:51.876070] INFO  [COMMON] schedule_one (ob_dag_scheduler.
cpp:2997) [1597][T1002_DagSchedu][T1002][YB427F000001-0006032C0D448715-0-0] [lt=16]
schedule one task(task=0x7fa9264c8390, priority="PRIO_COMPACTION_HIGH", group id=0,
total_running_task_cnt=6, running_task_cnts_[priority]=6, low_limits_[priority]=6,
up_limits_[priority]=6, task->get_dag()->get_dag_net()=NULL)
```

5.5 T1002_MINI_MERGE

5.5.1 线程介绍

这个线程主要负责执行在 T1002_DagScheduler 中调度的任务。

5.5.2 日志流程

从完整日志中筛选出对应的任务标识 task_id，我们可以清楚地看到总共进行了 3 个任务调度。这里将日志分成了以下 3 个部分。

5.5.2.1 ObTabletMergePrepareTask

Prepare 任务：主要涉及一些初始化工作和检查项，为后续的任务做准备。

```
    [2023-08-18 06:44:51.420180] INFO  [SERVER] add_task (ob_sys_task_stat.
cpp:142) [1597][T1002_DagSchedu][T1002][Y0-0000000000000000-0-0] [lt=9] succeed
to add sys task(task={start_time:1692341091420175, task_id:YB427F000001-
0006032C0D448715-0-0, task_type:3, svr_ip:"127.0.0.1:2882", tenant_
id:1002, is_cancel:false, comment:"info="DAG_MINI_MERGE";ls_id=1001;tablet_
id=200001;compaction_scn=0;extra_info="merge_type="MINI_MERGE"";"})
    [2023-08-18 06:44:51.420192] INFO  [COMMON] schedule_one (ob_dag_scheduler.
cpp:2997) [1597][T1002_DagSchedu][T1002][YB427F000001-0006032C0D448715-0-0] [lt=12]
schedule one task(task=0x7fa9264c8080, priority="PRIO_COMPACTION_HIGH", group id=0,
total_running_task_cnt=6, running_task_cnts_[priority]=6, low_limits_[priority]=6,
```

```
up_limits_[priority]=6, task->get_dag()->get_dag_net()=NULL)
    ...
    [2023-08-18 06:44:51.421833] INFO    [STORAGE.COMPACTION] process (ob_
tablet_merge_task.cpp:976) [1561][T1002_MINI_MERG][T1002][YB427F000001-
0006032C0D448715-0-0] [lt=20] succeed to init merge ctx(task={this:0x7fa9264c8080,
type:15, status:2, dag:{ObIDag:{this:0x7fa95f358b20, type:0, name:"MINI_
MERGE", id:YB427F000001-0006032C0D448715-0-0, dag_ret:0, dag_status:2, start_
time:1692341091420191, running_task_cnt:1, indegree:0, consumer_group_id:0,
hash:7887337314793470841}, param:{merge_type:"MINI_MERGE", merge_version:0,
ls_id:{id:1001}, tablet_id:{id:200001}, report_:null, for_diagnose:false,
is_tenant_major_merge:false, need_swap_tablet_flag:false}, compat_mode:0,
ctx:{sstable_version_range:{multi_version_start:1, base_version:0, snapshot_
version:1692341091113671451}, scn_range:{start_scn:{val:1, v:0}, end_
scn:{val:1692341091275445526, v:0}}}})
```

5.5.2.2 ObTabletMergeTask

Merge 任务：该任务的重点在于写入宏块，将多版本的记录融合成一条记录，以实现数据的整理和合并。

```
    [2023-08-18 06:44:51.421879] INFO    [COMMON] schedule_one (ob_dag_scheduler.
cpp:2997) [1597][T1002_DagSchedu][T1002][YB427F000001-0006032C0D448715-0-0] [lt=8]
schedule one task(task=0x7fa9264c81b0, priority="PRIO_COMPACTION_HIGH", group_id=0,
total_running_task_cnt=6, running_task_cnts_[priority]=6, low_limits_[priority]=6,
up_limits_[priority]=6, task->get_dag()->get_dag_net()=NULL)
    ...
    ...
    [2023-08-18 06:44:51.875958] INFO    [STORAGE.COMPACTION] process
(ob_tablet_merge_task.cpp:1555) [1595][T1002_MINI_MERG][T1002]
[YB427F000001-0006032C0D448715-0-0] [lt=25] merge macro blocks ok(idx_=0,
task={this:0x7fa9264c81b0, type:1, status:2, dag:{ObIDag:{this:0x7fa95f35
8b20, type:0, name:"MINI_MERGE", id:YB427F000001-0006032C0D448715-0-0, dag_
ret:0, dag_status:2, start_time:1692341091420191, running_task_cnt:1, indegree:0,
consumer_group_id:0, hash:7887337314793470841}, param:{merge_type:"MINI_MERGE",
merge_version:0, ls_id:{id:1001}, tablet_id:{id:200001}, report_:null, for_
diagnose:false, is_tenant_major_merge:false, need_swap_tablet_flag:false}, compat_
mode:0, ctx:{sstable_version_range:{multi_version_start:1, base_version:0,
snapshot_version:1692341091113671451}, scn_range:{start_scn:{val:1, v:0}, end_
scn:{val:1692341091275445526, v:0}}}}})
```

5.5.2.3 ObTabletMergeFinishTask

Finish 任务：主要负责生成新的 MINI SSTable 并释放相关 MemTable。

```
    [2023-08-18 06:44:51.876070] INFO    [COMMON] schedule_one (ob_dag_scheduler.
cpp:2997) [1597][T1002_DagSchedu][T1002][YB427F000001-0006032C0D448715-0-0] [lt=16]
schedule one task(task=0x7fa9264c8390, priority="PRIO_COMPACTION_HIGH", group_id=0,
total_running_task_cnt=6, running_task_cnts_[priority]=6, low_limits_[priority]=6,
up_limits_[priority]=6, task->get_dag()->get_dag_net()=NULL)
```

```
    ...
    [2023-08-18 06:44:51.876907] INFO  [STORAGE.COMPACTION] create_sstable
(ob_tablet_merge_ctx.cpp:344) [1589][T1002_MINI_MERG][T1002][YB427F000001-
0006032C0D448715-0-0] [lt=50] succeed to merge sstable(param={table_key:{tablet_
id:{id:200001}, column_group_idx:0, table_type:"MINI", scn_range:{start_scn:{val:1,
v:0}, end_scn:{val:1692341091275445526, v:0}}}, sstable_logic_seq:0, schema_
version:1692341087064224, ...
    ...
    [2023-08-18 06:44:51.889896] INFO  [STORAGE] release_memtables (ob_i_memtable_
mgr.cpp:164) [1589][T1002_MINI_MERG][T1002][YB427F000001-0006032C0D448715-0-0]
[lt=6] succeed to release memtable(ret=0, i=1, scn={val:1692341091275445526, v:0})
    [2023-08-18 06:44:51.889938] INFO  [STORAGE.COMPACTION] process (ob_
tablet_merge_task.cpp:1209) [1589][T1002_MINI_MERG][T1002][YB427F000001-
0006032C0D448715-0-0] [lt=12] sstable merge finish(ret=0, merge_info={is_
inited:true, sstable_merge_info:{tenant_id:1002, ls_id:{id:1001}, tablet_
id:{id:200001}, compaction_scn:1692341091275445526, merge_type:"MINI_MERGE",
merge_cost_time:454652, merge_start_time:1692341091421154, merge_finish_
time:1692341091875806, dag_id:YB427F000001-0006032C0D448715-0-0, occupy_
size:63203471, new_flush_occupy_size:63203471, original_size:75545791, compressed_
size:62951855, macro_block_count:31, multiplexed_macro_block_count:0, new_micro_
count_in_new_macro:3823, multiplexed_micro_count_in_new_macro:0, total_row_
count:333312, incremental_row_count:333312,
    ...
```

最终，DAG 任务执行完毕后，相关任务会被清除，标志着数据冻结和转储流程成功执行。

```
    [2023-08-18 06:44:51.890015] INFO  [COMMON] finish_dag_ (ob_dag_scheduler.
cpp:2563) [1589][T1002_MINI_MERG][T1002][YB427F000001-0006032C0D448715-0-0]
[lt=19] dag finished(dag_ret=0, runtime=469823, dag_cnt=9, dag_cnts_[dag.get_
type()]=9, &dag=0x7fa95f358b20, dag={ObIDag:{this:0x7fa95f358b20, type:0,
name:"MINI_MERGE", id:YB427F000001-0006032C0D448715-0-0, dag_ret:0, dag_
status:3, start_time:1692341091420191, running_task_cnt:0, indegree:0, consumer_
group_id:0, hash:7887337314793470841}, param:{merge_type:"MINI_MERGE", merge_
version:0, ls_id:{id:1001}, tablet_id:{id:200001}, report_:null, for_
diagnose:false, is_tenant_major_merge:false, need_swap_tablet_flag:false}, compat_
mode:0, ctx:{sstable_version_range:{multi_version_start:1, base_version:0,
snapshot_version:1692341091113671451}, scn_range:{start_scn:{val:1, v:0}, end_
scn:{val:1692341091275445526, v:0}}})
    [2023-08-18 06:44:51.890035] INFO  [SERVER] del_task (ob_sys_task_stat.
cpp:171) [1589][T1002_MINI_MERG][T1002][YB427F000001-0006032C0D448715-0-0]
[lt=18] succeed to del sys task(removed_task={start_time:1692341091420175, task_
id:YB427F000001-0006032C0D448715-0-0, task_type:3, svr_ip:"127.0.0.1:2882",
tenant_id:1002, is_cancel:false, comment:"info="DAG_MINI_MERGE";ls_id=1001;tablet_
id=200001;compaction_scn=0;extra_info="merge_type="MINI_MERGE"";"})
```

6 从 Oracle 迁移至 OceanBase 后存储过程语法报错问题的诊断

作者：余振兴

6.1 背景信息

客户反馈一个存储过程从 Oracle 迁移到 OceanBase Oracle 模式后，执行时报语法错误。报错信息如下：

```
call pro_table_demo('t_cc_demo', to_char(sysdate, 'yyyy-mm-dd'));

报错信息在 p17_db_log 中，报错信息：
-5001 ; ORA-00900: You have an error in your SQL syntax; check the manual that
corresponds to your OceanBase version for the right syntax to use near ')  when
matched then update set a.REMINDER_COUNT=b.REMINDER_COUNT,a.EXT_CUST_NO1' at line 1
```

6.2 问题诊断

对这类报语法错误的 SQL，通常的诊断方式是执行一遍，获取该 SQL 的 trace_id，从日志中获取实际传入变量后的真实 SQL，再进行排查与判断。

6.2.1 获取该存储过程的 trace 信息

具体操作步骤如下：

(1) 执行 set ob_enable_trace_log=on。

(2) 执行问题 SQL。

(3) 执行 show trace。

(4) 执行 show trace 后，会得到 trace_id。

(5) 用这个 trace_id 去查询 gv$sql_audit 表以获取 svr_ip 值，从而得到实际运行该 SQL 的 Observer 服务器 IP 地址。

(6) 在对应 IP 地址的主机上执行 grep trace_id /home/admin/oceanbase/log/observer.log。

6.2.2 定位报错语句

基于获取的 trace log 信息，结合报错位点，找到实际报错的 SQL 语句。例如用报错信息中的 when matched then update set a.REMINDER_COUNT=b.REMINDER_COUNT

部分进行匹配，得到以下 SQL（SQL 做了字段精简）。

```
merge into t_cc_demo a using
    (select REMINDER_COUNT,...,ELECTRICALPIN_EMPLOY_NAME
    from t_cc_demo@dblink_demo
    where lastupt_dttm >= to_date('2023-02-16','yyyy-mm-dd')
    and lastupt_dttm<to_date('2023-02-16','yyyy-mm-dd')+1) b on ()
when matched then
    update set a.REMINDER_COUNT=b.REMINDER_COUNT,...,a.ELECTRICALPIN_EMPLOY_NAME=b.ELECTRICALPIN_EMPLOY_NAME
when not matched then
    insert (REMINDER_COUNT,...,ELECTRICALPIN_EMPLOY_NAME) values (b.REMINDER_COUNT,...,b.ELECTRICALPIN_EMPLOY_NAME)
```

6.2.3 对比报错

手工执行获取的 SQL，观测是否有相同的报错。经过验证，出现的报错与存储过程执行时相同，基本确定是由该 SQL 导致。下面开始针对该 SQL 做进一步诊断。

6.2.4 分析报错原因

可以看到，该 SQL 条件中 ON 后的括号匹配关联条件为空，因此初步判断报错是由此处的条件缺失导致，但还需要进一步分析存储过程中的逻辑进行判断。

```
merge into t_cc_demo a using (select REMINDER_COUNT,...ELECTRICALPIN_EMPLOY_NAME
from t_cc_demo@dblink_demo
where lastupt_dttm >= to_date('2023-02-16','yyyy-mm-dd')
and lastupt_dttm<to_date('2023-02-16','yyyy-mm-dd')+1) b
on ()  -- 存在问题的点，关联条件不存在
when matched then update set
-- 以下的部分省略
```

6.2.5 分析存储过程中的定义

完整的存储过程定义如下：

```
create or replace procedure pro_table_demo(p_par_table  in varchar2,
                            archive_date in varchar2) is
... 存储过程较长，省略部分无关代码

    -- 取表所有字段
    cursor c_column is
      select t.column_name
        from user_tab_columns t
       where t.table_name = upper(p_par_table);

    -- 取表除主键外的字段
    cursor c_not_pkey is
      select t.column_name
```

```
          from user_tab_columns t
       where t.table_name = upper(p_par_table)
         and t.column_name not in
             (select col.column_name
                from user_constraints con, user_cons_columns col
               where con.constraint_name = col.constraint_name
                 and con.constraint_type = 'P'
                 and col.table_name = upper(p_par_table));
    --取表的主键
    cursor c_pkey is
      select col.column_name
        from user_constraints con, user_cons_columns col
       where con.constraint_name = col.constraint_name
         and con.constraint_type = 'P'
         and col.table_name = upper(p_par_table);
  begin
    for c1 in c_column loop
      v_column        := v_column || c1.column_name || ',';
      v_column_insert := v_column_insert || 'b.' || c1.column_name || ',';
    end loop;
    v_column1       := substr(v_column, 0, length(v_column) - 1);
    v_column_insert1 := '(' || substr(v_column_insert,
                                      0,
                                      length(v_column_insert) - 1) || ') ';
    for c2 in c_not_pkey loop
      v_column_update := v_column_update || 'a.' || c2.column_name || '=b.' ||
                         c2.column_name || ',';
    end loop;
    v_column_update1 := substr(v_column_update,
                               0,
                               length(v_column_update) - 1);
    for c3 in c_pkey loop
      v_column_pkey := v_column_pkey || 'b.' || c3.column_name || '=a.' ||
                       c3.column_name || ' and ';
    end loop;
    v_column_pkey1 := '(' ||
                              substr(v_column_pkey, 0, length(v_column_pkey) - 5) || ') ';
    v_sql_str      := 'merge into ' || upper(p_par_table) || ' a ' ||
                      'using (select ' || v_column1 || ' from ' ||
                      upper(p_par_table) ||
                      '@dblink_demo where lastupt_dttm >= ' ||
                      'to_date('''  || archive_date || ''',''yyyy-mm-dd'')' ||
                      ' and lastupt_dttm< ' ||
                              'to_date('''  || archive_date || ''',''yyyy-mm-dd'')+1' ||
                      ') b on ';
    v_sql_str1     := v_column_pkey1 || ' when matched then update set ';
```

```
            v_sql_str3      := v_column_update1 || ' when not matched then insert (';
            v_sql_str2      := v_column1 || ') values ' || v_column_insert1;

            execute immediate v_sql_str || v_sql_str1 || v_sql_str3 || v_sql_str2;
            v_all_cnt := sql%rowcount;
            commit;
            --统计变动的记录数
            v_ins_cnt := 0;
            v_upd_cnt := 0;
            v_del_cnt := 0;
            v_step_tm := v_step_tm || 'step1=' ||
                        round((sysdate - v_end_tm) * 24 * 60 * 60) || '秒';
            v_end_tm  := sysdate;
        end pro_table_demo;
```

6.2.5.1 分析1

结合报错的位点，可以知道问题主要出现在v_sql_str定义的SQL结尾以及v_sql_str1定义的开头部分，v_sql_str1开头部分拼接的SQL存在异常。于是，进一步分析v_sql_str1的具体获取方式。

```
    v_sql_str       := 'merge into ' || upper(p_par_table) || ' a ' ||
                       'using (select ' || v_column1 || ' from ' ||
                       upper(p_par_table) ||
                       '@dblink_demo where lastupt_dttm >= ' ||
                       'to_date(''' || archive_date || ''',''yyyy-mm-dd'')' ||
                       ' and lastupt_dttm< ' ||
                       'to_date(''' || archive_date || ''',''yyyy-mm-dd'')+1' ||
                       ') b on ';
    v_sql_str1      := v_column_pkey1 || ' when matched then update set ';
    ---- 省略部分无关代码
    execute immediate v_sql_str || v_sql_str1 || v_sql_str3 || v_sql_str2;
```

6.2.5.2 分析2

v_sql_str变量的具体值是v_column_pkey1变量定义的，而v_column_pkey1变量引用的是v_column_pkey变量的定义。继续往上追溯：

```
    for c3 in c_pkey loop
      v_column_pkey := v_column_pkey || 'b.' || c3.column_name || '=a.' ||
                       c3.column_name || ' and ';
    end loop;
    v_column_pkey1 := '(' ||
                     substr(v_column_pkey, 0, length(v_column_pkey) - 5) || ') ';
```

6.2.5.3 分析3

v_column_pkey是由游标c_pkey定义的SQL获取的，因此找到游标的SQL定义进行分析。

```
cursor c_pkey is
  select col.column_name
    from user_constraints con, user_cons_columns col
   where con.constraint_name = col.constraint_name
     and con.constraint_type = 'P'
     and col.table_name = upper(p_par_table);
```

6.2.6 具体分析定位后的 SQL 语句

套入具体的表名对该游标 SQL 进行查询，发现返回值为空，获取不到该表的主键信息。

```
-- 无记录返回
select col.column_name
    from user_constraints con, user_cons_columns col
    where con.constraint_name = col.constraint_name
    and con.constraint_type = 'P'
    and col.table_name = upper('t_cc_demo');

Empty set (1.35 sec)
```

6.2.6.1 分析 1

查询该表的所有约束条件，发现该表不包含 constraint_type = 'P' 的主键约束，但包含一个对 SRT_ID 字段的唯一键及非空约束，而且从 constraint_name 中的 PK_t_cc_demo 约束名判断，该字段确实为该表的主键（见图 10）。

```
select col.column_name,constraint_type,con.constraint_name
    from user_constraints con, user_cons_columns col
    where con.constraint_name = col.constraint_name
    and col.table_name = upper('t_cc_demo');
```

```
+-----------------+-----------------+----------------------------------------------+
| COLUMN_NAME     | CONSTRAINT_TYPE | CONSTRAINT_NAME                              |
+-----------------+-----------------+----------------------------------------------+
| SRT_ID          | U               | PK_T_CC_L0_SRT                               |
| SRT_ID          | C               | T_CC_L0_SRT_OBNOTNULL_1669630398067668       |
| SRT_ORIGNAL_ID  | C               | T_CC_L0_SRT_OBNOTNULL_1669630398067707       |
| SRT_PROFILE_ID  | C               | T_CC_L0_SRT_OBNOTNULL_1669630398067718       |
+-----------------+-----------------+----------------------------------------------+
4 rows in set (0.401 sec)
```

图 10

6.2.6.2 分析 2

横向对比 Oracle 中该表的约束信息，得到图 11。可以看到 Oracle 侧 SRT_ID 字段确实存在主键类型约束，但在 OceanBase 侧转为了唯一键约束。

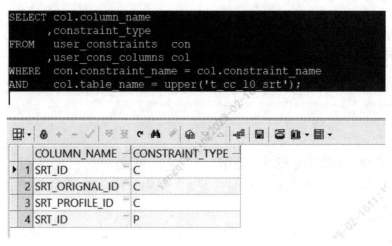

图 11

6.2.7 小结

(1) 由于迁移后游标获取主键字段时,匹配不到约束类型为 P 的字段,以至于后续 SQL 拼接出现条件为空的情况,从而导致 SQL 报语法错误。

(2) 该问题主要是由于 OMS 迁移时将部分分区表的主键转换为唯一键导致,具体的转换原因和逻辑参考 8.4 节。

6.3 修复方式

将获取主键的 SQL 调整为取唯一键约束类型,并且约束名称以 PK 开头(排除其他唯一键的干扰)。

```
-- 取表的主键(修改前)
cursor c_pkey is
    select col.column_name
    from user_constraints con, user_cons_columns col
    where con.constraint_name = col.constraint_name
        and con.constraint_type = 'P'
        and col.table_name = upper(p_par_table);

-- 取表的主键(修改后)
cursor c_pkey is
    select col.column_name
    from user_constraints con, user_cons_columns col
    where con.constraint_name = col.constraint_name
        and con.constraint_type in ('U','P')
        and con.constraint_name like 'PK%'
        and col.table_name = upper(p_par_table);
```

6.4 知识扩展

在 Oracle 中，分区表是堆表结构，数据和索引分开，分区键可以不是主键或者主键的一部分；在 OceanBase 中，分区表是索引组织表，分区键必须是主键或主键的一部分。当 Oracle 侧分区键不是主键或主键一部分时，为了在 OceanBase 侧能成功构建分区表，OMS 会对主键约束进行转换将其改为唯一性约束，以便能正常创建分区表。以下是验证哪些分区表会进行主键转换的示例：

```sql
-- ------------------- Oracle 侧表结构
-- 1. 主键就是分区键
CREATE TABLE "T_PARTKEY_IS_PK" (
    "ACT_ID" NUMBER(10,0) NOT NULL,
    "SRT_ID" NUMBER(10,0),
    "SRT_ORIGNAL_ID" NUMBER(10,0),
    "CRT_DTTM" DATE,
    "LASTUPT_DTTM" DATE,
    CONSTRAINT "PK_T_PARTKEY_IS_PK" PRIMARY KEY ("CRT_DTTM")
)
PARTITION BY RANGE ("CRT_DTTM")
(
        PARTITION "P201512" VALUES LESS THAN (TO_DATE(' 2016-01-01 00:00:00', 'SYYYY-MM-DD HH24:MI:SS', 'NLS_CALENDAR=GREGORIAN')),
        PARTITION "PMAX" VALUES LESS THAN (MAXVALUE)
);

-- 2. 主键不是分区键
CREATE TABLE "T_PARTKEY_NOT_PK" (
    "ACT_ID" NUMBER(10,0) NOT NULL,
    "SRT_ID" NUMBER(10,0),
    "SRT_ORIGNAL_ID" NUMBER(10,0),
    "CRT_DTTM" DATE,
    "LASTUPT_DTTM" DATE,
    CONSTRAINT "PK_T_PARTKEY_NOT_PK" PRIMARY KEY ("ACT_ID")
)
PARTITION BY RANGE ("CRT_DTTM")
(
        PARTITION "P201512" VALUES LESS THAN (TO_DATE(' 2016-01-01 00:00:00', 'SYYYY-MM-DD HH24:MI:SS', 'NLS_CALENDAR=GREGORIAN')),
        PARTITION "PMAX" VALUES LESS THAN (MAXVALUE)
);

-- 3. 主键是分区键的一部分，且分区键是主键多列中的第一列
CREATE TABLE "T_PARTKEY_IS_FIRST_COLUMNS_PK" (
    "ACT_ID" NUMBER(10,0) NOT NULL,
    "SRT_ID" NUMBER(10,0),
```

```sql
        "SRT_ORIGNAL_ID" NUMBER(10,0),
        "CRT_DTTM" DATE,
        "LASTUPT_DTTM" DATE,
        CONSTRAINT "PK_T_PARTKEY_IS_FIRST_COLUMNS" PRIMARY KEY ("CRT_DTTM","ACT_ID")
    )
    PARTITION BY RANGE ("CRT_DTTM")
    (
            PARTITION "P201512" VALUES LESS THAN (TO_DATE(' 2016-01-01 00:00:00', 'SYYYY-MM-DD HH24:MI:SS', 'NLS_CALENDAR=GREGORIAN')),
            PARTITION "PMAX" VALUES LESS THAN (MAXVALUE)
    );

    -- 4. 主键是分区键的一部分, 且分区键不是主键多列中的第一列
    CREATE TABLE "T_PARTKEY_NOT_FIRST_COLUMNS_PK" (
        "ACT_ID" NUMBER(10,0) NOT NULL,
        "SRT_ID" NUMBER(10,0),
        "SRT_ORIGNAL_ID" NUMBER(10,0),
        "CRT_DTTM" DATE,
        "LASTUPT_DTTM" DATE,
         CONSTRAINT "PK_T_PARTKEY_NOT_FIRST_COLUMNS" PRIMARY KEY ("ACT_ID","CRT_DTTM")
    )
    PARTITION BY RANGE ("CRT_DTTM")
    (
            PARTITION "P201512" VALUES LESS THAN (TO_DATE(' 2016-01-01 00:00:00', 'SYYYY-MM-DD HH24:MI:SS', 'NLS_CALENDAR=GREGORIAN')),
            PARTITION "PMAX" VALUES LESS THAN (MAXVALUE)
    );


    -- -------------------- Oracle 侧约束状态
    col table_name for a30
    col COLUMN_NAME for a10
    col CONSTRAINT_NAME for a30
    col CONSTRAINT_TYPE for a10
    SELECT CON.TABLE_NAME,
           COL.COLUMN_NAME,
           CON.CONSTRAINT_NAME,
           CON.CONSTRAINT_TYPE
      FROM USER_CONSTRAINTS CON, USER_CONS_COLUMNS COL
     WHERE CON.CONSTRAINT_NAME = COL.CONSTRAINT_NAME
       AND CON.CONSTRAINT_TYPE = 'P'
       AND CON.TABLE_NAME LIKE 'T_PARTKEY%'
     ORDER BY CON.TABLE_NAME, COL.POSITION;

    TABLE_NAME                                              COLUMN_NAM CONSTRAINT_
NAME                        CONSTRAINT
    ------------------------------  ---------- ------------------------------ ----------
```

```
T_PARTKEY_IS_FIRST_COLUMNS_PK    ACT_ID       PK_T_PARTKEY_IS_FIRST_COLUMNS      P
T_PARTKEY_IS_FIRST_COLUMNS_PK    CRT_DTTM     PK_T_PARTKEY_IS_FIRST_COLUMNS      P
T_PARTKEY_IS_PK                  CRT_DTTM     PK_T_PARTKEY_IS_PK                 P
T_PARTKEY_NOT_FIRST_COLUMNS_PK   CRT_DTTM     PK_T_PARTKEY_NOT_FIRST_COLUMNS     P
T_PARTKEY_NOT_FIRST_COLUMNS_PK   ACT_ID       PK_T_PARTKEY_NOT_FIRST_COLUMNS     P
T_PARTKEY_NOT_PK                 ACT_ID       PK_T_PARTKEY_NOT_PK                P

6 rows selected.

-- -------------------------------- 通过 OMS 迁移到 OB 侧的约束状态
SELECT CON.TABLE_NAME,
       COL.COLUMN_NAME,
       CON.CONSTRAINT_NAME,
       CON.CONSTRAINT_TYPE
FROM USER_CONSTRAINTS CON, USER_CONS_COLUMNS COL
WHERE CON.CONSTRAINT_NAME = COL.CONSTRAINT_NAME
AND CON.CONSTRAINT_NAME NOT LIKE '%OMS_ROWID'
AND CON.CONSTRAINT_NAME NOT LIKE '%OBNOTNULL%'
ORDER BY CON.TABLE_NAME, COL.POSITION;
```

TABLE_NAME	COLUMN_NAME	CONSTRAINT_NAME	CONSTRAINT_TYPE
T_PARTKEY_IS_FIRST_COLUMNS_PK	ACT_ID	PK_T_PARTKEY_IS_FIRST_COLUMNS	P
T_PARTKEY_IS_FIRST_COLUMNS_PK	CRT_DTTM	PK_T_PARTKEY_IS_FIRST_COLUMNS	P
T_PARTKEY_IS_PK	CRT_DTTM	PK_T_PARTKEY_IS_PK	P
T_PARTKEY_NOT_FIRST_COLUMNS_PK	CRT_DTTM	PK_T_PARTKEY_NOT_FIRST_COLUMNS	P
T_PARTKEY_NOT_FIRST_COLUMNS_PK	ACT_ID	PK_T_PARTKEY_NOT_FIRST_COLUMNS	P
T_PARTKEY_NOT_PK	ACT_ID	PK_T_PARTKEY_NOT_PK	U

```
6 rows in set (0.16 sec)

-- [INFO] [CONVERT] CONSTRAINT "PK_T_PARTKEY_NOT_PK" PRIMARY KEY ("ACT_ID") -> CONSTRAINT "PK_T_PARTKEY_NOT_PK" UNIQUE ("ACT_ID")
```

6.5 结论

当 Oracle 侧的主键不包含分区键时，OMS 会将 Oracle 侧的主键改为唯一键，但保持约束名一致且 OMS 在做表结构迁移时，会给出存在转换的注释信息。

对于存储过程的报错或者 SQL 执行报错，均可使用以上方法获取实际的 SQL 执行的 trace id 进行日志诊断，得到具体报错的 SQL 进行分析。

7 Join 估行不准选错执行计划该如何优化？

作者：胡呈清

数据库版本：OceanBase 3.2.3.3。

7.1 问题描述

一个 Join 查询，关联字段包含组合主键中的第 1、2、4 个字段，执行 Nested-Loop Join 时，被驱动表只能匹配主键的前两个字段，成本 cost1 较低，但实际效率不高；并且驱动表的扇出 n（也就是输出行数）估行比实际小很多。在计算总成本时：

$$\text{Join 总成本} \approx （\text{驱动表成本} + n*cost1）$$

在本案例中驱动表成本是固定的，执行计划中 n 的估算值只有 5000，但实际值达 60 万，cost1=154。计算成本时，n*cost1 比实际值小很多，优化器最终选择了 Nested-Loop Join，如果被驱动表可以匹配主键的全部字段，效率是很高的，但这里由于只能匹配前两个字段，效率较低，导致整个查询耗时非常长。

7.2 分析过程

7.2.1 分析执行计划

问题 SQL 如下（执行耗时 500s 以上）：

```
select
count(*) from
(
SELECT
JGBM AS QYDJID,
SEGMENT3 AS FNUMBER,
PERIOD_NAME AS SSQJ,
...
FROM
(
SELECT
...
FROM
DC_ACCOUNTBALANCE_TEMP A,
DEF_ACCOUNTCONFIG B,
DC_ACCOUNT C,
NVAT_ACCANDTAXIDMAPFORP07 D,
```

```
        BI_CHOICEOFUNIT E
    WHERE
    A.SEGMENT1 = D.ZTJGBM
    AND D.SBDWID = E.SBDWID
    AND B.JGBM = E.DEPTCODE
    AND B.YXQSNY <= (
    substr(A.PERIOD_NAME, 4, 6) || substr(A.PERIOD_NAME, 1, 2)
    )
    AND (
    substr(A.PERIOD_NAME, 4, 6) || substr(A.PERIOD_NAME, 1, 2)
    ) <= B.YXJZNY
    AND C.QYDJID = B.SYZT
    AND C.FNUMBER = A.SEGMENT3
    AND C.ACCOUNTYEAR = substr(A.PERIOD_NAME, 4, 6)
    AND a.period_name = '10-2023'
    ) SUB
    GROUP BY
    JGBM,
    SEGMENT3,
    PERIOD_NAME
    ) X
    left join DC_ACCOUNTBALANCE A
    ON (
    A.SSQJ = X.SSQJ
    AND A.QYDJID = X.QYDJID
    AND A.FNUMBER = X.FNUMBER
    );
```

执行计划如下（多余信息已删除），结合 SQL 内容进行解读：

(1)X 表是 A、B、C、D、E 等 5 张表关联的结果，将其与 A 表进行关联查询。从执行计划看，主要成本集中在 X 表，因此先执行 X 部分，确认是否慢在这部分。执行耗时只有 5 秒，结果有 61 万行，但执行计划估算的行数只有 5123。

(2)X 部分执行很快，慢在 A 部分。因为是 Nested-Loop Join，A 作为被驱动表会循环查询 61 万次（batch_join=false），每次查询走主键，执行计划的 13 号 算 子 中 range_key([A.SSQJ(0x7eb5a42ec400)], [A.QYDJID(0x7eb5a42ed840)], [A.DATAUSE(0x7ec8f84434e0)], [A.FNUMBER(0x7eb5a42eec80)]), range(MIN ; MAX) 部分信息说明索引里有 4 个字段，但是 range_cond([A.SSQJ(0x7eb5a42ec400) = ?(0x7ec8f8451e20)], [A.QYDJID(0x7eb5a42ed840) = ?(0x7ec8f8452950)]) 这部分表示只能用索引的前两个字段，这会是导致执行慢的原因吗？此外，有个信息可以提供佐证：A：table_rows：32310843, physical_range_rows：391, logical_range_rows：391, 优化器估算 A 表每次查询需要扫描 391 行，这个效率确实不高。

(3)在估算 Nested-Loop Join 的总成本时，计算逻辑是驱动表的成本＋驱动表的扇出 × 被驱动表查询一次的成本。这个 SQL 中驱动表的扇出（5123）比实际值（61 万）小很多，估算出的总成本比实际小很多。

```
========================================================================
|ID|OPERATOR                    |NAME                              |EST. ROWS|COST   |
------------------------------------------------------------------------
|0 |SCALAR GROUP BY             |                                  |1        |3947739|
|1 | NESTED-LOOP OUTER JOIN     |                                  |5123     |3947543|
|2 |  SUBPLAN SCAN              |X                                 |5123     |3154937|
|3 |   HASH GROUP BY            |                                  |5123     |3154861|
|4 |    HASH JOIN               |                                  |5123     |3149203|
|5 |     TABLE SCAN             |C                                 |81314    |31453  |
|6 |     HASH JOIN              |                                  |63573    |2940900|
|7 |      HASH JOIN             |                                  |1898     |35447  |
|8 |       TABLE SCAN           |D(IDX_ACCANDTAXIDMAPFORP07_CMB1)  |2011     |778    |
|9 |       HASH JOIN            |                                  |1736     |32462  |
|10|        TABLE SCAN          |E(IDX_BI_CHOICEOFUNIT_CMB1)       |1704     |660    |
|11|        TABLE SCAN          |B                                 |29154    |11277  |
|12|      TABLE SCAN            |A(IDX_DC_ACCOUNTBALANCE_TEMP_TEST)|639387   |2468263|
|13|  TABLE SCAN                |A                                 |1        |154    |
========================================================================

Outputs & filters:
-------------------------------------
...
13 - output([remove_const(1)(0x7ec8f846ba40)]), filter([A.
FNUMBER(0x7eb5a42eec80) = ?(0x7ec8f8453480)]),
    access([A.FNUMBER(0x7eb5a42eec80)]), partitions(p0),
    is_index_back=false, filter_before_indexback[false],
    range_key([A.SSQJ(0x7eb5a42ec400)], [A.QYDJID(0x7eb5a42ed840)],
[A.DATAUSE(0x7ec8f84434e0)], [A.FNUMBER(0x7eb5a42eec80)]), range(MIN ; MAX),
    range_cond([A.SSQJ(0x7eb5a42ec400) = ?(0x7ec8f8451e20)],
[A.QYDJID(0x7eb5a42ed840) = ?(0x7ec8f8452950)])

Used Hint:
...

Optimization Info:
-------------------------------------
...
A:table_rows:32310843, physical_range_rows:391, logical_range_rows:391, index_
back_rows:0, output_rows:0, est_method:local_storage, optimization_method=cost_
based, avaiable_index_name[DC_ACCOUNTBALANCE],...
```

7.2.2 分析表的统计信息

上一节我们分析得出：X 部分查询很快，慢在 A 表查询，要查询 61 万次。A 表查询时使用了主键的前两个字段。因此需要分析一下 A 表的统计信息，看看主键的 4 个字段的 NDV 分别是多少，结果如下：

(1)SSQJ、QYDJID 两个字段的 NDV 并不高，每组值的重复次数可以通过统计信息估算：32310843/(85×972)=391。这就是执行计划中的 physical_range_rows:391，意思就是每次查询大概要扫 391 行数据，如果只执行一次是没问题的，但这个 SQL 里需要执行 61 万次，总耗时就大了。

(2) 另外 SQL 中关联字段包含了主键的 3 个字段，不在条件里的第 3 个字段 DATAUSE 实际值都为 1。从逻辑上来看，SQL 中加上 "AND A.DATAUSE = 1" 条件，结果不会变，这样做的好处是 A 表查询时可以使用主键的所有字段，每次只需要扫 1 行数据，效率会高很多。另一种更好的方式是主键中去掉 DATAUSE 字段，不过 OceanBase 不支持修改主键。

```
-- 查询
select column_name,num_distinct from all_tab_col_statistics where table_name='DC_ACCOUNTBALANCE';
-- 结果
column_name          num_distinct
SSQJ                 85
QYDJID               972
DATAUSE              1
FNUMBER              2616
```

7.2.3 改写

7.2.3.1 方法 1：加 "AND A.DATAUSE = 1"

加上这个条件后，SQL 耗时从 500 秒降到 8 秒。执行计划如下，A 表每次只需要扫描 1 行：

```
==========================================================================
|ID|OPERATOR                   |NAME                               |EST. ROWS|COST    |
--------------------------------------------------------------------------
|0 |SCALAR GROUP BY            |                                   |1        |3214924 |
|1 | NESTED-LOOP OUTER JOIN    |                                   |5123     |3214729 |
|2 |  SUBPLAN SCAN             |X                                  |5123     |3154937 |
|3 |   HASH GROUP BY           |                                   |5123     |3154861 |
|4 |    HASH JOIN              |                                   |5123     |3149203 |
|5 |     TABLE SCAN            |C                                  |81314    |31453   |
|6 |     HASH JOIN             |                                   |63573    |2940900 |
|7 |      HASH JOIN            |                                   |1898     |35447   |
|8 |       TABLE SCAN          |D(IDX_ACCANDTAXIDMAPFORP07_CMB1)   |2011     |778     |
|9 |       HASH JOIN           |                                   |1736     |32462   |
|10|        TABLE SCAN         |E(IDX_BI_CHOICEOFUNIT_CMB1)        |1704     |660     |
|11|        TABLE SCAN         |B                                  |29154    |11277   |
|12|      TABLE SCAN           |A(IDX_DC_ACCOUNTBALANCE_TEMP_TEST) |639387   |2468263 |
|13|  TABLE GET                |A                                  |1        |11      |
==========================================================================
Outputs & filters:
...
13 - output([remove_const(1)(0x7eb91646c790)]), filter(nil),
     access([A.SSQJ(0x7eb91646b730)]), partitions(p0),
     is_index_back=false,
```

```
        range_key([A.SSQJ(0x7eae68cec980)], [A.QYDJID(0x7eae68ceddc0)],
[A.DATAUSE(0x7eae68cf05d0)], [A.FNUMBER(0x7eae68cef200)]), range(MIN ; MAX),
        range_cond([A.DATAUSE(0x7eae68cf05d0) = 1(0x7eae68cefeb0)],
[A.SSQJ(0x7eae68cec980) = ?(0x7eb916451ce0)], [A.QYDJID(0x7eae68ceddc0) =
?(0x7eb916452810)], [A.FNUMBER(0x7eae68cef200) = ?(0x7eb916453340)])
      ...
      Optimization Info:
      -------------------------------------
      A:table_rows:32310843, physical_range_rows:1, logical_range_rows:1, index_
back_rows:0, output_rows:1, est_method:local_storage, optimization_method=rule_
based, heuristic_rule=unique_index_without_indexback
```

7.2.3.2 方法 2：加 Hint 走 Hash Join

前面我们分析 A 表查询只能使用主键索引的前 2 个字段，效率不高，在这种情况下可以看一下 Hash Join 的执行效率，加 Hint " /*+ leading(X A) use_hash(A) */"，耗时只要 40 秒。执行计划如下，结合前面的分析进行解读：被驱动表 A 除了关联条件没有其他条件，因此要做全表扫描，成本很高，所以总成本也很高，并且显然比 Nested-Loop Join 的成本高。在没有 Hint 干预的情况下，优化器会选 Nested-Loop Join。

```
=====================================================================
|ID|OPERATOR              |NAME                              |EST. ROWS|COST     |
---------------------------------------------------------------------
|0 |SCALAR GROUP BY       |                                  |1        |52828380 |
|1 | HASH OUTER JOIN      |                                  |5123     |52828184 |
|2 |  SUBPLAN SCAN        |X                                 |5123     |3154937  |
|3 |   HASH GROUP BY      |                                  |5123     |3154861  |
|4 |    HASH JOIN         |                                  |5123     |3149203  |
|5 |     TABLE SCAN       |C                                 |81314    |31453    |
|6 |     HASH JOIN        |                                  |63573    |2940900  |
|7 |      HASH JOIN       |                                  |1898     |35447    |
|8 |       TABLE SCAN     |D(IDX_ACCANDTAXIDMAPFORP07_CMB1)  |2011     |778      |
|9 |       HASH JOIN      |                                  |1736     |32462    |
|10|        TABLE SCAN    |E(IDX_BI_CHOICEOFUNIT_CMB1)       |1704     |660      |
|11|        TABLE SCAN    |B                                 |29154    |11277    |
|12|       TABLE SCAN     |A(IDX_DC_ACCOUNTBALANCE_TEMP_TEST)|639387   |2468263  |
|13|  TABLE SCAN          |A                                 |32310843 |12497986 |
=====================================================================
```

7.3 总结

这是一个很典型的问题，如果 Join 时关联表太多，容易选错执行计划。原因是估算驱动表的扇出很容易产生误差，尤其 Join 的结果作为驱动表时，这相当于要估算 Join 的结果有多少行，误差会更大。而优化器在估算 Nested-Loop Join 算法的成本时，驱动表的扇出对计算结果影响很大，也就是说 Nested-Loop Join 的成本估算结果很容易产生误差，所以容易选错执行计划。

8 Oracle 中部分不兼容对象迁移到 OceanBase 的处理方式

作者：余振兴

8.1 背景介绍

在进行国产化改造过程中，我们需要将 Oracle 数据库迁移到 OceanBase（Oracle 模式）数据库，虽然 OceanBase 对于 Oracle 兼容性已经足够好，但依旧还有一些特殊语法或对象需要单独处理，下面介绍笔者对遇到的一些不完全兼容对象的处理逻辑。

8.2 Oracle 中 LOB 类数据迁移到 OceanBase 时的处理逻辑

Oracle 中 CLOB 和 BLOB 类型均可达到 4GB 大小（以 Oracle 11.2 为例），而 OceanBase 数据库当前版本（3.2.3.x）所支持的大对象数据类型的信息如表 1 所示。

表 1

类型	BLOB	CLOB
长度	变长	变长
自定义长度上限（字符）	48MB	48MB
字符集	BINARY	与租户字符集一致

考虑到从 Oracle 迁移到 OceanBase，如果涉及 LOB 类字段，可能会存在 LOB 数据大于 48MB 时数据丢失的问题，需要提前发现这类数据并进行处理。

8.2.1 找到 Oracle 中 LOB 数据最大长度

我们可以做一个实验生成 CLOB 及 BLOB 类型数据，使用 Oracle 自带的 DBMS_LOB 包获取对应类型的最大值。

8.2.1.1 构建包含 LOB 类型的数据表

代码如下：

```
CREATE TABLE t_lob(
c_ID NUMBER,
c_clob CLOB,
c_blob BLOB
);
```

8.2.1.2 创建数据存储过程

随机插入 100 条记录到 t_lob 表。

```
CREATE OR REPLACE PROCEDURE insert_random_lob_data AS
BEGIN
DECLARE
l_random_string VARCHAR2(10000);
l_random_blob BLOB;
BEGIN
FOR i IN 1..100 LOOP
l_random_string := dbms_random.string('U', dbms_random.value(1, 10000));
dbms_lob.createtemporary(l_random_blob, TRUE);
dbms_lob.writeappend(l_random_blob, LENGTH(l_random_string), utl_raw.cast_to_raw(l_random_string));

INSERT INTO t_lob(c_ID, c_clob, c_blob)
VALUES(i, l_random_string, l_random_blob);

dbms_lob.freetemporary(l_random_blob);
END LOOP;
COMMIT;
END;
END;
/
```

8.2.1.3 查询 t_lob 该表中 CLOB 和 BLOB 字段的最大值

代码如下（见图 12）：

```
SELECT MAX(DBMS_LOB.GETLENGTH(C_CLOB)) AS LONGEST_CLOB,
MAX(DBMS_LOB.GETLENGTH(C_BLOB)) AS LONGEST_BLOB
FROM T_LOB;
```

图 12

8.2.2 获取整个数据库中 LOB 字段值较大的清单

排除系统用户，获取 LOB 字段清单后，再基于清单中的 LOB 字段单独分析其最大值（见图 13）。

```
SELECT COL.OWNER,
COL.TABLE_NAME,
```

```sql
    COL.COLUMN_NAME,
    COL.DATA_TYPE,
    COL.AVG_COL_LEN,
    COL.CHAR_LENGTH,
    TAB.NUM_ROWS
FROM DBA_TABLES TAB, DBA_TAB_COLUMNS COL
WHERE TAB.OWNER = COL.OWNER
AND TAB.TABLE_NAME = COL.TABLE_NAME
AND COL.DATA_TYPE IN ('CLOB', 'BLOB')
AND COL.OWNER NOT IN ('SYS', 'SYSTEM')
AND COL.OWNER IN
(SELECT USERNAME FROM DBA_USERS WHERE ACCOUNT_STATUS = 'OPEN')
AND COL.TABLE_NAME NOT LIKE 'BIN%';
```

	OWNER	TABLE_NAME	COLUMN_NAME	DATA_TYPE	AVG_COL_LEN	CHAR_LENGTH	NUM_ROWS
1	ZHENXING	TABLE_CLOB	R1	CLOB		0	
2	ZHENXING	T_LOB	C_BLOB	BLOB		0	
3	ZHENXING	T_LOB	C_CLOB	CLOB		0	

图 13

8.3 Oracle 中 DISABLE 约束在 OMS 迁移过程中的处理逻辑

对 Oracle 中的约束类非表对象做一致性校验时,发现部分约束在 OMS 迁移完成后丢失了,需要分析其丢失的原因。

8.3.1 问题分析

从 OMS 界面中获取 DDL 的语句可以看到有 2 个 WARN,且类型是 DISCARD,表示 OMS 判断其是 DISABLE 状态的约束,直接选择了舍弃。

```
    -- [WARN] [DISCARD] CONSTRAINT "PK_T_PARTKEY_IS_PK" PRIMARY KEY ("CRT_DTTM")
DISABLE NOVALIDATE -> [NULL]
    -- [WARN] [DISCARD] CHECK ("ACT_ID" IS NOT NULL) DISABLE NOVALIDATE -> [NULL]
    CREATE TABLE "T_PARTKEY_IS_PK" (
    "ACT_ID" NUMBER(10,0),
    "SRT_ID" NUMBER(10,0),
    "SRT_ORIGNAL_ID" NUMBER(10,0),
    "CRT_DTTM" DATE,
    "LASTUPT_DTTM" DATE
    )
```

8.3.2 问题结论

Oracle 侧处于 DISABLE 状态的约束通过 OMS 迁移时会被舍弃,不会在 OceanBase 侧创建。比对约束对象时,需要额外注意 Oracle 端约束是否处于 DISABLE 状态。这种情况本身对业务和功能没有影响。

8.3.3 校验约束时提前排除状态为 DISABLE 的约束

可以通过以下语句观测源端 Oracle 约束的状态（见图 14）。

```sql
-- 手工将T_PARTKEY_IS_PK表的约束都disable
ALTER TABLE ZHENXING.T_PARTKEY_IS_PK DISABLE NOVALIDATE CONSTRAINT PK_T_PARTKEY_IS_PK;
ALTER TABLE ZHENXING.T_PARTKEY_IS_PK DISABLE CONSTRAINT SYS_C0011109;

SELECT OWNER,
TABLE_NAME,
CONSTRAINT_NAME,
CONSTRAINT_TYPE,
INDEX_NAME,
STATUS
FROM DBA_CONSTRAINTS
WHERE OWNER = 'ZHENXING'
AND TABLE_NAME = 'T_PARTKEY_IS_PK';
```

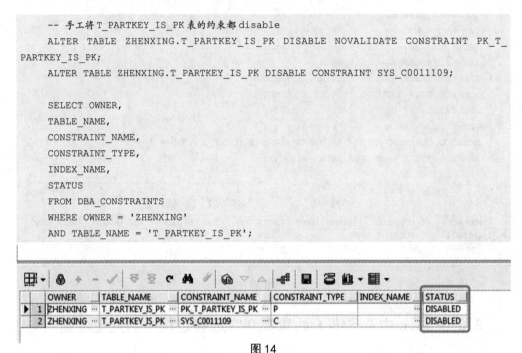

图 14

8.4 Oracle 中分区表迁移到 OceanBase 后，带有的自动分区属性丢失

自动分区属性是 Oracle 11g 的特性，可以用 INTERVAL 语法基于天、月、年自动创建分区。Oracle 中的分区表通过 OMS 迁移到 OceanBase 后，自动分区属性会丢失，导致当分区未自动创建时新增数据无法写入分区表，引发报错。

8.4.1 问题分析

从 OMS 界面获取的 DDL 语句显示，有 1 个 WARN 信息，且类型是 DISCARD，表示 OMS 判断其不完全兼容，故直接选择了舍弃掉。

```
-- OMS 迁移表结构时记录的 WARN 信息，表示自动分区属性由于不兼容会自动被舍弃
[WARN] [DISCARD]   INTERVAL (NUMTOYMINTERVAL (1,'MONTH')) -> [NULL]
```

8.4.2 问题结论

所以，在 Oracle 迁移到 OceanBase 前，需要把 Oracle 端存在自动分区属性的表提前找出来，避免由于迁移到 OceanBase 后分区为未自动创建导致数据无法插入而报错。并且找出这类分区后，先在 Oracle 端创建足够的多分区，避免迁移过程中源端分区数增加而比对不一致的情况。最后，记录清单，告知业务开发人员，待后续用其他方式定期

生成新分区。

8.4.3 如何找出 Oracle 中自动分区的表

8.4.3.1 Oracle 侧模拟自动分区

代码如下：

```
-- 创建基于天的自动分区表
SQL> create table interval_sales (
prod_id number(6),
time_id date)
partition by range (time_id)
INTERVAL(NUMTOYMINTERVAL(1, 'MONTH'))
(partition p1 values less than (to_date('2015-01-01','yyyy-mm-dd')));

-- 查询当前分区，默认生成 1 个定义好的分区
SQL> SELECT TABLE_NAME, PARTITION_NAME
FROM USER_TAB_PARTITIONS
WHERE TABLE_NAME = 'INTERVAL_SALES';

TABLE_NAME                      PARTITION_NAME
------------------------------  ------------------------------
INTERVAL_SALES                  P1

-- 插入数据（不在默认分区内）
SQL> INSERT INTO INTERVAL_SALES VALUES(001, TO_DATE('2015-02-01', 'yyyy-mm-dd'));

-- 自动生成了新分区
TABLE_NAME                      PARTITION_NAME
------------------------------  ------------------------------
INTERVAL_SALES                  P1
INTERVAL_SALES                  SYS_P221

-- 单独查看该分区数据（验证数据确实存在新分区）
SQL> SELECT * FROM INTERVAL_SALES PARTITION(SYS_P221);

PROD_ID TIME_ID
---------- ---------
     1 01-FEB-15
```

8.4.3.2 统计 Oracle 侧有哪些表是自动分区的表

代码如下（见图 15）：

```
/*
PARTITION_COUNT: Number of partitions in the table. For interval partitioned
tables, the value of this column is always 1048575.
*/
SELECT T1.OWNER,
```

```
    T1.TABLE_NAME,
    T1.INTERVAL,
    T1.PARTITIONING_TYPE,
    T1.PARTITION_COUNT,
    T1.SUBPARTITIONING_TYPE        AS SUB_TYPE,
    T1.SUBPARTITIONING_KEY_COUNT SUB_COUNT,
    T1.STATUS
    FROM DBA_PART_TABLES T1
    WHERE 1 = 1
    AND TABLE_NAME NOT LIKE 'BIN%'
    AND (INTERVAL IS NOT NULL OR PARTITION_COUNT = 1048575);
```

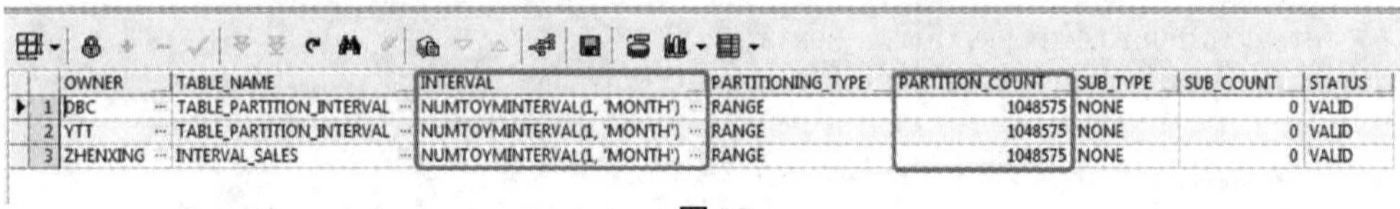

	OWNER	TABLE_NAME	INTERVAL	PARTITIONING_TYPE	PARTITION_COUNT	SUB_TYPE	SUB_COUNT	STATUS
1	DBC	TABLE_PARTITION_INTERVAL	NUMTOYMINTERVAL(1, 'MONTH')	RANGE	1048575	NONE	0	VALID
2	YTT	TABLE_PARTITION_INTERVAL	NUMTOYMINTERVAL(1, 'MONTH')	RANGE	1048575	NONE	0	VALID
3	ZHENXING	INTERVAL_SALES	NUMTOYMINTERVAL(1, 'MONTH')	RANGE	1048575	NONE	0	VALID

图 15

8.5 总结

本文分析了 3 种 Oracle 对象和 OceanBase 对象不兼容时的处理方法以及在迁移前通过提前统计发现问题的操作方式，在迁移前发现这类问题能有效避免迁移过程中报错的问题。

9 一个关于 NOT IN 子查询的 SQL 优化案例

作者：胡呈清

数据库版本：OceanBase 3.2.3.3

9.1 问题描述

前段时间遇到一个慢 SQL，NOT IN 子查询被优化器改写成 NESTED-LOOP ANTI JOIN，但是被驱动表无法使用索引而进行全表扫描，执行耗时 16 秒。这个 SQL 如下：

```
SELECT AGENT_ID, MAX(REL_AGENT_ID)
FROM T_LDIM_AGENT_UPREL
WHERE AGENT_ID NOT IN (select AGENT_ID
    from T_LDIM_AGENT_UPREL
    where valid_flg = '1')
group by AGENT_ID;
```

简略执行计划如下:

```
----------------------------------------------------------------------------------
|ID|OPERATOR              |NAME                                        |EST. ROWS|COST     |
----------------------------------------------------------------------------------
|0 |MERGE GROUP BY        |                                            |146      |62970523 |
|1 | NESTED-LOOP ANTI JOIN|                                            |149      |62970511 |
|2 |  TABLE SCAN          |T_LDIM_AGENT_UPREL(I_LDIM_AGENT_UPREL_AGENT_ID)|27760 |10738    |
|3 |  MATERIAL            |                                            |13880    |11313    |
|4 |   SUBPLAN SCAN       |VIEW1                                       |13880    |11115    |
|5 |    TABLE SCAN        |T_LDIM_AGENT_UPREL                          |13880    |10906    |
----------------------------------------------------------------------------------
```

9.2 问题分析

9.2.1 分析表结构和数据量

表结构如下,关联字段 AGENT_ID 是有索引的:

```
CREATE TABLE "T_LDIM_AGENT_UPREL" (
  "REL_AGENT_ID" NUMBER(22) CONSTRAINT "T_LDIM_AGENT_UPREL_
OBNOTNULL_1679987669730612" NOT NULL ENABLE,
  "AGENT_ID" NUMBER(22),
  "EMPLOYEE_ID" NUMBER(22),
  "EMP_PARTY_FULLNAME" VARCHAR2(60),
  "GRP_ID" NUMBER(22),
  "GRP_PARTY_FULLNAME" VARCHAR2(255),
  "CS_ID" NUMBER(22),
  "CS_ORGAN_NAME" VARCHAR2(255),
  "CRT_DTTM" DATE,
  "LASTUPT_DTTM" DATE,
  "VALID_FLG" VARCHAR2(1),
  "VALID_DTTM" DATE,
  "INVALID_DTTM" DATE,
  CONSTRAINT "PK_T_LDIM_AGENT_UPREL" PRIMARY KEY ("REL_AGENT_ID")
);
CREATE INDEX "IDX_T_LDIM_AGENT_UPREL_CT" on "T_LDIM_AGENT_UPREL" ("CRT_DTTM")
GLOBAL ;
CREATE INDEX "IDX_T_LDIM_AGENT_UPREL_LT" on "T_LDIM_AGENT_UPREL" ("LASTUPT_
DTTM") GLOBAL ;
CREATE INDEX "I_LDIM_AGENT_UPREL_AGENT_ID" on "T_LDIM_AGENT_UPREL" ("AGENT_
ID") GLOBAL ;
```

数据量:T_LDIM_AGENT_UPREL 表一共 2.7 万行,子查询结果 3900 行。

9.2.2 判断直接原因

从执行计划、表结构和数据量来看,这个 SQL 效率低有两个原因:

(1)关联字段 AGENT_ID 有索引,但对被驱动表做查询时却使用全表扫描,效率必定低。为什么不使用索引？

(2) 既然被驱动表不使用索引，基于成本的比较，优化器为什么没有选择更高效的 HASH ANTI JOIN？

问题得一个一个看，先分析第二个问题。

9.2.3 使用 HINT 干预 JOIN 算法

使用如下 HINT 都不生效（并且尝试了 Outline Data 中的写法）：

```
/*+ use_hash(A B)*/
/*+ USE_HASH(@"SEL$1" ("VIEW1"@"SEL$1" )) */
/*+ NO_USE_NL_AGGREGATION */
```

执行计划显示 Used Hint 部分都为空，说明 HINT 无法生效，原因未知：

```
Used Hint:
-------------------------------------
/*+
*/


Outline Data:
-------------------------------------
/*+
BEGIN_OUTLINE_DATA
NO_USE_HASH_AGGREGATION(@"SEL$1")
LEADING(@"SEL$1" ("REPORT.A"@"SEL$1" "VIEW1"@"SEL$1" ))
USE_NL(@"SEL$1" ("VIEW1"@"SEL$1" ))
PQ_DISTRIBUTE(@"SEL$1" ("VIEW1"@"SEL$1" ) LOCAL LOCAL)
USE_NL_MATERIALIZATION(@"SEL$1" ("VIEW1"@"SEL$1" ))
INDEX(@"SEL$1" "REPORT.A"@"SEL$1" "I_LDIM_AGENT_UPREL_AGENT_ID")
FULL(@"SEL$2" "REPORT.B"@"SEL$2")
END_OUTLINE_DATA
*/
```

9.2.4 对比 Oracle 执行计划

注意：当 OceanBase 上看到的执行计划不符合预期，但又找不到原因时，可以通过对比 Oracle 的执行计划来进一步排查。

在 Oracle 上执行计划如下（这里得用 set autotrace on 的方式查看真实执行计划）：

(1) 可以使用 HASH ANTI JOIN，并且包含一个重要信息 HASH JOIN RIGHT ANTI NA（EXPLAIN 是看不到 NA 的）。

(2) 直接搜索就可以得到大概的解释，NA 即 Null-Aware Anti Join，这种反连接能够处理 NULL 值。下面展开讲。

```
SQL> set autotrace on
```

```
SQL> SELECT AGENT_ID, MAX(REL_AGENT_ID)
FROM T_LDIM_AGENT_UPREL
WHERE AGENT_ID NOT IN (select AGENT_ID
from T_LDIM_AGENT_UPREL
where valid_flg = '1')
group by AGENT_ID;  2    3    4    5    6

no rows selected

Execution Plan
----------------------------------------------------------
Plan hash value: 1033962367

--------------------------------------------------------------------------------
| Id  | Operation              | Name              | Rows  | Bytes | Cost (%CPU)| Time     |
--------------------------------------------------------------------------------
|   0 | SELECT STATEMENT       |                   |     9 |   171 |   276   (2)| 00:00:04 |
|   1 |  HASH GROUP BY         |                   |     9 |   171 |   276   (2)| 00:00:04 |
|*  2 |   HASH JOIN RIGHT ANTI NA|                 |  9672 |  179K |   275   (2)| 00:00:04 |
|*  3 |    TABLE ACCESS FULL   | T_LDIM_AGENT_UPREL|  3886 | 31088 |   137   (1)| 00:00:02 |
|   4 |    TABLE ACCESS FULL   | T_LDIM_AGENT_UPREL| 28098 |  301K |   137   (1)| 00:00:02 |
--------------------------------------------------------------------------------

Predicate Information (identified by operation id):
---------------------------------------------------
2 - access("AGENT_ID"="AGENT_ID")
3 - filter("VALID_FLG"='1')
```

9.2.5 NULL 值与 NOT IN

为了更好地说明 NULL 值对 NOT IN 的影响，下面举个简单的例子：

```
create table t1(a number,b varchar2(50),c varchar2(50) not null);
insert into t1 values(1,'aaa','aaa'),(2,'bbb','bbb'),(3,'ccc','ccc'),(4,NULL,'ddd');
commit;
```

只要 NOT IN 后面的子查询或者常量集合有 NULL 值出现，则整个 SQL 的执行结果就会为 NULL：

```
obclient [TESTUSER]> select * from t1 where b not in('aaa',NULL);
Empty set (0.004 sec)

obclient [TESTUSER]> select tt.b from t1 tt where tt.a=4;
+------+
| B    |
+------+
| NULL |
+------+
1 row in set (0.007 sec)
```

```
obclient [TESTUSER]> select t.* from t1 t where b not in(select tt.b from t1
tt where tt.a=4);
Empty set (0.005 sec)
```

NOT EXISTS 对 NULL 值不敏感，这意味着 NULL 值对 NOT EXISTS 的执行结果不会有什么影响：

```
obclient [TESTUSER]> select t.* from t1 t where not EXISTS (select tt.b from
t1 tt where t.b=tt.b and tt.a=4);
+------+------+-----+
| A    | B    | C   |
+------+------+-----+
|    1 | aaa  | aaa |
|    2 | bbb  | bbb |
|    3 | ccc  | ccc |
|    4 | NULL | ddd |
+------+------+-----+
4 rows in set (0.005 sec)
```

IN 对 NULL 值也不敏感：

```
obclient [TESTUSER]> select * from t1 where b in('aaa',NULL);
+------+------+-----+
| A    | B    | C   |
+------+------+-----+
|    1 | aaa  | aaa |
+------+------+-----+
1 row in set (0.004 sec)

obclient [TESTUSER]> select t.* from t1 t where b in(select tt.b from t1 tt
where tt.a<5);
+------+------+-----+
| A    | B    | C   |
+------+------+-----+
|    1 | aaa  | aaa |
|    2 | bbb  | bbb |
|    3 | ccc  | ccc |
+------+------+-----+
3 rows in set (0.002 sec)
```

结合 Null-Aware Anti Join，我们可以得到如下结论：NOT IN 和 <>ALL 对 NULL 值敏感，这意味着 NOT IN 后面的子查询或者常量集合一旦有 NULL 值出现，则整个 SQL 的执行结果就会为 NULL。

所以一旦相关的连接列上出现了 NULL 值（实际上只会判断字段是否有 NOT NULL 约束），此时 Oracle 如果还按照通常的 ANTI JOIN 的逻辑来处理（和 INNER

JOIN 的处理逻辑类似，差别在于只返回不满足关联条件的结果，而 INNER JOIN 对 NULL 值是不敏感的），得到的结果就不对了。

为了解决 NOT IN 和 <>ALL 对 NULL 值敏感的问题，Oracle 推出了改良的 ANTI JOIN（Oracle 11g 新增了参数 _OPTIMIZER_NULL_AWARE_ANTIJOIN，默认为 true），这种反连接能够处理 NULL 值，Oracle 称其为 Null-Aware Anti Join（在真实的执行计划中显示为 XX ANTI NA）。

9.2.6 小结

到这里我们能解释一个问题：为什么 OceanBace 不能使用 HASH ANTI JOIN？

原因是关联字段 AGENT_ID 没有 NOT NULL 约束，由于 NOT IN 对 NULL 敏感，不能使用普通的 ANTI JOIN，否则遇到 NULL 结果将不正确。Oracle 11g 推出的 Null-Aware ANTI JOIN 可以处理 NULL 敏感的场景，但是 OceanBace 3.x 还没有这个功能，因此不能使用 HASH ANTI JOIN，OB4.x 版本将推出 _OPTIMIZER_NULL_AWARE_ANTIJOIN 参数，和 Oracle 保持一致。

9.3 优化建议

既然 NOT IN 对 NULL 敏感，就可以从两个方向优化。先和业务确认 NOT IN 子查询结果集中是否可能出现 NULL，如果不会，就进一步确认关联字段 AGENT_ID 是否会有 NULL 值。如果不会，则下面三种方式任选其一，最佳选择是方法 1，最符合开发规范。

(1) 为 AGENT_ID 字段加上 NOT NULL 约束，这样优化器就可以使用 HASH ANTI JOIN 了。

(2) NOT EXISTS 对 NULL 值不敏感，因此可以将 NOT IN 改写为（或者也可以改写成 LEFT JOIN WHERE xx IS NULL 这种 ANTI JOIN 语法）：

```
SELECT AGENT_ID, MAX(REL_AGENT_ID)
FROM T_LDIM_AGENT_UPREL t1
WHERE NOT EXISTS (select AGENT_ID
from T_LDIM_AGENT_UPREL t2
where t1.agent_id=t2.agent_id and valid_flg = '1')
group by AGENT_ID;
```

改写后的执行计划使用了 HASH RIGHT ANTI JOIN，执行耗时只要 50ms。

```
==================================================================
|ID|OPERATOR                   |NAME    |EST. ROWS|COST  |
------------------------------------------------------------------
|0 |HASH GROUP BY              |        |146      |46828 |
|1 | HASH RIGHT ANTI JOIN|     |        |149      |46697 |
|2 |  SUBPLAN SCAN             |VIEW1   |13880    |11115 |
```

```
|3 |   TABLE SCAN         |T2                                     |13880  |10906|
|4 |   TABLE SCAN         |T1(I_LDIM_AGENT_UPREL_AGENT_ID)|27760  |10738|
======================================================================================
```

(3) 为父查询、子查询都加上 AND AGENT_ID is NOT NULL 条件，也可以让优化器使用 HASH ANTI JOIN：

```
SELECT AGENT_ID, MAX(REL_AGENT_ID)
FROM T_LDIM_AGENT_UPREL
WHERE AGENT_ID NOT IN (select AGENT_ID
from T_LDIM_AGENT_UPREL
where valid_flg = '1' and AGENT_ID is not null )
and AGENT_ID is not null
group by AGENT_ID;
```

执行计划如下：

```
|ID|OPERATOR                |NAME                                              |EST. ROWS|COST |
------------------------------------------------------------------------------------------------
|0 |HASH GROUP BY           |                                                  |146      |47472|
|1 | HASH RIGHT ANTI JOIN|                                                  |149      |47341|
|2 |  SUBPLAN SCAN          |VIEW1                                             |13880    |11173|
|3 |   TABLE SCAN           |T_LDIM_AGENT_UPREL                                |13880    |10965|
|4 |  TABLE SCAN            |T_LDIM_AGENT_UPREL(I_LDIM_AGENT_UPREL_AGENT_ID)|27760  |11324|
```

9.4 答疑

问题 1：HASH JOIN 只能用于关联条件的等值查询，不支持连接条件是大于、小于、不等于和 LIKE 的场景。为什么 NOT IN、NOT EXISTS 可以使用 HASH ANTI JOIN？

NOT IN、NOT EXISTS 子查询和 WHERE t1.a!=t2.a 看起来相似，但其实语义是不一样的，下面的例子可以说明。NOT IN 的语义其实是"如果存在相等的值，则丢弃外表结果"，因此本质上 NOT IN 的实现方式还是做等值查找，所以 HASH ANTI JOIN 的实现和 HASH JOIN 本质一样，只是在返回结果时做了相反的判断。

```
obclient [TESTUSER]> select * from t1 t join t1 tt on t.a!=tt.a;
+------+------+------+------+------+------+
| A    | B    | C    | A    | B    | C    |
+------+------+------+------+------+------+
|    1 | aaa  | aaa  |    2 | bbb  | bbb  |
|    1 | aaa  | aaa  |    3 | ccc  | ccc  |
|    1 | aaa  | aaa  |    4 | NULL | ddd  |
|    2 | bbb  | bbb  |    1 | aaa  | aaa  |
|    2 | bbb  | bbb  |    3 | ccc  | ccc  |
|    2 | bbb  | bbb  |    4 | NULL | ddd  |
```

```
|   3 | ccc  | ccc |   1 | aaa  | aaa  |
|   3 | ccc  | ccc |   2 | bbb  | bbb  |
|   3 | ccc  | ccc |   4 | NULL | ddd  |
|   4 | NULL | ddd |   1 | aaa  | aaa  |
|   4 | NULL | ddd |   2 | bbb  | bbb  |
|   4 | NULL | ddd |   3 | ccc  | ccc  |
+-----+------+-----+-----+------+------+
12 rows in set (0.005 sec)

obclient [TESTUSER]> select t.* from t1 t where a not in(select tt.a from t1 tt);
Empty set (0.005 sec)
```

还可以用 Oracle 的执行计划和优化报告来验证：

```
## 执行计划的 2 号算子 HASH JOIN RIGHT ANTI NA 有如下条件，这里能说明做的是等值查找
  2 - access("AGENT_ID"="AGENT_ID")

## 另外可以通过下面的方法查看优化器改写后的 SQL：
alter session set tracefile_identifier='10053c';
alter session set events '10053 trace name context forever,level 1';
执行 SQL；
alter session set events '10053 trace name context off';
cd /u01/oracle/diag/rdbms/repo/repo/trace
```

cat repo_ora_6702_10053c.trc 在 "Final query after transformations" 部分即为优化器改写后的 SQL，关联条件也是等值查询：

```
Final query after transformations:******* UNPARSED QUERY IS *******
SELECT "T_LDIM_AGENT_UPREL"."AGENT_ID" "AGENT_ID",MAX("T_LDIM_AGENT_
UPREL"."REL_AGENT_ID") "MAX(REL_AGENT_ID)" FROM "REPORT"."T_LDIM_AGENT_UPREL" "T_
LDIM_AGENT_UPREL","REPORT"."T_LDIM_AGENT_UPREL" "T_LDIM_AGENT_UPREL" WHERE "T_
LDIM_AGENT_UPREL"."AGENT_ID"="T_LDIM_AGENT_UPREL"."AGENT_ID" AND "T_LDIM_AGENT_
UPREL"."VALID_FLG"='1' GROUP BY "T_LDIM_AGENT_UPREL"."AGENT_ID"
kkoqbc: optimizing query block SEL$5DA710D3 (#1)
```

问题 2：为什么 OceanBase 可以使用 NESTED-LOOP ANTI JOIN？它能处理涉及 NULL 敏感的情况吗？怎么实现的？是否因为它的实现方式导致了对被驱动表只能全表扫描而不能走索引？

从结果来看，OceanBase 的 NESTED-LOOP ANTI JOIN 查询结果正确，能处理涉及 NULL 敏感的情况。

其实现方式可以从执行计划看出一些端倪：

```
|ID|OPERATOR        |NAME       |EST. ROWS|COST     |
---------------------------------------------------
|0 |MERGE GROUP BY  |           |146      |62970523 |
```

```
|1 | NESTED-LOOP ANTI JOIN|                                                    |149    |62970511|
|2 | TABLE SCAN          |T_LDIM_AGENT_UPREL(I_LDIM_AGENT_UPREL_AGENT_ID)|27760  |10738   |
|3 | MATERIAL            |                                                    |13880  |11313   |
|4 | SUBPLAN SCAN        |VIEW1                                               |13880  |11115   |
|5 | TABLE SCAN          |T_LDIM_AGENT_UPREL                                  |13880  |10906   |

Outputs & filters:
-------------------------------------
  0 - output([T_LDIM_AGENT_UPREL.AGENT_ID(0x7eeef19c3fe0)], [T_FUN_MAX(T_LDIM_
AGENT_UPREL.REL_AGENT_ID(0x7eeef19c50f0))(0x7eeef19c49e0)]), filter(nil),
      group([T_LDIM_AGENT_UPREL.AGENT_ID(0x7eeef19c3fe0)]), agg_func([T_FUN_MAX(T_
LDIM_AGENT_UPREL.REL_AGENT_ID(0x7eeef19c50f0))(0x7eeef19c49e0)])
  1 - output([T_LDIM_AGENT_UPREL.AGENT_ID(0x7eeef19c3fe0)], [T_LDIM_AGENT_UPREL.
REL_AGENT_ID(0x7eeef19c50f0)]), filter(nil),
      conds([(T_OP_OR, T_LDIM_AGENT_UPREL.AGENT_ID(0x7eeef19c3fe0) = VIEW1.
AGENT_ID(0x7eeef19ce070)(0x7eeef19ce360), (T_OP_IS, T_LDIM_AGENT_UPREL.
AGENT_ID(0x7eeef19c3fe0), NULL, 0)(0x7eeef19cf2e0), (T_OP_IS, VIEW1.AGENT_
ID(0x7eeef19ce070), NULL, 0)(0x7eeef19cfee0))(0x7eeef19cec00)]), nl_params_(nil),
batch_join=false
  2 - output([T_LDIM_AGENT_UPREL.AGENT_ID(0x7eeef19c3fe0)], [T_LDIM_AGENT_UPREL.
REL_AGENT_ID(0x7eeef19c50f0)]), filter(nil),
      access([T_LDIM_AGENT_UPREL.AGENT_ID(0x7eeef19c3fe0)], [T_LDIM_AGENT_UPREL.REL_
AGENT_ID(0x7eeef19c50f0)]), partitions(p0),
      is_index_back=false,
      range_key([T_LDIM_AGENT_UPREL.AGENT_ID(0x7eeef19c3fe0)], [T_LDIM_AGENT_UPREL.
REL_AGENT_ID(0x7eeef19c50f0)]), range(MIN,MIN ; MAX,MAX)always true
  3 - output([VIEW1.AGENT_ID(0x7eeef19ce070)]), filter(nil)
  4 - output([VIEW1.AGENT_ID(0x7eeef19ce070)]), filter(nil),
      access([VIEW1.AGENT_ID(0x7eeef19ce070)])
  5 - output([T_LDIM_AGENT_UPREL.AGENT_ID(0x7eeef1a609a0)]), filter([T_LDIM_
AGENT_UPREL.VALID_FLG(0x7eeef1a606b0) = ?(0x7eeef1a60c90)]),
      access([T_LDIM_AGENT_UPREL.VALID_FLG(0x7eeef1a606b0)], [T_LDIM_AGENT_UPREL.
AGENT_ID(0x7eeef1a609a0)]), partitions(p0),
      is_index_back=false, filter_before_indexback[false],
      range_key([T_LDIM_AGENT_UPREL.REL_AGENT_ID(0x7eeef1a821a0)]), range(MIN ; MAX)
always true
```

把1号NESTED-LOOP ANTI JOIN算子的Outputs&filters单独拿出来看：

```
  1 - output([T_LDIM_AGENT_UPREL.AGENT_ID(0x7eeef19c3fe0)], [T_LDIM_AGENT_UPREL.
REL_AGENT_ID(0x7eeef19c50f0)]), filter(nil),
      conds([(T_OP_OR, T_LDIM_AGENT_UPREL.AGENT_ID(0x7eeef19c3fe0) = VIEW1.
AGENT_ID(0x7eeef19ce070)(0x7eeef19ce360), (T_OP_IS, T_LDIM_AGENT_UPREL.
AGENT_ID(0x7eeef19c3fe0), NULL, 0)(0x7eeef19cf2e0), (T_OP_IS, VIEW1.AGENT_
ID(0x7eeef19ce070), NULL, 0)(0x7eeef19cfee0))(0x7eeef19cec00)]), nl_params_(nil),
batch_join=false
```

匹配条件是：

```
        where T_LDIM_AGENT_UPREL.AGENT_ID=VIEW1.AGENT_ID
        Or T_LDIM_AGENT_UPREL.AGENT_ID is NULL -- 判断父查询 AGENT_ID 是否为空，如果遇到
NULL 值，则剔除这行结果
        Or VIEW1.AGENT_ID is NULL -- 判断子查询结果集 AGENT_ID 是否为 NULL，如果遇到 NULL 值，
直接进入 JOIN_END 阶段，不返回任何数据
```

以上逻辑是可以正确处理 NULL 值的。

按照这个逻辑，即使加上 Or VIEW1.AGENT_ID IS NULL 条件，被驱动表依然是可以使用索引的，只有 IS NOT NULL 无法使用索引：

```
##SQL
select AGENT_ID from T_LDIM_AGENT_UPREL
where AGENT_ID='124253' or AGENT_ID is null;

## 执行计划
===============================================================================
|ID|OPERATOR   |NAME                                              |EST. ROWS|COST|
-------------------------------------------------------------------------------
|0 |TABLE SCAN|T_LDIM_AGENT_UPREL(I_LDIM_AGENT_UPREL_AGENT_ID)|1        |46  |
===============================================================================

Outputs & filters:
-------------------------------------
  0 - output([T_LDIM_AGENT_UPREL.AGENT_ID(0x7eef739f9120)]), filter(nil),
      access([T_LDIM_AGENT_UPREL.AGENT_ID(0x7eef739f9120)]), partitions(p0),
      is_index_back=false,
      range_key([T_LDIM_AGENT_UPREL.AGENT_ID(0x7eef739f9120)], [T_LDIM_AGENT_
UPREL.REL_AGENT_ID(0x7eef73a40830)]), range(124253,MIN ; 124253,MAX), (NULL,MIN ;
NULL,MAX),
      range_cond([T_LDIM_AGENT_UPREL.AGENT_ID(0x7eef739f9120) = ?(0x7eef739f7a50)
OR (T_OP_IS, T_LDIM_AGENT_UPREL.AGENT_ID(0x7eef739f9120), NULL, 0)(0x7eef739f86d0)
(0x7eef739f6dd0)])
```

按照经验，此时我们应该查看 Oracle 上 NESTED-LOOP ANTI JOIN NA 的处理逻辑，不过在 Oracle 上调不出这个执行计划，因此线索中断。

推断：目前 OceanBase 3.x 版本没有实现真正意义上的 NESTED-LOOP ANTI JOIN NA，但是 NESTED-LOOP ANTI JOIN 可以正确处理 NULL 敏感问题。OceanBase4.x 会实现 NESTED-LOOP ANTI JOIN NA，实现方式就是我们前面推理出的逻辑，也就是说 OceanBase 3.x 用的不是这一套逻辑，执行计划虽然这么显示，但实际上不是这样，对被驱动表匹配查询时就是要遍历全表，不能直接走索引匹配。

问题 3：加上 /*+ no_rewrite */ 后，采用 SUBPLAN FILTER 算子，尽管父查询显示可以使用索引，为什么执行效率还是较低？

加了 /*+ no_rewrite */ 的执行计划，执行耗时 7 秒，比原始 SQL（耗时 16 秒）快。从执行逻辑来看：

(1) 这是非相关子查询，每次重复执行的结果都是一样的，所以执行一次后将结果保存在参数集合 init_plan_idxs_([1]) 中，表示子查询只需要执行一次。

(2) 从参数中获取右边非相关子查询的结果，FILTER 被下推到左边计划中以执行父查询。注意，条件是 A.AGENT_ID!= ALL(subquery(1))，这里使用了不等于操作（!=），因此无法使用索引快速过滤数据，需要扫描整个索引，所以执行效率并不高。如果这里用的不是 NOT IN 而是 IN，则可以利用索引快速查找。

```
===================================================================
|ID|OPERATOR         |NAME                         |EST. ROWS|COST    |
-------------------------------------------------------------------
|0 |MERGE GROUP BY   |                             |3659     |58062035|
|1 | SUBPLAN FILTER  |                             |13880    |58061224|
|2 |  TABLE SCAN     |A(I_LDIM_AGENT_UPREL_AGENT_ID)|27760   |10738   |
|3 |  TABLE SCAN     |B                            |13880    |10906   |
===================================================================

Outputs & filters:
-------------------------------------
  0 - output([A.AGENT_ID(0x7ee843c44330)], [T_FUN_MAX(A.REL_AGENT_
ID(0x7ee843c45440))(0x7ee843c44d30)]), filter(nil),
      group([A.AGENT_ID(0x7ee843c44330)]), agg_func([T_FUN_MAX(A.REL_AGENT_
ID(0x7ee843c45440)(0x7ee843c44d30)])
  1 - output([A.AGENT_ID(0x7ee843c44330)], [A.REL_AGENT_ID(0x7ee843c45440)]),
filter([A.AGENT_ID(0x7ee843c44330) != ALL(subquery(1)(0x7ee843bf8e60))
(0x7ee843bf8470)]),
      exec_params_(nil), onetime_exprs_(nil), init_plan_idxs_([1])
  2 - output([A.AGENT_ID(0x7ee843c44330)], [A.REL_AGENT_ID(0x7ee843c45440)]),
filter(nil),
      access([A.AGENT_ID(0x7ee843c44330)], [A.REL_AGENT_ID(0x7ee843c45440)]),
partitions(p0),
      is_index_back=false,
      range_key([A.AGENT_ID(0x7ee843c44330)], [A.REL_AGENT_ID(0x7ee843c45440)]),
range(MIN,MIN ; MAX,MAX)always true
  3 - output([B.AGENT_ID(0x7ee843c41350)]), filter([B.VALID_FLG(0x7ee843c40c40) =
?(0x7ee843c40520)]),
      access([B.VALID_FLG(0x7ee843c40c40)], [B.AGENT_ID(0x7ee843c41350)]),
partitions(p0),
      is_index_back=false, filter_before_indexback[false],
      range_key([B.REL_AGENT_ID(0x7ee843cb5bb0)]), range(MIN ; MAX)always true
```

10 OceanBase 是如何关闭主备线程的？

作者：何文超

10.1 背景

在 MySQL 主备同步中，存在 stop slave；reset slave all 这样的命令来控制关闭主备线程，删除主备相关信息。那么在分布式的 OceanBase 中是否存在类似场景？所用的命令是否相同？如不同，又有什么区别？

10.1.1 说明

(1) 在 MySQL 中，主备同步是主备库同步，而在 OceanBase 中，类似场景存在于主备集群中。

(2) OceanBase 主备集群没有 stop slave; reset slave all 等命令，但有处理类似场景的机制。

下面来详细介绍 OceanBase 中的 "stop slave; reset slave all"。

10.1.2 环境准备

准备一套 OceanBase 主备集群。

10.2 OceanBase 中的 stop slave

下面通过几个实验来验证 clog 是如何影响 OceanBase 主备集群状态的。

10.2.1 实验 1：关闭 clog，集群是否可用？

关闭 clog 同步（在主集群 sys 租户上操作）：

```
MySQL [(none)]> alter system disable cluster synchronization 'hwc_cluster' cluster_id=1682755173;
    //hwc_clog：备集群名
    //cluster_id：备集群 ID
```

查看同步状态（在主集群 sys 租户上操作）：

```
MySQL [(none)]> select * from oceanbase.v$ob_standby_status\G;
*************************** 1. row ***************************
    cluster_id: 1682755173
```

```
cluster_name: hwc_cluster
cluster_role: PHYSICAL STANDBY
cluster_status: DISABLED
current_scn: 1688630508000921
rootservice_list: 10.186.64.63:2882:2881
redo_transport_options: ASYNC NET_TIMEOUT = 30000000
protection_level: MAXIMUM PERFORMANCE
synchronization_status: CLUSTER IS DISABLED
1 row in set (0.00 sec)


//synchronization_status: CLUSTER IS DISABLED
// 主备集群断开 clog 日志同步
```

结论：关闭 clog 同步，OceanBase 主备集群关闭。

10.2.2 实验 2：特殊情况下，新数据是否丢失?

验证当"主备集群 clog 同步断开时间 >clog 的保留时间"时，再次开启主备集群间的 clog 同步，新数据是否丢失。

修改 clog 保留天数为 1 天：

```
MySQL [(none)]> ALTER SYSTEM SET clog_expire_days=1;
```

1 天后，主集群插入新数据：

```
MySQL [lpp]> select * from test;
+------+------+
| c1   | c2   |
+------+------+
|    1 | eee  |
|    2 | eee  |
|    3 | eee  |
+------+------+
3 rows in set (0.01 sec)
MySQL [lpp]> insert into test(c1,c2) values(4,'ddd');
Query OK, 1 row affected (0.02 sec)
```

开启 clog 同步（在主集群 sys 租户上操作）：

```
MySQL [(none)]> alter system enable cluster synchronization 'hwc_cluster' cluster_id=1682755173;
  //hwc_clog：备集群名
  //cluster_id：备集群 ID
```

检查备集群是否同步到新数据（连接串需要添加 -c）：

```
MySQL [lpp]> select /*+READ_CONSISTENCY(WEAK) */ * from test;
+------+------+
| c1   | c2   |
+------+------+
```

```
|   1 | eee |
|   2 | eee |
|   3 | eee |
|   4 | ddd |
+------+------+
4 rows in set (0.00 sec)
```

结论：备集群同步到新数据。

原理：当开启主备集群 clog 同步时，会自动检测数据一致性，如发现数据不一致，会自动拉取基线数据进行同步。

10.2.3 实验 3：停止 clog 同步后，备集群是否可用？

clog 正常同步时，备集群查询数据（连接串需要添加 -c）：

```
MySQL [LPP]> select /*+READ_CONSISTENCY(WEAK) */ * from test;
+------+------+
| c1   | c2   |
+------+------+
|   1 | eee |
|   2 | eee |
|   3 | eee |
|   4 | ddd |
+------+------+
4 rows in set (0.00 sec)
```

停止 clog 同步，备集群查询数据（连接串需要添加 -c）：

```
MySQL [lpp]> select /*+READ_CONSISTENCY(WEAK) */ * from test;
ERROR 4012 (HY000): Timeout
```

结论：当停止 clog 同步时，备集群不可用。

10.3 OceanBase 中的 reset slave all

在 MySQL 中通过 reset slave all 删除主备相关信息，从库可以作为一个独立的库，可读可写。

在 OceanBase 中通过主备集群解耦来删除主备集群关系，具体步骤涉及生产环境，解耦步骤更为烦琐，此处不详细阐述，可参考官方文档。

10.4 OceanBase 与 MySQL 的区别

那么，OceanBase 主备集群与 MySQL 主备库，在关闭主备线程，删除主备相关信息上有哪些区别呢？

10.4.1 MySQL

(1) 命令操作位置：备库。

(2) 停止同步命令：stop slave。

(3) 删除主备关系：reset slave all。

(4) 当 binlog 同步断开，主节点日志过期时，重新打开日志同步，备库会丢数据。

(5) 当 binlog 同步断开时，备库可用。

10.4.2 OceanBase

(1) 命令操作位置：主集群。

(2) 停止同步命令：alter system disable cluster synchronization 'hwc_cluster' cluster_id=xxxxxxxxx。

(3) 删除主备关系：主备库解耦（较为烦琐，OCP V3.3.0 可以白屏化操作）。

(4) 当 clog 同步断开，主节点日志过期时，重新打开日志同步，备集群不会丢数据。

(5) 当 clog 同步断开，备集群不可用。

11 如何有效使用 outline 功能？

作者：胡呈清

为了防止某些 SQL 的执行计划发生变化，我们通常会创建 outline 来绑定执行计划。但是为什么在实际操作中，我们创建 outline 并验证时总会遇到不生效的情况呢？

阅读本文后，你将了解如何获取 SQL ID、创建 outline，以及如何正确验证 outline 的效果。

下面以一个实例来演示 outline 的正确使用方法。

11.1 实例演示

我们得知业务中 SELECT count(*) FROM ACT_HI_COMMENT; 这个 SQL 选错了索引，需要指定其使用 IDX_ACT_HI_COMMENT_02 这个索引。

11.1.1 获取 SQL ID

可以从 gv$sql_audit 中获取 SQL ID。

```
obclient [SYS]> select query_sql,sql_id,svr_ip,plan_id,tenant_id from gv$sql_
```

```
audit where query_sql like 'SELECT%count%ACT_HI_COMMENT%';
    +--------------------------------------------+--------------------------------
-+------------+--------+-----------+
    | QUERY_SQL                                  | SQL_ID
| SVR_IP     | PLAN_ID | TENANT_ID |
    +--------------------------------------------+--------------------------------
-+------------+--------+-----------+
    | SELECT count(*) FROM ACT_HI_COMMENT;       | AC1ED40EC4D5E1A9D75944216745063A
| 26.0.8.170 |  99968 |      1001 |
    | SELECT count(*) FROM ACT_HI_COMMENT        | 46815AF386F959D17293BCF931FEEAF1
| 26.0.8.170 |  99798 |      1001 |
    +--------------------------------------------+--------------------------------
-+------------+--------+-----------+
    2 rows in set (8.695 sec)
```

此时我们发现一个问题：结果中有两条 SQL（对应两个 SQL ID），区别是一个有分号，另一个没分号。哪个是业务下发的 SQL？答案是没有分号的那个。理由如下：

(1) 应用程序发起的 SQL 请求，以及 obclient 客户端执行的 SQL，在 gv$sql_audit 中都是没有分号的。

(2) 在 ODC 中执行的 SQL 无论加不加分号，在 gv$sql_audit 中都有分号。

11.1.2 创建 outline

我们用第二个 SQL ID 来绑定执行计划，注意这里要在对应的 schema 下创建 outline，常见的误区是在 sys 用户下创建，这是不会生效的：

```
    -- 创建 outline
    conn JTZJGL;
    create outline test_outline on '46815AF386F959D17293BCF931FEEAF1' using hint
/*+ index(ACT_HI_COMMENT IDX_ACT_HI_COMMENT_02) */;

    -- 查询 outline
    obclient [JTZJGL]> select OUTLINE_ID,DATABASE_NAME,OUTLINE_NAME,OUTLINE_
SQL,SQL_ID,OUTLINE_CONTENT from gv$outline;
    +-------------------+---------------+--------------+-------------+----------
--------------------+----------------------------------------------------------+
    | OUTLINE_ID        | DATABASE_NAME | OUTLINE_NAME | OUTLINE_SQL | SQL_ID
| OUTLINE_CONTENT                                          |
    +-------------------+---------------+--------------+-------------+----------
--------------------+----------------------------------------------------------+
    | 1100611139404781  | JTZJGL        | TEST_OUTLINE |             | 46815AF386F9
59D17293BCF931FEEAF1 | /*+index(ACT_HI_COMMENT IDX_ACT_HI_COMMENT_02) */ |
    +-------------------+---------------+--------------+-------------+----------
--------------------+----------------------------------------------------------+
    1 row in set (0.005 sec)
```

11.1.3 验证效果

11.1.3.1 误区

认为可以通过 explain 观察执行计划的方式来验证 outline 的效果,这是一个误区。explain 无法验证 outline 的效果(执行计划不会改变)。

```
obclient [JTZJGL]> explain SELECT count(*) FROM ACT_HI_COMMENT\G
*************************** 1\. row ***************************
Query Plan: ===================================================================
========
|ID|OPERATOR         |NAME                                      |EST. ROWS|COST |
-----------------------------------------------------------------------------
|0 |SCALAR GROUP BY|                                          |1        |89615|
|1 | TABLE SCAN      |ACT_HI_COMMENT(IDX_ACT_HI_COMMENT_01)|210872   |81567|
===================================================================

Outputs & filters:
-------------------------------------
0 - output([T_FUN_COUNT(*)]), filter(nil),
group(nil), agg_func([T_FUN_COUNT(*)])
1 - output([1]), filter(nil),
access([ACT_HI_COMMENT.TASK_ID_]), partitions(p0)

1 row in set (0.004 sec);
```

11.1.3.2 正确的方式

应该执行原始 SQL 来验证 outline 的效果。前面我们讲了 ODC 会自动给 SQL 加分号,所以只能在 obclient 客户端中执行 SQL,然后在 gv$plan_cache_plan_stat 视图中查询这个 SQL 对应的 OUTLINE_ID。

OUTLINE_ID 只要不为 -1 就说明生效了。从 OUTLINE_DATA 中也可以看出 SQL 使用了我们指定的 IDX_ACT_HI_COMMENT_02 索引:

```
-- 执行原始 SQL
obclient [JTZJGL]> SELECT count(*) FROM ACT_HI_COMMENT;
+----------+
| COUNT(*) |
+----------+
|   210872 |
+----------+
1 row in set (0.110 sec)

-- 然后在 gv$plan_cache_plan_stat 视图中查询这个 SQL 对应的 OUTLINE_ID,只要不为 -1 就说
明生效了
obclient [SYS]> select SQL_ID,PLAN_ID,QUERY_SQL,OUTLINE_ID,OUTLINE_DATA from
gv$plan_cache_plan_stat where sql_id='46815AF386F959D17293BCF931FEEAF1';
```

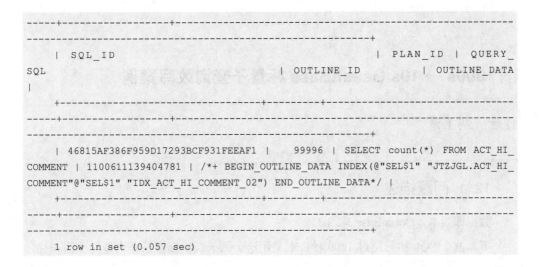

```
1 row in set (0.057 sec)
```

11.2 总结

(1) 建议始终使用 SQL ID 来创建 outline，因为 SQL 文本即使多一个空格都会导致 outline 不生效。

(2) 通过 gv$sql_audit 获取 SQL_ID，注意，业务 SQL 在 gv$sql_audit 中不会带有分号。

(3) 创建 outline 时需要登录业务租户，并在对应的 schema 下创建，不能在 sys 用户下创建。验证 outline 时的注意事项如下：

①用 explain 查看执行计划是不会改变的，不能用来验证 outline 是否生效。

②执行原始 SQL 时，执行的 SQL 文本需要从 gv$sql_audit 获取，并且不能有任何修改，不能美化 SQL 格式后执行。不能使用 ODC 执行 SQL，因为 ODC 会给 SQL 加上分号或者改写 SQL，这样会导致 SQL ID 变化，无法命中 outline。

③执行原始 SQL 后，可以通过 gv$plan_cache_plan_stat 检查目标 SQL 的 plan cache 状态，如果 outline_id 不为 -1，则说明命中了 outline。也可以通过 gv$plan_cache_plan_explain 查看验证时执行的 SQL 的真实执行计划（先通过 gv$sql_audit 获取 tenant_id、ip、port、plan_id 4 个元素），核对其是否和 outline 中定义的 hint 一致。

12 1000s → 10s OceanBase 标量子查询改写案例

作者：胡呈清

12.1 问题描述

数据库版本：OceanBase 3.2.3.3。

下面这个 SQL 执行超过 1000 秒！本文用这个例子谈谈标量子查询慢的原因和优化方法。

```
select
rq.processinstid processinstid,
rq.question_id questionId,
rq.question_no questionNo,
to_char(rq.rev_start_date, 'yyyy-MM-dd') revStartDate,
(
select
e.name
from
e
where
e.category_code = 'REV_SOURCE'
and e.code = rq.rev_source
) revSource,
(
select
e.name
from
e
where
e.category_code = 'QUESTION_TYPE'
and e.code = rq.question_type
) questionType,
rq.question_summary questionSummary,
rq.question_desc questionDesc,
to_char(rq.question_discover_date, 'yyyy-MM-dd') questionDiscoverDate,
rq.aud_project_type audProjectType,
(
select
d.dept_name
from
```

```sql
d
where
d.dept_id = rq.check_dept
) checkDept,
(
select
to_char(wm_concat(distinct(k.org_name)))
from
o,
k
where
o.question_id = rq.question_id
and o.ASC_ORG = k.org_id
and o.REFORM_TYPE = '0'
) ascOrg,
(
select
to_char(wm_concat(distinct(k.dept_name)))
from
o,
fnd_dept_t k
where
o.question_id = rq.question_id
and o.MAIN_REV_DEPT = k.dept_id
and o.REFORM_TYPE = '0'
) mainRevDept,
(
select
e.name
from
e
where
e.category_code = 'REV_FINISH_STATE'
and e.code = rq.rev_finish_state
) revFinishState,
to_char(rq.compliance_date, 'yyyy-MM-dd') complianceDATE
from
rq
left join REM_QUESTION_PLAN_T t on rq.question_id = t.question_id
left join fnd_org_t org on t.ASC_ORG = org.org_id
where
1 = 1
and rq.asc_org is null
and (
t.asc_org in (
select
f.org_id
from
```

```
f
where
f.org_type = 'G'
)
or rq.created_by_org in (
select
f.org_id
from
f
where
f.org_type = 'G'
)
)
and rq.company_type = 'G';
```

12.2 分析过程

执行计划如下：

```
==================================================================
|ID|OPERATOR              |NAME            |EST. ROWS|COST       |
------------------------------------------------------------------
|0 |SUBPLAN FILTER        |                |6283     |788388847  |
|1 | SUBPLAN FILTER       |                |6283     |1325483    |
|2 |  HASH OUTER JOIN     |                |8377     |210530     |
|3 |   TABLE SCAN         |RQ              |7966     |77932      |
|4 |   TABLE SCAN         |T               |152919   |59150      |
|5 |   TABLE SCAN         |F               |440      |2763       |
|6 |   TABLE SCAN         |F               |440      |2763       |
|7 |   TABLE SCAN         |E(SYS_C0011218) |1        |92         |
|8 |   TABLE SCAN         |E(SYS_C0011218) |1        |92         |
|9 |   TABLE GET          |D               |1        |46         |
|10| SCALAR GROUP BY      |                |1        |62483      |
|11|  NESTED-LOOP JOIN    |                |1        |62483      |
|12|   TABLE SCAN         |O               |1        |62468      |
|13|   TABLE GET          |K               |1        |28         |
|14| SCALAR GROUP BY      |                |1        |62483      |
|15|  NESTED-LOOP JOIN    |                |1        |62483      |
|16|   TABLE SCAN         |O               |1        |62468      |
|17|   TABLE GET          |K               |1        |27         |
|18|   TABLE SCAN         |E(SYS_C0011218) |1        |92         |
==================================================================
```

每个子算子的成本都不高，但总成本很高！

下面结合 SQL 语法语义进行解读。首先，这个 SQL 从语法上分为两部分：标量子查询，即投影部分的子查询；外部查询，即 FROM 子句的关联查询和子查询。

因此，这个 SQL 的执行逻辑是（也就是执行计划里的 0 号 SUBPLAN FILTER 算子）：先执行外部查询，得到结果集 r（执行计划中的 1~6 号算子）；再执行标量子查询，从结果集 r 中取一行数据，带到标量子查询中执行（执行计划中的 7~18 号算子）。重复上述操作，直到循环取完最后一行数据。

为了定位 SQL 到底慢在哪一步，我们继续拆解分析。

(1) 先拆出外部查询（即对应的 1~6 号算子部分）单独执行，很快得到结果为 13 万行，也就意味着所有标量子查询都需要执行 13 万次。

(2) 从执行计划来看，7、8、9、18 号算子对应的 4 个标量子查询都可以利用索引，效率较高。只保留外部查询和这 4 个标量子查询，执行耗时很短。

(3) 关键在于 10、14 号算子，其对应的 2 个标量子查询除了和外表关联外，本身内部还关联 o、k 这 2 张表，这两张表要做 13 万次关联，很明显效率会很低。

SQL 中 10、14 号算子对应的标量子查询如下，还可以再拆解 SQL，单独做一次 o、k 表的关联查询（如下**加粗**部分）要 200 毫秒：

```
select
xxx,
(
select
to_char(wm_concat(distinct(k.org_name)))
from
REM_QUESTION_PLAN_T o,
fnd_org_t k
where
o.question_id = rq.question_id
and o.ASC_ORG = k.org_id
and o.REFORM_TYPE = '0'
) ascOrg,
(
select
to_char(wm_concat(distinct(k.dept_name)))
from
REM_QUESTION_PLAN_T o,
fnd_dept_t k
where
o.question_id = rq.question_id
and o.MAIN_REV_DEPT = k.dept_id
and o.REFORM_TYPE = '0'
) mainRevDept,
xxx
from t(外部查询，结果有 13 万行);
```

12.3 结论

标量子查询的执行计划只能是循环嵌套连接，也就是 SUBPLAN FILTER 算子（等同于 NESTED-LOOP JOIN 执行逻辑），它的执行效率取决于两个因素：外部查询的结果集大小与子查询的效率。

因此，只有当外部查询结果集不大，并且子查询的关联字段有高效索引时，执行效率才高。如果关联字段没有索引，优化器也没法像 JOIN 语法一样使用 HASH JOIN 算子，执行效率很低。

在上面这个慢 SQL 中，有两个标量子查询不只和外表关联，它内部还有关联查询，所以即使关联字段有索引，子查询单次执行的效率也受限，再加上要执行 13 万次，耗时就长了。所以，这个 SQL 只能改写成 LEFT JOIN 来优化，这也是标量子查询的标准优化方法。

12.4 优化方案

这个 SQL 的标量子查询中有聚合函数，应该先以 GROUP BY 聚合再和外表关联。SQL（局部）改写如下：

```sql
with t1 as (
select
o.question_id,
to_char(wm_concat(distinct(k.org_name))) as org_name
from
REM_QUESTION_PLAN_T o,
fnd_org_t k
where
o.ASC_ORG = k.org_id
and o.REFORM_TYPE = '0'
group by
o.question_id
),
t2 as (
select
o.question_id,
to_char(wm_concat(distinct(k.dept_name))) as dept_name
from
REM_QUESTION_PLAN_T o,
fnd_dept_t k
where
o.MAIN_REV_DEPT = k.dept_id
and o.REFORM_TYPE = '0'
group by
```

```
    o.question_id
  )
  select
  xxx,
  t1.org_name as ascOrg,
  t2.dept_name as mainRevDept,
  xxx
  from t（外部查询，结果有13万行）
  left join t1 on t.question_id=t1.question_id
  left join t2 on t.question_id=t2.question_id;
```

改写后的执行计划如下（变成使用 HASH OUTER JOIN 算法），可以看到成本从 7.88 亿降到了 365 万，执行耗时降到 10 秒！

```
=================================================================
|ID|OPERATOR            |NAME            |EST. ROWS|COST    |
-----------------------------------------------------------------
|0 |SUBPLAN FILTER      |                |6318     |3653489 |
|1 | MERGE GROUP BY     |                |6318     |1636701 |
|2 |  SORT              |                |6318     |1632074 |
|3 |   SUBPLAN FILTER   |                |6318     |1613799 |
|4 |    HASH OUTER JOIN |                |8424     |492531  |
|5 |     HASH OUTER JOIN|                |8377     |331672  |
|6 |      MERGE OUTER JOIN|              |7966     |198317  |
|7 |       TABLE SCAN   |RQ              |7966     |77932   |
|8 |       SUBPLAN SCAN |T2              |2351     |119098  |
|9 |        MERGE GROUP BY|              |2351     |119062  |
|10|         SORT       |                |2352     |118658  |
|11|          HASH JOIN |                |2352     |113818  |
|12|           TABLE SCAN|K              |22268    |8614    |
|13|           TABLE SCAN|O              |76460    |60075   |
|14|      TABLE SCAN    |T               |152919   |59150   |
|15|     SUBPLAN SCAN   |T1              |76415    |118014  |
|16|      HASH JOIN     |                |76415    |116865  |
|17|       TABLE SCAN   |K               |7033     |2721    |
|18|       TABLE SCAN   |O               |76460    |60075   |
|19|    TABLE SCAN      |F               |440      |2763    |
|20|    TABLE SCAN      |F               |440      |2763    |
|21| TABLE SCAN         |E(SYS_C0011218) |1        |92      |
|22| TABLE SCAN         |E(SYS_C0011218) |1        |92      |
|23| TABLE GET          |D               |1        |46      |
|24| TABLE SCAN         |E(SYS_C0011218) |1        |92      |
=================================================================
```

13 日志盘过小也会导致创建租户失败？

作者：郑增权

13.1 背景

某客户出于节约资源的想法，将日志盘设置得比较小，大小约为集群内存规格的 1.5 倍。他创建租户时，在 CPU 和内存都充足的情况下，却出现报错："LOG_DISK resource not enough"。我们尝试复现问题并定位原因。

13.2 环境信息

架构：单节点集群。

版本：OceanBase4.2.1.4。

```
MySQL [oceanbase]> select svr_ip,status,build_version from __all_server;
+--------------+--------+------------------------------------------------------------------------------------------+
| svr_ip       | status | build_version                                                                            |
+--------------+--------+------------------------------------------------------------------------------------------+
| 10.186.64.61 | ACTIVE | 4.2.1.4_104010012024030714-c4f3400ad2839e337bc9dab5d1bfe1d01134a1d7(Mar  7 2024 14:32:22) |
+--------------+--------+------------------------------------------------------------------------------------------+
1 row in set (0.01 sec)
```

查看集群可分配的 CPU、内存、日志盘容量。

(1)MEM_CAPACITY：observer 进程可用的内存大小。

(2)LOG_DISK_CAPACITY：日志盘空间总大小，41.8GB。

(3)LOG_DISK_ASSIGNED：日志盘已分配大小，6GB。

(4) 剩余可分配 CPU 数量：18-2=16

(5) 剩余可分配内存大小：24-2=22GB

```
MySQL [oceanbase]> SELECT SVR_IP, SVR_PORT , ZONE , SQL_PORT , CPU_CAPACITY
, CPU_CAPACITY_MAX, CPU_ASSIGNED , CPU_ASSIGNED_MAX ,MEMORY_LIMIT/1024/1024/1024
MEMORY_LIMIT_GB , MEM_CAPACITY/1024/1024/1024 MEM_CAPACITY_GB, MEM_
```

```
ASSIGNED/1024/1024/1024 MEM_ASSIGNED_GB ,LOG_DISK_CAPACITY/1024/1024/1024 LOG_
DISK_CAPACITY_GB,LOG_DISK_ASSIGNED/1024/1024/1024 LOG_DISK_ASSIGNED_GB,LOG_DISK_
IN_USE/1024/1024/1024 LOG_DISK_IN_USE_GB  FROM GV$OB_SERVERS;
+--------------+----------+-------+----------+--------------+------------------
-+------------+------------------+-----------------+------------------+----------
-------+---------------------+--------------------+
|   SVR_IP     | SVR_PORT | ZONE  | SQL_PORT | CPU_CAPACITY | CPU_CAPACITY_
MAX | CPU_ASSIGNED | CPU_ASSIGNED_MAX | MEMORY_LIMIT_GB | MEM_CAPACITY_GB | MEM_
ASSIGNED_GB | LOG_DISK_CAPACITY_GB | LOG_DISK_ASSIGNED_GB | LOG_DISK_IN_USE_GB |
+--------------+----------+-------+----------+--------------+------------------
-+------------+------------------+-----------------+------------------+----------
-------+---------------------+--------------------+
| 10.186.64.61 |     2882 | zone1 |     2881 |           18 |                18
|            2 |                2 |  30.000000000000 |  24.000000000000 |
2.000000000000 |     41.875000000000 |       6.000000000000 |     0.125000000000
|
+--------------+----------+-------+----------+--------------+------------------
-+------------+------------------+-----------------+------------------+----------
-------+---------------------+--------------------+
1 row in set (0.00 sec)
```

可以看到，当前集群仅 sys 租户占用了 2C2G（表示 CPU 数量为 2，内存为 2GB，下同）的资源。

```
MySQL [oceanbase]> SELECT a.tenant_name,a.tenant_id,b.name unit_config,c.name
pool_name,b.max_cpu,b.min_cpu,MEMORY_SIZE/1024/1024/1024 as MEMORY_SIZE
    -> FROM
    -> OCEANBASE.DBA_OB_TENANTS a,
    -> OCEANBASE.DBA_OB_UNIT_CONFIGS b,
    -> OCEANBASE.DBA_OB_RESOURCE_POOLS c
    -> WHERE a.tenant_id=c.tenant_id
    -> AND b.unit_config_id = c.unit_config_id
    -> ORDER BY a.tenant_id desc;
+-------------+-----------+-------------------------------+-----------+--------
---+---------+----------------+
| tenant_name | tenant_id | unit_config                   | pool_name | max_cpu
| min_cpu | MEMORY_SIZE    |
+-------------+-----------+-------------------------------+-----------+--------
---+---------+----------------+
| sys         |         1 | config_sys_zone1_twoctwog_xio | sys_pool  |       2
|       2 | 2.000000000000 |
+-------------+-----------+-------------------------------+-----------+--------
---+---------+----------------+
1 row in set (0.01 sec)
```

13.3 问题复现及疑问

13.3.1 创建资源单元

尝试创建 1 个规格为 4C12G 的资源单元。

```
MySQL [oceanbase]> CREATE RESOURCE UNIT mem_test_unit MEMORY_SIZE = '12G',MAX_CPU = 4, MIN_CPU = 4;
Query OK, 0 rows affected (0.02 sec)
```

13.3.2 创建资源池

出现报错：LOG_DISK resource not enough。

```
MySQL [oceanbase]> CREATE RESOURCE POOL pool_evan UNIT='mem_test_unit', UNIT_NUM=1, ZONE_LIST=('zone1');
ERROR 4733 (HY000): zone 'zone1' resource not enough to hold 1 unit. You can check resource info by views: DBA_OB_UNITS, GV$OB_UNITS, GV$OB_SERVERS.
    server '"10.186.64.61:2882"' LOG_DISK resource not enough
```

问题 1：剩余资源为 16C22G，为何创建一个 4C12G 的资源池会失败？

问题 2：报错信息提到 LOG_DISK，日志盘容量与内存规格存在何种关联？

13.4 日志盘大小与租户内存大小的关系

尝试新建规格为 1C1G 的租户，分析租户内存大小与日志盘容量分配的规律。

(1) 新建规格为 1C1G 的租户。

(2)LOG_DISK_ASSIGNED_GB 增长至 9GB，相较之前增加了 3GB。

```
MySQL [oceanbase]> CREATE RESOURCE UNIT unit_1g MEMORY_SIZE = '1G',MAX_CPU = 1, MIN_CPU = 1;
Query OK, 0 rows affected (0.01 sec)
MySQL [oceanbase]> CREATE RESOURCE POOL pool_1g UNIT='unit_1g', UNIT_NUM=1, ZONE_LIST=('zone1');
Query OK, 0 rows affected (0.02 sec)
MySQL [oceanbase]> CREATE TENANT IF NOT EXISTS tenant_1g
    ->              PRIMARY_ZONE = 'zone1',
    ->              RESOURCE_POOL_LIST=('pool_1g')
    ->              set OB_TCP_INVITED_NODES='%';
Query OK, 0 rows affected (26.05 sec)
MySQL [oceanbase]>
MySQL [oceanbase]> SELECT SVR_IP, SVR_PORT , ZONE , SQL_PORT , CPU_CAPACITY ,CPU_ASSIGNED ,MEMORY_LIMIT/1024/1024/1024 MEMORY_LIMIT_GB , MEM_CAPACITY/1024/1024/1024 MEM_CAPACITY_GB, MEM_ASSIGNED/1024/1024/1024 MEM_ASSIGNED_GB ,LOG_DISK_CAPACITY/1024/1024/1024 LOG_DISK_CAPACITY_GB,LOG_DISK_ASSIGNED/1024/1024/1024 LOG_DISK_ASSIGNED_GB  FROM GV$OB_SERVERS;
```

```
+---------------+----------+-------+----------+--------------+--------------+---------
------+----------------+------------------+----------------------+----------------------+
| SVR_IP        | SVR_PORT | ZONE  | SQL_PORT | CPU_CAPACITY | CPU_ASSIGNED | MEMORY_
LIMIT_GB | MEM_CAPACITY_GB | MEM_ASSIGNED_GB | LOG_DISK_CAPACITY_GB | LOG_DISK_ASSIGNED_GB |
+---------------+----------+-------+----------+--------------+--------------+---------
------+----------------+------------------+----------------------+----------------------+
| 10.186.64.61  |     2882 | zone1 |     2881 |           18
|          3 | 30.000000000000 | 24.000000000000 |      3.000000000000 |     41.875000000000 |
9.000000000000  |
+---------------+----------+-------+----------+--------------+--------------+---------
------+----------------+------------------+----------------------+----------------------+
1 row in set (0.00 sec)
```

为租户每分配 1GB 内存，则对应分配 3GB 日志盘容量。

13.5 代入本文初始环境中计算

计算本文初始背景集群剩余内存规格：

```
(LOG_DISK_CAPACITY_GB - LOG_DISK_ASSIGNED_GB) / 3 = (41.875 - 6.00) / 3 ≈ 11.958 GB
```

即集群剩余可用的内存上限为 11.958GB，取整为 11GB。

13.6 验证

13.6.1 释放资源

删掉租户 tenant_1g 和对应的 RESOURCE POOL，释放资源。

```
MySQL [oceanbase]> DROP TENANT tenant_1g;
Query OK, 0 rows affected (35.04 sec)
MySQL [oceanbase]> DROP RESOURCE POOL pool_1g;
Query OK, 0 rows affected (0.01 sec)
MySQL [oceanbase]> SELECT SVR_IP, SVR_PORT , ZONE , SQL_PORT , CPU_
CAPACITY ,CPU_ASSIGNED ,MEMORY_LIMIT/1024/1024/1024 MEMORY_LIMIT_GB , MEM_
CAPACITY/1024/1024/1024 MEM_CAPACITY_GB, MEM_ASSIGNED/1024/1024/1024 MEM_
ASSIGNED_GB ,LOG_DISK_CAPACITY/1024/1024/1024 LOG_DISK_CAPACITY_GB,LOG_DISK_
ASSIGNED/1024/1024/1024 LOG_DISK_ASSIGNED_GB  FROM GV$OB_SERVERS;
+---------------+----------+-------+----------+--------------+--------------+---------
------+----------------+------------------+----------------------+----------------------+
| SVR_IP        | SVR_PORT | ZONE  | SQL_PORT | CPU_CAPACITY | CPU_ASSIGNED | MEMORY_
LIMIT_GB | MEM_CAPACITY_GB | MEM_ASSIGNED_GB | LOG_DISK_CAPACITY_GB | LOG_DISK_ASSIGNED_GB |
+---------------+----------+-------+----------+--------------+--------------+---------
------+----------------+------------------+----------------------+----------------------+
| 10.186.64.61  |     2882 | zone1 |     2881 |           18
```

```
|                2 | 30.000000000000 | 24.000000000000 |  2.000000000000 |    41.875000000000 |
6.000000000000 |
    +---------------+------------+--------+----------+----------------+----------------+------
--------+----------------+-----------------+----------------------+
    1 row in set (0.00 sec)
```

13.6.2 重新创建

创建一个规格为 16C11G 的租户。

```
    MySQL [oceanbase]> CREATE RESOURCE UNIT unit_11g MEMORY_SIZE = '11G',MAX_CPU = 16, MIN_CPU = 16;
    Query OK, 0 rows affected (0.01 sec)
    MySQL [oceanbase]> CREATE RESOURCE POOL pool_11g UNIT='unit_11g', UNIT_NUM=1, ZONE_LIST=('zone1');
    Query OK, 0 rows affected (0.02 sec)

    MySQL [oceanbase]> CREATE TENANT IF NOT EXISTS tenant_11g
    ->              PRIMARY_ZONE = 'zone1',
    ->              RESOURCE_POOL_LIST=('pool_11g')
    ->              set OB_TCP_INVITED_NODES='%';
    Query OK, 0 rows affected (25.99 sec)
    MySQL [oceanbase]> SELECT SVR_IP, SVR_PORT , ZONE , SQL_PORT , CPU_CAPACITY ,CPU_ASSIGNED ,MEMORY_LIMIT/1024/1024/1024 MEMORY_LIMIT_GB , MEM_CAPACITY/1024/1024/1024 MEM_CAPACITY_GB, MEM_ASSIGNED/1024/1024/1024 MEM_ASSIGNED_GB ,LOG_DISK_CAPACITY/1024/1024/1024 LOG_DISK_CAPACITY_GB,LOG_DISK_ASSIGNED/1024/1024/1024 LOG_DISK_ASSIGNED_GB  FROM GV$OB_SERVERS;
    +---------------+----------+-------+----------+--------------+--------------+---------------------+-----------------+-----------------+----------------------+---------------------+
    | SVR_IP        | SVR_PORT | ZONE  | SQL_PORT | CPU_CAPACITY | CPU_ASSIGNED | MEMORY_LIMIT_GB | MEM_CAPACITY_GB | MEM_ASSIGNED_GB | LOG_DISK_CAPACITY_GB | LOG_DISK_ASSIGNED_GB |
    +---------------+----------+-------+----------+--------------+--------------+---------------------+-----------------+-----------------+----------------------+---------------------+
    | 10.186.64.61  |     2882 | zone1 |     2881 |           18 |           18 | 30.000000000000 | 24.000000000000 | 13.000000000000 |     41.875000000000 | 39.000000000000 |
    +---------------+----------+-------+----------+--------------+--------------+---------------------+-----------------+-----------------+----------------------+---------------------+
    1 row in set (0.01 sec)
```

13.7 疑问解答

下面我们来解答前面提出的两个疑问。

问题 1：剩余资源为 16C22G，为何创建一个 4C12G 的资源池会失败？

(1) 由于日志盘规格为 41.875GB，且 sys 租户已占用 6GB 日志盘份额，经前文的计算可得：集群剩余可用的内存上限为 11GB。

(2) 所创建租户的内存规格超过 11GB，会因申请不到对应份额的日志盘容量而引发报错 "LOG_DISK resource not enough"。

问题 2：报错信息提到 LOG_DISK，日志盘容量与内存规格存在何种关联？

LOG_DISK_SIZE 默认值为内存规格值的 3 倍，最小值为 2GB。

13.8 建议

日志盘大小尽量设置为内存上限的 3 或 4 倍（在生产环境中至少是 3 倍），避免因日志盘不足导致集群已有的内存无法进行分配。

14 Oracle 模式竟然可以使用 Repeatable Read？

作者：任仲禹

14.1 背景

你看到文章标题可能会有疑惑，OceanBase 的 Oracle 模式只支持 2 种隔离级别：读已提交（Read Committed，RC）与可串行化（Serializable），为什么还讨论在 Oracle 模式下使用可重复读（Repeatable Read，RR）隔离级别这个问题？起因是交付时客户提出如下疑问：

"我的 Java 应用通过 oceanbase-jdbc 访问 OceanBase Oracle 模式，业务上想实现 MySQL 中的可重复读效果（即事务内 2 次相同查询看到的数据是不变的），所以我将会话设置为只读 conn.setReadOnly(true)，但程序运行结果不符合预期。"

乍一听没完全理解，沟通后我才梳理清楚：①客户了解到 set transaction read only; 命令可以实现可重复读的效果。②所以应用中配置了 conn.setReadOnly(true) 想实现此效果。

根据官网原文：设置 ReadOnly 时，不推荐执行 set session transaction readonly，推荐使用 Connection.setReadOnly(xx) 接口。

此时，还剩两个问题：

(1) OceanBase 中的命令 set transaction read only; 为何能实现可重复读效果？

(2) 配置 conn.setReadOnly(true) 是否正确，若不正确，该如何配置？

14.2 分析

涉及的环境如下：

(1) OceanBase：323bp10hotfix5

(2) JDBC 客户端：oceanbase-jdbc 2.4.3

14.2.1 为何 set transaction read only 能实现可重复读效果？

在 OceanBase 中，只读事务中的所有查询都引用了数据库的同一份快照，从而提供多表、多查询、读取一致的视图。所以，在只读事务内 2 次相同查询所看到的数据是一致的，也就实现了可重复读的效果。这对于多用户更新相同的表且运行多个查询时非常有用，也满足客户的业务需求。

14.2.2 配置 conn.setReadOnly 是否正确？

通过截取程序运行堆栈分析，客户环境用的是 Hikari 连接池，调用路径从下往上为 Hikari → OceanBase-client → setReadOnly → setSessionReadOnly（见图 16）。

```
Thread [main] (Suspended (entry into method setSessionReadOnly in AbstractMastersListener))
    MastersFailoverListener(AbstractMastersListener).setSessionReadOnly(boolean, Protocol) line: 326
    MastersFailoverListener.switchReadOnlyConnection(Boolean) line: 297
    FailoverProxy.invoke(Object, Method, Object[]) line: 207
    $Proxy173.setReadonly(boolean) line: not available
    OceanBaseConnection.setReadOnly(boolean) line: 996
    HikariProxyConnection(ProxyConnection).setReadOnly(boolean) line: 397
    HikariProxyConnection.setReadOnly(boolean) line: not available
    NativeMethodAccessorImpl.invoke0(Method, Object, Object[]) line: not available [native method]
    NativeMethodAccessorImpl.invoke(Object, Object[]) line: 62
    DelegatingMethodAccessorImpl.invoke(Object, Object[]) line: 43
    Method.invoke(Object, Object...) line: 498
    JdbcMethodInvocation.invoke(Object) line: 34
    YuspConnection(WrapperAdapter).replayMethodsInvocation(Object) line: 57
    YuspConnection(AbstractConnectionAdapter).createConnectionsT(String, DataSource, int) line: 102
    YuspConnection(AbstractConnectionAdapter).getConnections(String, int) line: 84
    YuspConnection.prepareStatement(String) line: 36
```

图 16

最终，setSessionReadOnly 调用的是 set session transaction read only 命令（见图 17）。

```
protected void setSessionReadOnly(boolean readOnly, Protocol protocol) throws SQLException {
    if (protocol.versionGreaterOrEqual(paramInt1:5, paramInt2:6, paramInt3:5)) {
        logger.info(paramString:"SQL node [{}], conn={}] is now in {} mode.", new Object[] { protocol
            .getHostAddress().toString(),
            Long.valueOf(protocol.getServerThreadId()), readOnly ? "read-only" : "write" });
        protocol.executeQuery("SET SESSION TRANSACTION " + (readOnly ? "READ ONLY" : "READ WRITE"));
    }
}
```

图 17

按照笔者过往使用数据库的经验,"set transaction read only"等同于"set session transaction read only",[session] 只是默认值而已(MySQL 就是这样的)。

在 Oracle 模式中做实验测试一下。

14.2.2.1 set transaction read only

具体见表 2。

表 2

时序	Session1	Session2
T1	create table a (id number); insert into a values(1); commit; select * from a; // 返回记录 (1)	
T2		set transaction read only; select * from a; // 返回记录 (1)
T3	insert into a values(222); commit; select * from a; // 返回记录 (1),(222)	
T4		select * from a; // 返回记录 (1)

14.2.2.2 set session transaction read only

具体见表 3。

表 3

时序	Session1	Session2
T1	create table a (id number); insert into a values(1); commit; select * from a; // 返回记录 (1)	
T2		set session transaction read only; select * from a; // 返回记录 (1)
T3	insert into a values(222); commit; select * from a; // 返回记录 (1),(222)	
T4		select * from a; // 返回记录 (1),(222)

通过测试可知，两者的语义和效果是不一样的。虽然都能实现在其作用范围内只读，但从 OceanBase 支持人员处了解到，二者还是有以下区别：

（1）作用范围不同。

① set transaction read only 仅作用于当前事务，一旦该事务结束（Commit 或 Rollback），该设置就失效，会话中后续事务不会继承该设置。

② set session transaction read only 则影响整个会话中的所有事务，一旦设置，会话中接下来的所有事务都会被设置为只读模式，直到会话结束或重新设置为可读写模式。

（2）引用数据库快照的区别。

①仅 set transaction 命令开启的只读事务才能引用数据库的快照（继而通过读取一致性视图以获得可重复读的效果）。

② set seesion transaction 命令无法使只读事务获得快照。

14.3 正确的配置方式

既然 conn.setReadOnly（set session transaction read only）无法实现预期效果，那如何实现 set transaction read only 命令带来的可重复读效果呢？

OceanBase 产研团队提供了一个方法：要达到效果，除了需要设置 conn.setReadOnly(True)，还需在 JDBC option 中添加以下参数。

14.3.1 oracleChangeReadOnlyToRepeatableRead=True

该参数在 oceanbase-client 2.4.7 中引入，目的是实现可重复读（实质是快照读）的效果。应用配置完成后，确实有效果。截至本文发布时，官网查不到该参数的详细说明，我们先结合源码看一下它的实现方式（见图 18）：

图 18

对于 setReadOnly，在满足了 isOracleMode = true 和 oracleChangeReadOnlyToRepeatableRead = true 情况下，将会把会话的隔离级别设置为 setTransactionIsolation(readOnly ? 4 : 2)。

在本例的场景下，readOnly 的值被设置为 True，那么传递给 setTransactionIsolation 方法的值就是 4（见图 19）。

```java
public void setTransactionIsolation(int level) throws SQLException {
    cmdPrologue();
    this.lock.lock();
    try {
        lockLogger.debug(paramString:"AbstractQueryProtocol.setTransactionIsolation locked");

        if (this.transactionIsolationLevel == level) {
            return;
        }
        String query = "SET SESSION TRANSACTION ISOLATION LEVEL";
        switch (level) {
            case 1:
                query = query + " READ UNCOMMITTED";
                break;
            case 2:
                query = query + " READ COMMITTED";
                break;
            case 4:
                query = query + " REPEATABLE READ";
                break;
            case 8:
                query = query + " SERIALIZABLE";
                break;
            default:
                throw new SQLException("Unsupported transaction isolation level");
        }
        executeQuery(query);
        this.transactionIsolationLevel = level;
    } finally {
        this.lock.unlock();
        lockLogger.debug(paramString:"AbstractQueryProtocol.setTransactionIsolation unlocked");
    }
}
```

图 19

由上可知，当值为 4 时，JDBC 将会传递 16.3.2 节所示的 SQL 给后端 OBServer。

14.3.2 SET SESSION TRANSACTION ISOLATION LEVEL REPEATABLE READ

前文提到 Oracle 模式仅支持 RC 和 Serializable，那该命令发到 OceanBase 上的行为是怎样的？继续测试（见表 4）。

表4

时序	Session1	Session2
T1	create table a (id number); insert into a values(1); commit; select * from a;　　// 返回记录 (1)	
T2		SET SESSION TRANSACTION ISOLATION LEVEL REPEATABLE READ; select * from a;　　// 返回记录 (1)
T3	insert into a values(222); commit; select * from a;　　// 返回记录 (1),(222)	
T4		select * from a;　　// 返回记录（1） show variables like 'transaction_isolation'\G // 返回记录 VARIABLE_NAME: transaction_isolation VALUE:REPEATABLE-READ

结果是 OBoracle 可以实现可重复读的效果，且通过客户端命令查询到当前会话被设置为 RR。

14.4 RR 和 Serializable

最后再简单说明一下，官网提到 OceanBase 的 MySQL 模式支持 3 种隔离级别（RC、RR、Serializable），Oracle 模式支持 2 种（RC、Serializable）。但是实际在 OceanBase 数据库中只实现了 2 种隔离级别，即读已提交（RC）和可串行化（Serializable）。

(1) 当用户指定 RR 隔离级别时，实际使用的是 Serializable。也就是说，OceanBase 数据库的 RR 隔离级别更加严格，不会出现幻读的异常情况。

(2) 但在底层实现上，OceanBase 数据库的 Serializable 隔离级别实际上使用的是快照隔离（Snapshot Isolation，SI），不能保证严格的可串行化。

14.5 结论

应用通过 oceanbase-client 驱动访问 OceanBase Oracle 模式数据库时，要想实现 RR 的效果，除了需要设置 setReadOnly 为 True，还需要满足：

(1) oceanbase-client 的版本高于 2.3.8。

(2) JDBC Option 中配置 oracleChangeReadOnlyToRepeatableRead=True。在 OceanBase Oracle 模式数据库中，会话可以被设置为 RR 隔离级别，但只是会话变量显示为 RR，实际底层实现上用的是 SI。

15 ActionDB 扩展 OceanBase GIS 能力：新增 ST_PointN 函数

作者：ActionDB 团队

在江苏省某行政单位的 ActionDB 项目中，由于该系统强依赖于地图，涉及大量坐标处理，而 OceanBase 原生几何属性函数（geometry property function）无法满足需求，需要在地理信息系统（geographic information system，简称 GIS）功能中增加 ST_PointN 函数。

面对这一挑战，ActionDB 技术团队深入解析了 OceanBase 中已有的 GIS 函数，包括从表达式解析到算法调用的完整逻辑链路。为了实现 ST_PointN 函数，技术团队需在现有基础上注册 ST_PointN 相关元信息，并开发地理（GEO）函数的对应实现，确保 ST_PointN 函数的无缝集成与高效运行。

15.1 什么是 ActionDB

ActionDB 作为一款卓越的企业级分布式数据库，其设计核心依托于 OceanBase 的开源内核，辅以爱可生在开源数据库领域的深厚积累与技术专长，荣获原厂的正式授权及内核级技术支持。

ActionDB 集 OceanBase 的稳健性与高性能于一身，更进一步强化了与 MySQL 的兼容性，融合爱可生独有的安全特性与用户友好的运维管理工具，缔造了更高品质、更全面的数据库解决方案。

ActionDB 的 MySQL 8.0 协议具有全面兼容能力，辅以基于 MySQL binlog 的双向复制技术，为业务系统与下游数据平台提供了安全无虞、无缝迁移的完美方案，确保数据迁移的零风险与无感知。

15.2 ST_PointN 函数介绍

ST_PointN 函数用于在给定的几何对象中提取第 N 个点，常用于几何对象分析和地理信息系统（GIS）。ST_PointN 函数接受一个几何对象（如线或多边形）和一个索引 N，返回该几何对象的第 N 个点。该函数的主要作用是帮助用户从复杂的几何对象中提取具体的点,以便进行进一步的地理分析或处理。

常见的应用场景如下:

(1) 道路和路线分析:分析交通路线时,提取路线中的特定点进行详细分析或优化。

(2) 环境监测:用于从多边形或线形区域中提取监测点,从而对环境数据进行更精确的分析。

(3) 城市规划:从复杂的多边形中提取特定点以帮助规划人员做出决策。

(4) 导航系统:通过提取路径中的关键点,优化导航指引和路径规划。

15.3 函数功能实现

15.3.1 如何添加注册信息?

添加 ST_PointN 函数注册信息时,需要在 OceanBase 已有的 GIS 相应信息后追加内容。

(1) 定义函数名:在头文件 ob_name_def.h 中添加函数名称,以便在其他地方使用和引用。

(2) 定义函数 ID:为函数分配一个唯一的 ID,每个 GIS 系统函数(T_FUN_SYS_ST_xx)的 ID 不重复。

(3) 添加序列化信息:所有的函数与表达式需要在 ObExpr::EvalFunc g_expr_eval_functions 注册,用于函数信息的序列化。

(4) 注册函数实现类:在工厂方法 register_expr_operators 中注册函数实现类,使其可在执行 SQL 查询时被识别并使用。

(5) 添加函数与 GIS 算法的对应信息:由 ob_geo_func_register 维护函数与 GIS 算法的对应关系,添加相应信息(见图 20)。

```
template <>
struct ObGeoFunc<ObGeoFuncType::PointN>
{
    typedef ObGeoFuncPointN gis_func;
};
```

图 20

15.3.2 如何实现 ST_PointN 函数?

ST-PointN 函数功能的实现包含两部分:

(1) ObExprSTPointN 类。表达式处理的入口,负责 GIS 算法参数的有效性检查,并准备 GIS 上下文。

在 SQL 表达式的执行阶段，位于 src/sql/engine/expr 目录下的各模块负责处理各种不同类型的 SQL 表达式（包括 GIS 函数）。ObExprSTPointN 类主要在算法执行前检查输入参数的有效性，并准备好执行 GIS 算法所需的上下文环境。

(2)ObExprSTPointN 算法实现。一旦 ObExprSTPointN 类完成参数检查后，它将调用 ObGeoFunc::geo_func::eval 方法。这会触发模板分发逻辑，最终执行具体的算法模块 ObGeoFuncPointNImpl，实现 ST_PointN 的功能。

15.3.2.1 添加函数实现类

代码如图 21 所示。

```
class ObExprSTPointN : public ObFuncExprOperator
{
public:
  explicit ObExprSTPointN(common::ObIAllocator &alloc);
  virtual ~ObExprSTPointN();
  virtual int calc_result_type1(ObExprResType &type,
                                ObExprResType &type1,
                                ObExprResType &type2,
                                common::ObExprTypeCtx &type_ctx) const override;
  static int eval_st_PointN(const ObExpr &expr, ObEvalCtx &ctx, ObDatum &res);
  virtual int cg_expr(ObExprCGCtx &expr_cg_ctx,
                      const ObRawExpr &raw_expr,
                      ObExpr &rt_expr) const override;
private:
  DISALLOW_COPY_AND_ASSIGN(ObExprSTPointN);
...
};
```

图 21

添加接口说明：

(1)calc_result_type1：检查传递的参数类型和数量是否正确。calc_result_type 族函数以后缀识别不同的参数个数，如 calc_result_type1 为一个函数参数。

(2)eval_st_PointN：检查所生成的 GIS 对象是否合法，检查 GIS 元数据与 GIS 上下文。调用 ObGeoTypeUtil 工具类的各个检查接口。设置 GIS 函数执行所需的 GIS 对象与 GIS 上下文（gis_ctx），并检查 GIS 相关对象是否有效。验证完成后，最后会调用 ObGeoFunc<ObGeoFuncType::PointN>::geo_func::eval(gis_context, result) 进入模板函数的分发逻辑，最终选择 ObGeoFuncPointNImpl 模块中匹配的 eval 函数。

15.3.2.2 添加算法实现

实现 ponitN 的 GIS 处理的核心逻辑。

(1) 注册支持的入参 GIS 子类型（如 linestring 和 multi_point）（见图 22）。

```
// geometrycollection
OB_GEO_CART_BINARY_FUNC_BEGIN(ObGeoFuncPointNImpl, ObWkbGeomCollection, ObWkbGeomCollection, bool)
OB_GEO_CART_BINARY_FUNC_GEO1_BEGIN(ObGeoFuncPointNImpl, ObWkbGeomCollection, bool)

// cases use disjoint (multi point cases)
OB_GEO_CART_BINARY_FUNC_BEGIN(ObGeoFuncPointNImpl, ObWkbGeomMultiPoint, ObWkbGeomPolygon, bool)
OB_GEO_CART_BINARY_FUNC_BEGIN(ObGeoFuncPointNImpl, ObWkbGeomMultiPoint, ObWkbGeomMultiPolygon, bool)
OB_GEO_CART_BINARY_FUNC_BEGIN(ObGeoFuncPointNImpl, ObWkbGeomPolygon, ObWkbGeomMultiPoint, bool)
```

图 22

(2) 定义 eval 族函数（见图 23）。

```
// ----- ObGeoFuncPointNImpl -----
class ObGeoFuncPointNImpl : public ObIGeoDispatcher<bool, ObGeoFuncPointNImpl>
{
public:
  ObGeoFuncPointNImpl();
  virtual ~ObGeoFuncPointNImpl() = default;

  // template for unary
  OB_GEO_UNARY_FUNC_DEFAULT(bool, OB_ERR_GIS_INVALID_DATA);
  OB_GEO_TREE_UNARY_FUNC_DEFAULT(bool, OB_ERR_GIS_INVALID_DATA);
  OB_GEO_CART_TREE_FUNC_DEFAULT(bool, OB_ERR_NOT_IMPLEMENTED_FOR_CARTESIAN_SRS);
  OB_GEO_GEOG_TREE_FUNC_DEFAULT(bool, OB_ERR_NOT_IMPLEMENTED_FOR_GEOGRAPHIC_SRS);

  // template for binary
  template <typename GeoType1, typename GeoType2>
  struct EvalWkbBi ...

  // default case for geography (calc using nonpoint_strategy)
  template <typename GeoType1, typename GeoType2>
  struct EvalWkbBiGeog ...

private:
  // geometry collection
  template <typename CollectonType>
  static int eval_pointN_geometry_collection(const ObGeometry *g1,
                                             const ObGeoEvalCtx &context,
                                             bool &result)
```

图 23

(3) 自定义的 pointN 函数。

Boost.Geometry 没有直接提供 pointN 函数，但可以通过访问几何体的内部结构实现类似功能。此方法适用于大多数 Boost.Geometry 提供的几何类型（如 linestring 和 multi_point）。

15.4 方案优势

15.4.1 量身定制，满足特定需求

通过添加自定义 GIS 函数，如 ST_PointN，可以根据具体应用需求量身定制数据库功能，确保数据库能够精准地满足业务需求，提高数据处理的灵活性和效率。

15.4.2 提高性能和效率

自定义 GIS 函数可以根据实际应用场景进行性能优化，比通用的开源解决方案更高效。这样可以减少复杂 GIS 操作的计算时间，提高查询性能。

15.4.3 增强系统可扩展性

自定义函数为系统增加了新的功能模块，使得 OceanBase 更加多样化和强大，能够支持更广泛的数据操作需求和业务场景。

15.4.4 提升用户体验

用户可以直接在数据库中定制函数，替代复杂的业务 GIS 操作，而不需要借助外部工具或进行额外的开发工作，简化了开发流程，提高了用户体验和开发效率。

15.4.5 数据安全性和一致性

自定义 GIS 函数在数据库内部实现，减少了数据传输过程中的安全隐患，保证了数据处理的一致性和安全性。

15.4.6 适应快速变化的业务需求

自定义函数使得数据库可以快速响应和适应不断变化的业务需求，无需等待开源社区的更新或第三方工具的升级，显著提高了开发和部署速度。

15.5 总结

我们在 OceanBase 中添加并实现了 ST_PointN 函数，具体步骤包括定义函数名和 ID、注册函数、参数类型检查、执行上下文设置、模板函数分发和具体算法实现等。

最终，通过调用自定义的 pointN 函数，实现了 OceanBase 中 GIS 功能的扩展。

ST_PointN 函数的实现，进一步丰富了 ActionDB 的 GIS 功能矩阵，提升了其在地理信息处理领域的应用价值与竞争力。ActionDB 本着依托开源回馈开源的初衷，本文讨论的 GIS 功能已经提交给 OceanBase 社区版，相信在下一个版本中就可以使用了。

16 ActionOMS 具备的延迟智能诊断功能

作者：`ActionOMS 团队`

16.1 案例背景

某客户需要将 Oracle 的数据同步到 OceanBase（MySQL 模式），涉及 OceanBase 的数据同步/迁移的工作就需要用到 OMS 了。

OceanBase 数据迁移服务（OceanBase migration service，OMS）是 OceanBase 数据库一站式数据传输和同步的产品。它支持多种关系型数据库、消息队列与 OceanBase 数据库之间的数据复制，是集数据迁移、实时数据同步和增量数据订阅于一体的数据传输服务，OMS 帮助低风险、低成本、高效率地实现 OceanBase 的数据流通，助力构建安全、稳定、高效的数据复制架构。

16.1.1 数据同步需要注意些什么

首先，保证数据不丢失、两端一致。其次，要提高数据同步的效率。

这样对于业务侧来说，目标端数据库的数据一直是最新的，数据同步的意义也就在这里。

16.1.2 数据同步问题诊断痛点

目前，OMS 在数据同步时如果出现性能波动或故障，只能展示同步的整体延迟和迁移流量、PRS 等指标值。如果要进一步定位延迟增加、数据同步阻塞等情况的具体原因，则需要用户在 OMS 服务器上手动执行诊断命令。

手动执行诊断命令的过程较为烦琐，展示效果不够直观，同时问题处理建议也常常缺乏足够精确的数据支撑。

16.2 ActionOMS

ActionOMS 是基于 OMS 本身的强大能力，结合爱可生公司在数据库及周边工具上多年的开发经验、对数据迁移/同步过程的深刻理解与丰富运维经验而推出的定制化版本。

16.2.1 官方授权

ActionOMS 是由 OceanBase 向爱可生进行了全部代码授权，可对 OMS 问题进行源码解释并修复，同时可以接受定制化开发的 OMS 版本。

16.2.2 版本介绍

2024 年 7 月中旬，ActionOMS 发布了 4.24.07.0 版本。该版本新增了智能性能诊断和智能故障诊断功能，采用量化推导，自顶向下（Top-Down）的思路，从进程开始，逐层深入，细化到具体工作线程、缓存队列等内部机制，最终定位性能问题。

智能诊断功能不需要手动介入，自动采集各项性能指标，从系统层面分析指标异常值，最终展现给用户精准的延迟故障点以及对应的调整方案。

16.3 实践案例

客户要求对数据同步过程中影响性能指标的原因进行排查，根据全面细致的监控数据给出延迟问题处理方案。

为了满足客户需求，ActionOMS 重塑了延迟诊断逻辑，支持自动化采集各组件的性能指标，并结合 SRE 常用的诊断思路，提供系统性、精准的诊断结果和优化方案，使用户在页面上能以更直观的方式了解问题原因并进行调整（见图 24）。

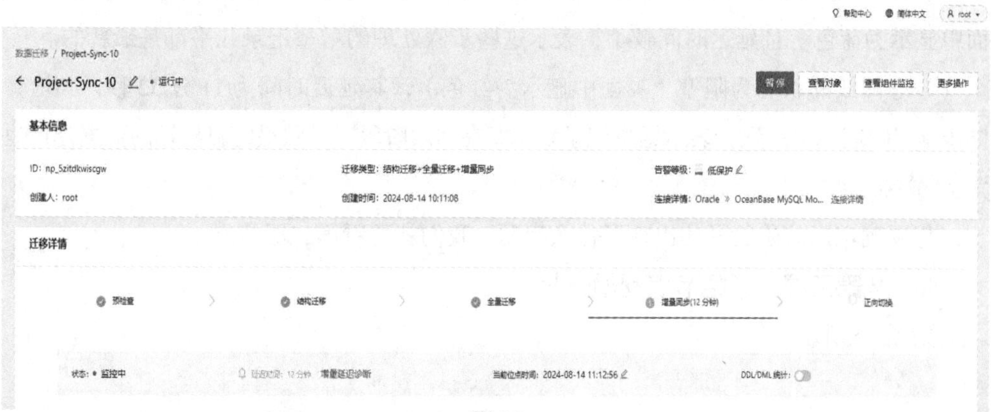

图 24

针对图 24 中出现的增量同步达到了 12 分钟，并且延迟时间还在一直增加的情况，通过新增的增量延迟诊断功能，我们可以看到数据同步过程中关键节点的延迟情况、出现问题的节点以及对应的处理建议（见图 25）。

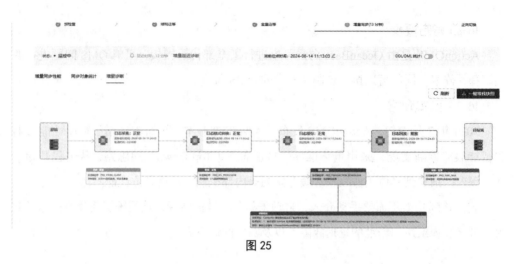

图 25

结构说明如下。

16.3.1 第一层：两端同步关键节点

第一层展示的是两端数据同步中比较关键的步骤及其对应的组件或进程。

显示当前节点处理增量记录的两个时间：

(1) 延迟时间 = 当前时间 – 节点处理的最新增量记录的时间。

(2) 最新的指标时间：当前诊断所依据的指标的生成时间。

从图 24 中可知，日志采集、日志格式转换、日志缓存等节点的状态都为正常（界面中显示为蓝色）且延迟时间较小，表示这些步骤处理的增量记录几乎都是最新的。

日志回放的状态为阻塞（界面中显示为红色），其延迟时间为 11 分 52 秒，和增量同步延迟时间基本一致，表示增量记录都阻塞在回放阶段。导致最终目标端的数据延迟也为 12 分钟左右。

想要确定回放阶段具体出现了什么问题，我们就要看第二层节点。

16.3.2 第二层：红色节点问题分析

如图 26 所示。

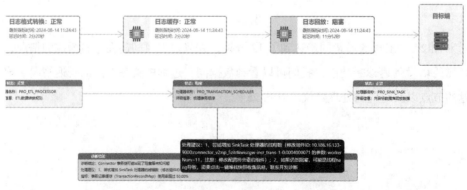

图 26

第二层展示的是第一层中状态为阻塞且标记为红色的子节点（第一层中不同的步骤会包含不同的第二层节点）。

该层显示问题节点工作处理器信息，包括：当前处理器名称、作用，以及状态信息，当状态为阻塞时，显示阻塞信息。

第一层日志回放出现了阻塞，那么第二层显示了日志回放步骤中更为具体的流程：store 接收 & 拼接事务→ ETL 数据转换→梳理事务顺序→向目标端回放数据。

其中，梳理事务顺序的节点状态为阻塞（该节点负责梳理增量记录中的事务关系，对于不相关事务可以并行回放，而相关事务由于彼此之间的依赖性，目前只能顺序回放）。

诊断结论显示，事务链可能出现了阻塞等未知问题，依据的指标为事务记录缓存的使用率超过 50%，对应的处理建议也给出了需要调整的参数和值。

16.4 处理效果

ActionOMS 针对系统给到的处理建议调整参数后，可以看到迁移流量和 RPS 均得到 2~3 倍提升（见图 27）。

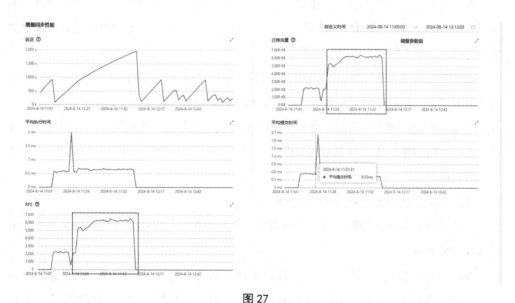

图 27

在 12 点左右，增量数据完成同步并且延迟回落（见图 28）。

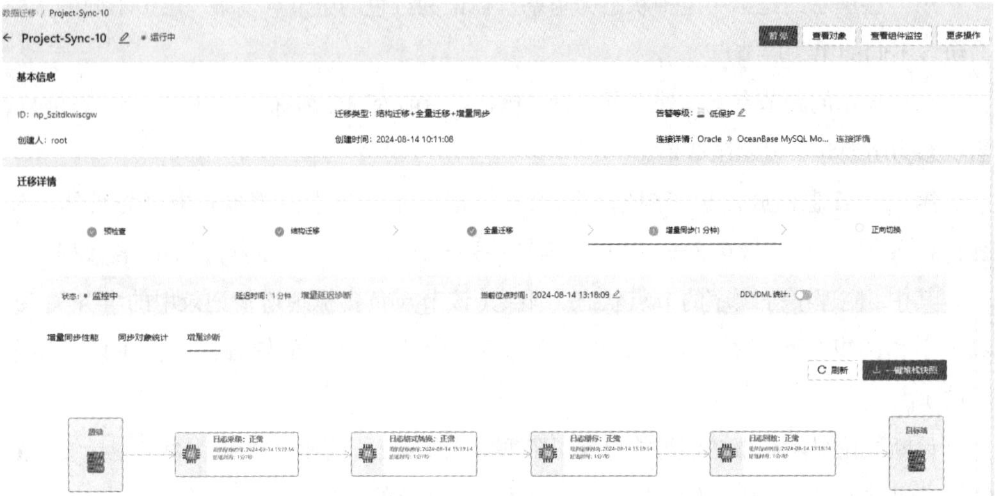

图 28

16.5 功能详解

16.5.1 自动化采集各项性能指标

端到端数据同步需求看起来很简单，不就是从源端库把数据拿到，再回放到目标端库吗？实际上实现起来还是相对烦琐的。其中包含如下过程：①源端数据抽取；②数据解析转换；③数据缓存；④目标端回放。

每个过程可能由不同的进程来负责，而每个进程又包含了大量的工作线程和缓存等。每个环节出现问题都有可能导致延迟增加、目标端数据不更新。

对于故障诊断来说，首先必须实现的就是采集同步过程中涉及的所有工作线程、缓存等各项性能指标，这一点也是后续诊断的基石，没有全部的性能指标，也就无法用全局视角看待问题。

16.5.2 从系统层面分析指标异常值

在全面获取各项性能指标后，往往需要结合故障相关的方法论才能得到更可靠的结论以及调整方案。

ActionOMS 依托爱可生公司多年在数据库及周边工具一线运维工作中得到的方法论，结合自顶向下（Top-Down）的诊断思路，从整体上衡量工作线程的利用率，通过分析如 _thread_used_1m（每分钟工作线程使用率）、_thread_rps_1min（每分钟工作线程吞吐量）、_queue_depth（队列深度）、_thread_idling_1min（每分钟线程空转率）等关键指标，识别是否存在性能问题。

在特定场景下，借助这些指标识别问题，比如：

(1) 线程调度是否均匀？程序会自动计算同组内指标值的离散值用于反映调度问题。

(2) 线程是否存在卡住的情况？通过线程使用率和吞吐量指标结合分析的方式，对于线程使用率相较于 CPU 使用率计算方式做了扩展，可识别线程虽然让出，但对于程序来说仍然卡住的情况。

(3) 在 ORACLE-RAC 中多实例中流量不均衡会导致延迟升高？由于每个实例都会独立产生 REDO LOG，ActionOMS 为了保持增量变更的有序性，对 RAC 架构这种多日志流来源场景的事务结果会进行合并排序。通过合并排序的线程空转率来判定是否存在因流量不均匀带来的延迟高问题。

16.5.3 给用户展现精准的延迟故障点

在数据同步过程中，一个节点出现问题，可能会导致上游节点的指标值都出现异常，对于诊断系统来说，溯源同等重要。

ActionOMS 采用自右到左思路，往往最下游出现的问题才是关键问题，首先暴露下游问题，解决后再重新诊断，逐步解决所有问题（见图 29）。

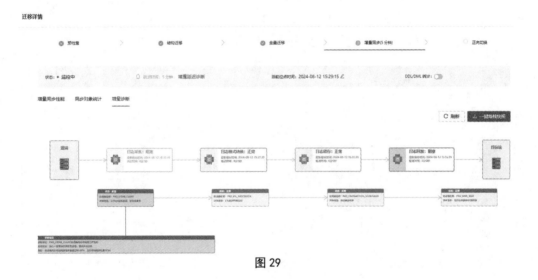

图 29

16.5.4 补充说明

考虑到现场场景的复杂性，现有的诊断方法论可能覆盖不了所有情况。为了能更方便地收集现场信息交由研发人工诊断，ActionOMS 在页面提供"一键堆栈快照"按钮，可一键收集增量过程中全部进程堆栈、火焰图等信息。后续诊断的方法论也会逐渐丰富。

为了满足自动采集以及诊断的需求，同时保证不会对被采集和诊断的系统造成明显的性能损耗，在指标采集和指标诊断的过程中，默认选择每分钟采集一次的频率并采用异步方式处理整个过程。

16.6 总结

对于一个数据同步系统来说，随着业务的变化，业务的数据流量特征也会随之变化。没有人能保证数据同步能在各种特征的流量中都平稳进行。

提供自动的诊断功能是为了实时监控和分析系统的各项性能指标，及时识别和处理潜在问题，确保系统在面对复杂、多变的业务流量时仍能稳定运行。它不仅能帮助系统管理员快速定位问题源头，还能提供相应的调整建议，以便在问题发生时迅速采取措施，避免数据同步延迟对业务造成更大的影响。

ActionOMS 不仅提升了数据同步系统的健壮性和可靠性，还大大降低了运维人员的工作负担，避免了因为未知问题导致的长时间排查过程，使得整个数据同步过程更加高效、稳定。

04 一问一实验（ChatDBA）篇

"一问一实验"是爱可生开源社区的王牌专栏。每一期通过一个数据库问题对应一组实验的形式，简明易懂地讲解数据库基础知识。自 2020 年连载至今已发布 50 篇，我们在前几册《大智小技》中有部分刊载。专栏问世以来，深受行业读者们的好评！真正成为了 DBA 们爱读、好读的技术内容。

2024 年 5 月 1 日，"一问一实验"全新归来，以下是对专栏作者的采访。

问题：为什么会断更这么久？

黄炎：冠冕堂皇地说，就是被其他更好玩的事情吸引了注意力。大模型的技术突破在技术生产力上提供了巨大的可能性，对这些可能性的探索占用了非常多的时间。

所幸我们找到了将 AI 和专栏合并发展的方式：我们在公司内部建立了 ChatDBA（给 DBA 用的智能辅助系统，负责辅助 DBA 进行故障诊断处理等工作）来协助日常工作；从第 51 期开始，专栏的主要作者更换成了 ChatDBA，并由人类专家对专栏质量进行点评。

问题：会给读者们带来什么新的体验？

黄炎：ChatDBA 使用了多种常见和不常见的技术（它不单纯是 LLM+RAG），这个复杂的结构主要想让 ChatDBA 脱离"搜索引擎"这个功能层次。在故障诊断处理中能进行引导/指导，对知识的使用更精确，思考模式更接近人类。我们希望由 AI 续写的专栏跟人类专家写的专栏没有差异（包括技术逻辑的正确性、知识内容的详略程度、接近人类的程度等方面）。

问题：这次的 AI 版（ChatDBA）和之前的版本有什么区别？

黄炎：随着 ChatDBA 后续的迭代，专栏的内容会逐渐产生一些变化。比如，单纯的故障诊断→深度的原理解析；单纯的数据库级别的分析→整合操作系统和网络等多方面知识；单纯的事实推理→技术猜想＋验证；单纯的解决问题→形成工具；短链推理→长链推理；现象推理→源码解释……

这些变化都会依赖于大模型本身的逻辑能力提高（希望国内外的大模型公司推出技术逻辑性更强的模型）以及 ChatDBA 的架构进化。

希望在专栏中向大家提供更多好玩的知识。

本篇共收录了 8 篇文章，通过 8 个不同的问题，从不同的使用场景出发，展现 ChatDBA 的各方面能力。

读者可访问爱可生开源社区视频号搜索相关内容，查看每篇文章对应的操作视频（见图 1）。

图 1

1 MySQL 日志报错定位

1.1 问题

在 MySQL 日志中发现有大量报错，可能是什么原因造成的？

报错信息：

```
    2024-04-26 23:27:06 [ERROR] /data/3306/base/5.7.26/bin/mysqld: The table '/data/3306/tmp/#sql_3d157_11' is full
    2024-04-27 00:11:04 [ERROR] /data/3306/base/5.7.26/bin/mysqld: The table '/data/3306/tmp/#sql_3d157_1' is full
```

1.2 实验

1.2.1 将问题丢给 ChatDBA

问题如图 2 所示。

图 2

有了 ChatDBA 加持，遇到问题就先别自己费劲了，扔给 ChatDBA 先看看它怎么说。

图 3 左侧为流程分析画布，展示 ChatDBA 对此问题的排查逻辑；右侧为互动区域。

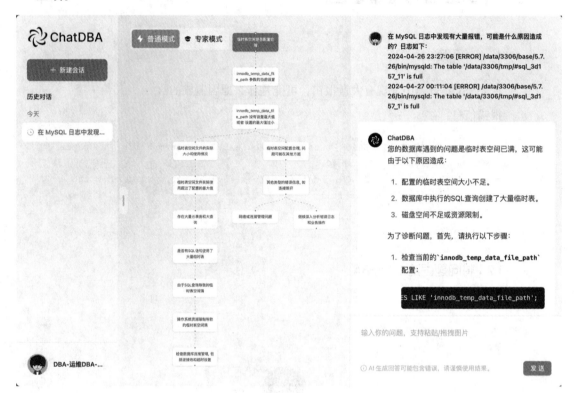

图 3

从图 4 左边的画布区域和右侧的对话区域可知，造成该问题的原因可能有以下几点：
- 配置的临时表空间大小不足。
- 数据库中执行的 SQL 查询创建了大量临时表。
- 磁盘空间不足或资源限制。

> 在 MySQL 日志中发现有大量报错，可能是什么原因造成的？日志如下：
> 2024-04-26 23:27:06 [ERROR] /data/3306/base/5.7.26/bin/mysqld: The table '/data/3306/tmp/#sql_3d157_11' is full
> 2024-04-27 00:11:04 [ERROR] /data/3306/base/5.7.26/bin/mysqld: The table '/data/3306/tmp/#sql_3d157_1' is full

ChatDBA

> 您的数据库遇到的问题是临时表空间已满，这可能由于以下原因造成：

1. 配置的临时表空间大小不足。
2. 数据库中执行的SQL查询创建了大量临时表。
3. 磁盘空间不足或资源限制。

为了诊断问题，首先，请执行以下步骤：

1. 检查当前的 `innodb_temp_data_file_path` 配置：

   ```
   SHOW VARIABLES LIKE 'innodb_temp_da
   ```

2. 查看 `/data/3306/tmp/` 目录的磁盘使用情况：

   ```
   df -h /data/3306/tmp/
   ```

3. 提供最近的慢查询日志，以便分析可能的资源消耗大的查询。

根据返回的信息，我们可以采取相应的解决方案：

1. 如果配置的临时表空间过小，可以通过修改 `my.cnf` 配置文件来增加大小，例如：

   ```
   service mysql stop
   vi /etc/mysql/my.cnf
   # 添加或更新以下配置
   innodb_temp_data_file_path=ibtmp1:
   service mysql start
   ```

2. 如果是由于特定查询导致的临时表过多，需要优化相关SQL，例如使用 `EXPLAIN` 和 `SHOW CREATE TABLE` 来分析，并调整查询或索引设计。

请先执行上述步骤，然后根据结果进行相应的调整或提供更多信息，以便进一步分析和解决。

图 4

人类专家评论区

点评：AI 识别了日志，明确知道是临时表空间的问题，而不是普通表空间的问题。

接下来，我们根据 ChatDBA 推荐的执行步骤查看一下相关配置。

1.2.2 ChatDBA 协助问题排查

首先，根据推荐的步骤检查参数的配置情况如何（见图 5）。

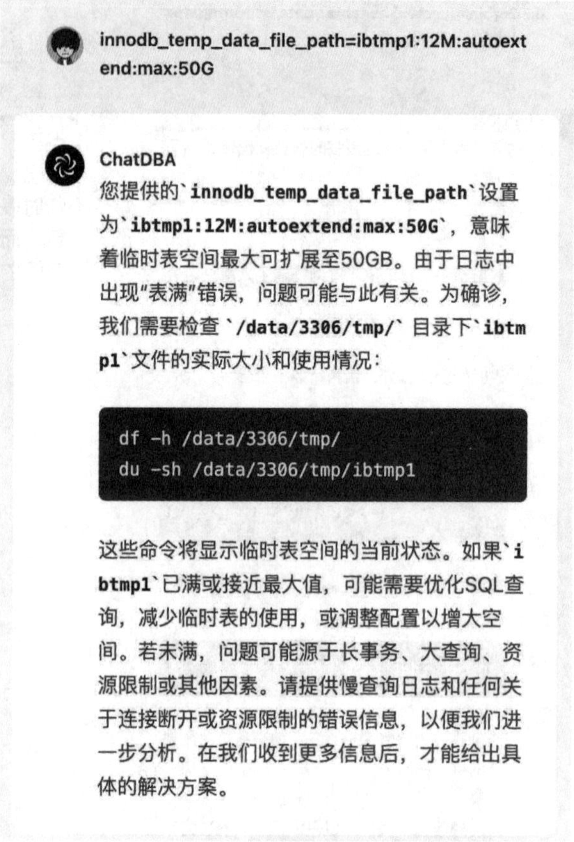

图 5

接下来，把从系统中查询到的配置信息提供给 ChatDBA。

在这里，ChatDBA 意识到日志表满的报错可能与这个配置有关，但是还不能确定，所以希望我们提供对应的临时表实际使用情况。执行 ChatDBA 给出的两条命令后可知，MySQL 的临时表空间（ibtmp1）实际已经占用了 50GB（见图 6）。

04 一问一实验（ChatDBA）篇

ibtmp1已经使用了50G

ChatDBA
问题的关键在于临时表空间ibtmp1已达到其配置的最大值50G，导致数据库无法创建更多临时表，从而触发"表满"错误。以下是问题的诊断和解决方案：

1. **问题诊断**：
   - 临时表空间满是因为实际使用达到了配置的最大限制50G。
   - 错误日志中的"表满"报错与临时表空间不足有关。
   - 大量的长事务和大查询可能是导致临时表大量使用的元凶。

2. **收集信息**：
   - 获取慢查询日志：`tail -n 100 /var/log/mysql/slow_query_log.log`
   - 检查运行的事务：`SHOW FULL PROCESSLIST`
   - 连接池状态：`SHOW GLOBAL STATUS LIKE 'Aborted%';` 和 `SHOW VARIABLES LIKE 'wait_timeout';`
   - 长时间未关闭的连接：`SELECT * FROM information_schema.processlist WHERE Time > 300;`

3. **解决方案**：
   - **优化SQL**：分析慢查询日志，优化占用资源的查询，添加合适索引，调整应用程序代码。
   - **扩展临时表空间**：增加`innodb_temp_data_file_path`配置，如设置为`ibtmp1:12M:autoextend:max:100G`，然后重启MySQL服务。
   - **监控和管理事务**：检查并结束长事务，避免资源过度消耗。
   - **优化连接管理**：检查连接池配置，确保不会出现资源耗尽的情况。

请注意，临时增加表空间大小是短期对策，长期解决方案应侧重于优化SQL查询和事务管理。在执行任何更改前，请确保有数据备份，以防不测。如果您需要进一步帮助，比如分析慢查询日志，我将乐意提供协助。

人类专家评论区

点评：这里更好的诊断方式，是找到"占用了大量临时表空间的长事务"。

图 6

347

可以看到，ChatDBA 已经基本确认问题是参数配置导致的，但是同时也指出长事务、长查询才有可能是导致问题的真正原因。

最后，ChatDBA 给出的解决方案是优化对应的 SQL 语句，但是在当前的业务场景下无法直接修改 SQL，所以我们选择通过修改参数的方式来临时处理。

1.2.3 ChatDBA 给出解决方案

ChatDBA 提供了扩展临时表空间的具体步骤，调整 innodb_temp_data_file_path 参数的配置（见图 7）。

图 7

1.2.4 实验总结

```
2024-04-26 23:27:06 [ERROR] /data/3306/base/5.7.26/bin/mysqld: The table '/data/3306/tmp/#sql_3d157_11' is full
```

针对上述报错的问题，我们看到了三个排查方向：参数配置过小；磁盘空间不足；SQL 语句创建了大量的临时表。

ChatDBA 给出的解决方向是先看配置情况（在我们日常工作中也可以参考这个案例的排查思路），先用查询的方法拿到配置信息快速定位问题，然后通过调整参数的方法临时解决问题，最后才是深度追踪问题的根源。

在这个案例中，因为用户的业务场景中有大量的长事务、长查询，所以导致对临时表空间占用率很高,需要配合业务方针对性地优化 SQL 语句才是解决此问题的最终办法。

1.3 问问 ChatGPT-4o

我们也将相同的问题发给了 ChatGPT-4o，让我们看看效果如何（见图 8）。

点评：
1. ChatGPT 给出的问题可能原因较为全面。

图 8（a）

- 如果使用的是 InnoDB 存储引擎，InnoDB 也会使用临时表空间。如果临时表空间不足，可以通过增加 `innodb_temp_data_file_path` 配置来扩展临时表空间。

5. 系统资源限制：
- 检查操作系统对单个进程的文件打开数或其他资源限制，确保 MySQL 有足够的系统资源可用。

> 2. 在未确定原因的情况下直接给出了操作步骤会导致用户不知道该如何进行下一步操作。

要解决这个问题，可以按照以下步骤进行排查和处理：

1. 检查磁盘空间：

如果磁盘空间不足，可以删除不必要的文件或扩展磁盘容量。

2. 增加 MySQL 临时表大小限制，在 MySQL 配置文件（通常是 `my.cnf` 或 `my.ini`）中添加或修改以下配置：

修改后重启 MySQL 服务以使配置生效。

3. 如果需要调整 InnoDB 临时表空间，可以在 MySQL 配置文件中添加或修改以下配置：

修改后重启 MySQL 服务以使配置生效。

通过上述步骤，你应该能够排查并解决 MySQL 日志中出现的 `The table is full` 错误。如果问题依然存在，可以提供更多的日志信息或系统配置细节，以便进行进一步的分析。

> 磁盘空间足够。在我的配置中tmp_table_size=32M，max_heap_table_size=32M，innodb_temp_data_file_path = ibtmp1:12M:autoextend:max:50G

如果磁盘空间足够且你的配置已经设置得比较合理，那么以下几个方面可能是导致问题的原因和可能的解决方法：

1. 临时表使用过多：
- 尽管你的 `tmp_table_size` 和 `max_heap_table_size` 已经设置为 32M，但某些查询可能会创建大量的临时表，导致内存中的临时表转移到磁盘上的临时表。这时即使磁盘空间足够，临时表的大小限制仍然会引发问题。
- 考虑进一步增大 `tmp_table_size` 和 `max_heap_table_size`，例如设置为 128M 或更大，根据你的内存容量调整：

> 点评：
> 3. 在用户提供了系统信息的情况下，ChatGPT并没有理解用户的输入，反而认为相关配置较为合理。

2. 大数据量操作：
- 检查是否有大量数据操作（例如复杂查询、排序、大量数据导入/导出等），这些操作可能会生成非常大的临时表。
- 尝试优化这些查询，减少临时表的生成。使用索引、分片查询等方式优化性能。

3. 配置 innodb_temp_data_file_path：

图 8（b）

- 虽然你的配置已经允许最大 50G 的临时表空间，但某些特定的操作可能会导致临时表空间使用过快。
- 监控实际的 InnoDB 临时表空间使用情况，可以通过以下查询查看当前使用情况：

4. 系统资源限制：

- 检查操作系统的资源限制，确保 MySQL 有足够的资源。你可以使用以下命令查看当前进程的资源限制：

5. 文件系统问题：

- 检查 MySQL 的临时目录所在的文件系统是否存在问题，例如，某些文件系统的碎片化可能导致空间利用效率低。
- 可以尝试将 MySQL 的临时目录（`tmpdir`）设置到另一个性能更好的磁盘或分区。

6. MySQL 版本问题：

- 确保使用的是 MySQL 的最新稳定版本，因为某些版本可能存在已知的 bug 或性能问题。
- 如果有可能，升级 MySQL 到最新的 5.7 版本或 8.0 版本。

调整配置后，重启 MySQL 服务以使新的配置生效：

通过上述步骤的排查和调整，你应该能找到并解决 MySQL 临时表满的问题。如果问题依然存在，可以提供更多的查询和操作细节，以便进行进一步的分析。

图 8（c）

可以看到这类通用的大语言模型偏好是给用户一个大而全的内容，但是缺少具体的操作指向性与问题理解能力，导致用户在输入信息后没有办法确定下一步应该做什么，从而真正解决问题。

2 MySQL 半同步复制频繁报错

2.1 问题：MySQL 的 error log 频繁报错

在 DBA 日常工作场景中，经常会遇到业务方的如下求助（见图 9）：

图 9

从图 10 中可以看到错误日志里面出现了很多的报错信息:

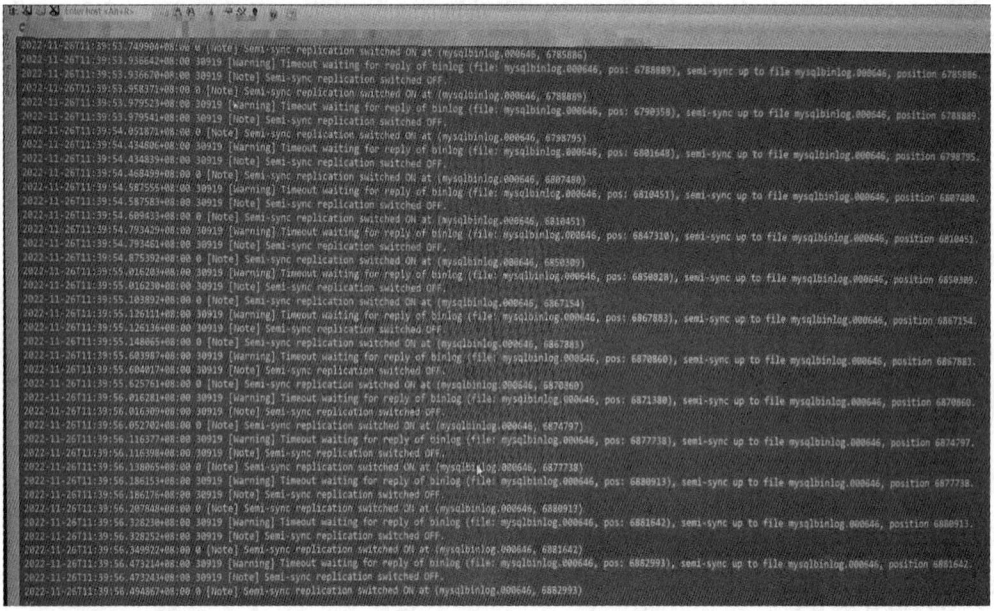

图 10

2.2 实验

2.2.1 将问题丢给 ChatDBA

问题如图 11 所示。

图 11

我们先把这个问题丢给 ChatDBA，让它看一下具体出了什么问题（见图 12）。

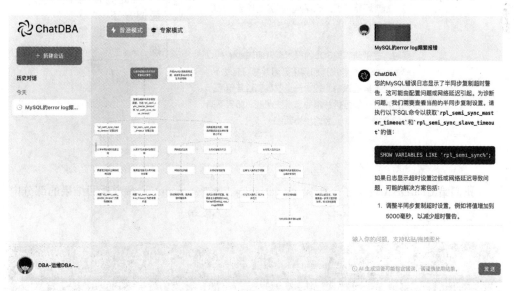

图 12

从左侧的问题逻辑图可以看到，ChatDBA 通过分析图片中的报错信息，认为最有可能的原因是参数配置的问题，其次是网络延迟或者写入压力比较大。

2.2.2 ChatDBA 协助问题排查

根据对应的排查逻辑，ChatDBA 首先希望获取对应的参数配置（见图 13）。

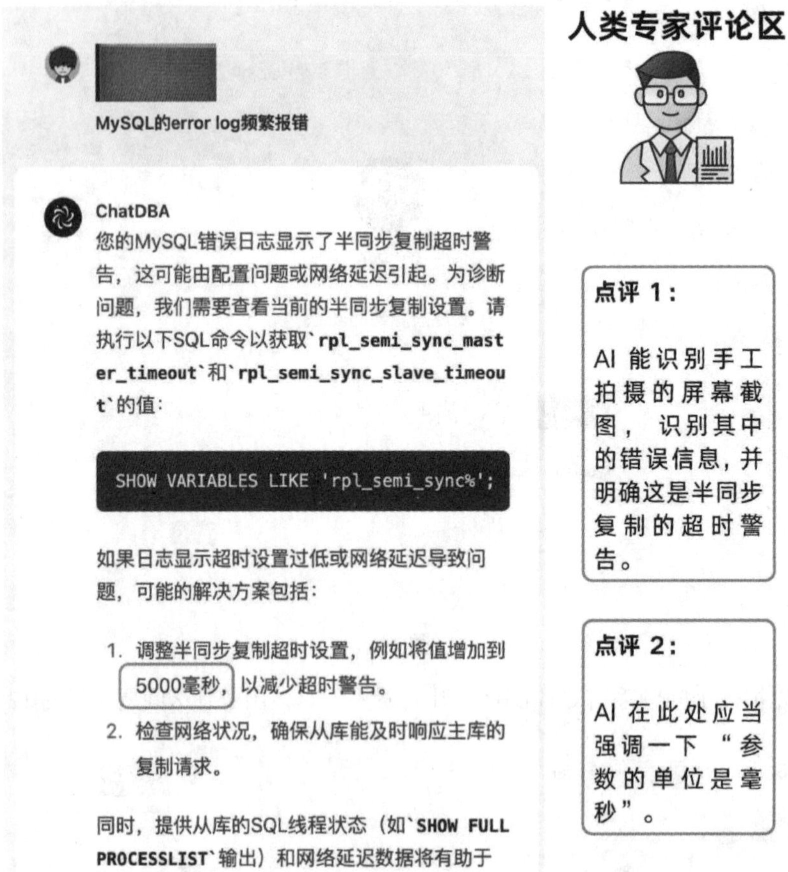

图 13

我们给 ChatDBA 发送了一张 show variables like '%timeout%' 命令返回结果的部分内容截图（见图 14）。

图 14

2.2.3 ChatDBA 给出解决方案

解决方案如图 15 所示。

人类专家评论区

SHOW VARIABLES LIKE 'rpl_semi_sync%';的输出如图

ChatDBA
根据您的问题，数据库出现半同步复制超时警告可能是由于`rpl_semi_sync_master_timeout`设置过低，其值仅为10，这导致主库等待从库确认的时间非常短暂。为了确认问题并提供最佳解决方案，我们需要以下关键信息：

1. 从库的`rpl_semi_sync_slave_timeout`值，以检查两边的超时设置是否匹配：

   `SHOW VARIABLES LIKE 'rpl_semi_sync`

2. 从库的SQL线程状态，以评估其处理复制请求的能力：

   `SHOW FULL PROCESSLIST;`

3. 主从服务器之间的网络延迟数据，以判断是否存在网络问题：

   `使用ping命令检查主从服务器之间的往返时间`

在等待这些信息的同时，您可以先尝试调整`rpl_semi_sync_master_timeout`的值，例如设置为5000毫秒，以减少超时警告：

`SET GLOBAL rpl_semi_sync_master_timeout`

同时，也建议将`rpl_semi_sync_slave_timeout`设置为相同的值，以保持一致。调整后，观察一段时间，看是否改善了情况。如果问题持续，可能需要进一步调查网络延迟或从库性能问题。如果您需要帮助执行上述步骤或解释结果，请告知。

点评 3：

通过"SHOW FULL PROCESSLIST"很难评估从库对复制的处理能力。
此处应当增加AI这方面的知识。

点评 4：

AI 已经意识到参数设置值极低，应当是对参数的时间单位理解有误。
此处应当将调整参数作为首要的解决方案。但AI对此处的处理，应该是高估了人类对参数的理解。

图 15

ChatDBA 识别出其中超时参数的配置是 10ms，这个值对于此场景而言过小了，所以需要调整为更大的值，这个小问题就顺利解决了。

2.2.4 实验总结

从日志的报错信息可以看到，MySQL 半同步复制在不断切换状态，并且多次遇到等待 binlog 回复超时的情况，此问题大概的原因有以下几种。

（1）配置问题

原因：MySQL 的半同步复制配置不当，导致超时设置过短。

解决方法：检查半同步复制相关的配置参数，例如 rpl_semi_sync_master_timeout，适当增加超时时间。

（2）网络延迟或不稳定

原因：主从服务器之间的网络不稳定会导致 binlog 传输延迟或丢失。

解决方法：检查和优化网络连接，确保主从服务器之间的网络稳定性。

（3）从服务器负载过高

原因：从服务器的负载过高，无法及时处理主服务器发送的 binlog。

解决方法：检查从服务器的性能，可能需要增加资源或优化查询。

（4）主服务器问题

原因：主服务器在生成 binlog 或发送 binlog 时出现问题。

解决方法：检查主服务器的性能和错误日志，确保其运行正常。

配置与网络情况相对比较好排查，可以从这两处入手快速排查问题原因，再逐步深入研究。

2.3 问问 Gemini

我们也将相同的问题发给了 Gemeni，让我们看看效果如何（见图 16）。

04 一问一实验（ChatDBA）篇

MySQL的error log频繁报错

人类专家评论区

◆ 根据您提供的图片信息，MySQL的error log频繁报错的原因是**半同步复制（semi-sync replication）出现问题**。

半同步复制是一种MySQL复制模式，它可以保证主服务器和从服务器的数据同步延迟在毫秒级。在半同步复制模式下，主服务器在更新数据之前，会先将更新内容发送给从服务器。从服务器收到更新内容之后，会进行校验，并向主服务器发送确认消息。如果主服务器没有收到确认消息，则会超时并重试。

在您提供的error log中，可以看到以下内容：

```
2022-11-26T11:39:53.958371+08:00 8 [Note] Semi-sync replication switched
2022-11-26T11:39:53.979523+08:00 30919 [Warning] Timeout waiting for re
2022-11-26T11:39:54.434839+08:00 10919 [Note] Semi-sync replication swi
2022-11-26T11:39:54.468499+08:00 0 [Note] Semi-sync replication switched
2022-11-26T11:39:54.587583+08:00 10919 [Note] Semi-sync replication swi
2022-11-26T11:39:54.609433+08:00 8 [Note] Semi-sync replication switched
2022-11-26T11:39:54.793461-00:00 30919 [Note] Semi-sync replication swi
2022-11-26T11:39:55.016230-08:00 30919 [Note] Semi-sync replication swi
2022-11-26T11:39:55.126136+08:00 30919 [Note] Semi-sync replication swi
2022-11-26T11:39:55.148065-08:00 e [Note] Semi-sync replication switched
2022-11-26T11:39:55.604017+08:00 30919 [Note] Semi-sync replication swi
2022-11-26T11:39:56.016309+06:00 30919 [Note] Semi-sync replication swi
2022-11-26T11:39:56.116390-08:00 30919 [Note] Semi-sync replication swi
2022-11-26T11:39:56.166176+08:00 30915 [Note] Semi-sync replication swi
2022-11-26T11:39:56.207848+08:00 8 [Note] Semi-sync replication switched
2022-11-26T11:39:56.320252+08:00
```

点评 1：

Genmini对图片的识别还是比较准确的。

可能的原因是什么？

显示草稿 ∨

◆ **MySQL半同步复制频繁报错的可能原因**

根据您提供的error log，MySQL半同步复制频繁报错的原因可能由以下几方面导致：

1. 主从服务器网络延迟过高

半同步复制要求主从服务器之间具有良好的网络连接，如果网络延迟过高，可能会导致主服务器在等待从服务器确认消息时超时，进而导致半同步复制被关闭。

2. 从服务器负载过高

如果从服务器负载过高，可能会导致其无法及时处理来自主服务器的更新消息，进而导致主服务器超时。

点评 2：

将半同步复制的解释作为提示，将知识补充给用户是一种比较好的体验，但缺少了具体的下一步排查计划。

图 16（a）

3. 主服务器或从服务器硬件故障

如果主服务器或从服务器出现硬件故障，也可能会导致半同步复制出现问题。

4. MySQL Bug

在某些情况下，MySQL Bug也可能导致半同步复制出现问题。

5. 其他原因

其他可能导致半同步复制频繁报错的原因还包括：

- 主服务器或从服务器的MySQL版本不兼容
- 主服务器或从服务器的MySQL配置错误
- 磁盘I/O瓶颈

如何排查和解决问题？

为了排查和解决MySQL半同步复制频繁报错的问题，建议您采取以下步骤：

1. 检查主从服务器之间的网络延迟

使用ping命令或其他网络诊断工具检查主从服务器之间的网络延迟。如果延迟过高，请采取措施降低延迟，例如升级网络带宽或优化网络配置。

2. 检查从服务器负载

检查从服务器的CPU、内存、I/O等资源的使用情况。如果负载过高，请采取措施降低负载，例如增加硬件资源或优化应用程序。

3. 检查主服务器和从服务器的硬件健康状况

使用硬件诊断工具检查主服务器和从服务器的硬件健康状况。如果发现任何硬件故障，请及时修复。

4. 检查MySQL版本和配置

确保主从服务器的MySQL版本兼容，并检查MySQL配置文件是否正确。

5. 检查MySQL Bug

搜索MySQL官方网站和社区论坛，了解是否存在已知的Bug会导致半同步复制出现问题。如果发现相关Bug，请升级MySQL版本或应用补丁。

6. 收集更多信息

如果以上步骤都无法解决问题，请收集更多信息，例如完整的error log、主服务器和从服务器的性能数据等，并寻求专业人士的帮助。

以下是一些额外的建议：

- 在主从服务器上启用MySQL的详细日志记录，以便更好地了解问题的原因。
- 使用MySQL的performance_schema功能收集有关半同步复制性能的数据。
- 在MySQL官方网站和社区论坛上寻求帮助。

希望以上信息对您有所帮助。

> **点评 3：**
>
> 给出的原因很全，但是缺少具体的指令以及下一步具体的行动计划。

图 16（b）

我们将这个问题分别发给了 ChatGPT-4o 与 Gemini，ChatGPT-4o 回答的效果与 ChatDBA 相当，但是 Gemini 的回答比较奇怪，虽然它也准确识别出对应图片的内容，但只是总结了一下原因，并没有做过多解释。当我们继续追问的时候发现它回答了多种原因，但是每个原因还是缺少具体指向性的操作。

3 MySQL 创建了用户却无法登录

3.1 问题：MySQL 创建了用户却无法登录

用户反馈自己进行了升级操作：从 MySQL 5.6 升级到 MySQL 5.7，并新建了一个 sky1@% 的用户，但是通过 socket 登录时却报错：Access denied。

3.2 实验

3.2.1 将问题丢给 ChatDBA

问题如图 17 所示。

图 17

我们先把这个问题丢给 ChatDBA，让它看一下具体出了什么问题（见图 18）。

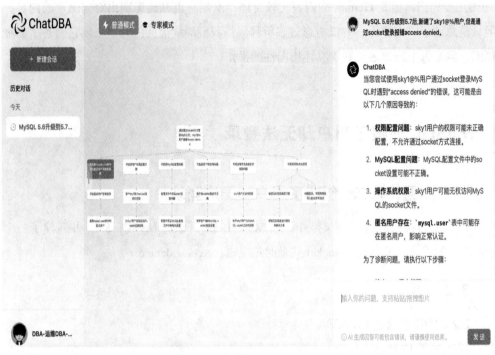

图 18

3.2.2 ChatDBA 协助问题排查

这个问题是用户虽然创建了名字为 sky1@% 的用户，但是却登录不上去，这时 ChatDBA 给出的可能原因有：

(1) 权限配置问题：没给这个用户配置 socket 登录方式。

(2) 数据库配置问题。

(3) 操作系统的权限问题。

(4) 匿名用户导致。

ChatDBA 提示，需要查询上面几处的具体信息（见图 19），同时也可以看到，此时 ChatDBA 推测了一种可能的解决方案，并给出了操作步骤。

MySQL 5.6升级到5.7后,新建了sky1@%用户,但是通过socket登录报错access denied。

ChatDBA

当您尝试使用sky1@%用户通过socket登录MySQL时遇到"access denied"的错误,这可能是由以下几个原因导致的:

1. **权限配置问题**:sky1用户的权限可能未正确配置,不允许通过socket方式连接。
2. **MySQL配置问题**:MySQL配置文件中的socket设置可能不正确。
3. **操作系统权限**:sky1用户可能无权访问MySQL的socket文件。
4. **匿名用户存在**:`mysql.user`表中可能存在匿名用户,影响正常认证。

为了诊断问题,请执行以下步骤:

1. **检查sky1用户权限**:

```
SELECT * FROM mysql.user WHERE User
```

2. **查找匿名用户**:

```
SELECT User, Host FROM mysql.user W
```

如果存在匿名用户,可能需要清理它们:

```
FLUSH PRIVILEGES;
DELETE FROM mysql.user WHERE User =
FLUSH PRIVILEGES;
```

3. **检查MySQL配置**:查看`my.cnf`或`my.ini`配置文件中的`socket`变量设置。
4. **确认操作系统权限**:确保sky1用户有权访问MySQL的socket文件。

请根据上述步骤提供相关结果,这将有助于我们确定问题所在并找到解决方案。如果清除匿名用户或调整权限后问题仍未解决,我们将进一步分析其他可能的原因。

图 19

3.2.3 ChatDBA 给出解决方案

解决方案见图 20。

人类专家评论区

用户sky1权限没有问题，用户信息表如下：

```
| user          | host      | authentication_string                        |
|---------------|-----------|----------------------------------------------|
| root          | localhost | *77664FA7C8BDAA068B86043F5D1087D474F6A950    |
| root          | centos    | *77664EA7C8BDAA068B86043F5D1087D474B6A950    |
| root          | ::1       | *77664EA7C8BDAA068B86043F5D1087D474B6A950b   |
| root          | 127.0.0.1 | *77664EA7C8BDAA068B86043F5D1087D474B6A950b   |
|               | localhost | NULL                                         |
|               | centos    | NULL                                         |
| repl          | %         | *ECE5887B232CECE4D3D90FFFAF543AB884D12AB     |
| mysql.session | localhost | *THISISNOTAVALIDPASSWORDTHATCANBEUSEDHEERE   |
| mysql.sys     | localhost | *THISISNOTAVALIDPASSWORDTHATCANBEUSEDHEERE   |
| dbsel         | %         | *ABE63EC4FB017F0E26B5086C9B35E7C45AC400C7    |
| sky           | %         | *D2207705EF335F527E04D9AEB1646AD0F1B4D84B    |
| sky1          | %         | *5F3ADC036F3AA756C121D6DF7F971E3C3CDE2DE22   |
```

ChatDBA

根据您的问题描述和之前的分析，问题的关键在于mysql.user表中的匿名用户可能导致sky1@%用户无法通过socket登录。为了解决这个问题，请按照以下步骤操作：

1. **删除匿名用户**：首先，确保您以具有足够权限（如root）的角色登录MySQL。然后，执行以下SQL命令：

```
DELETE FROM mysql.user WHERE User =
FLUSH PRIVILEGES;
```

这将删除所有Host为localhost和centos的匿名用户。在操作前，请务必备份数据。

点评 1：

AI 没有意识人类在此处需要个逻辑，应当给出 MySQL 升级+匿名用户和问题之间的关系。

图 20（a）

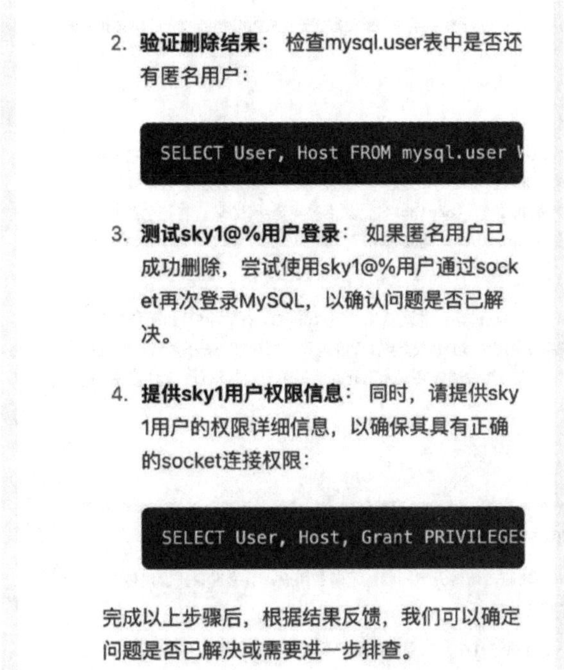

2. **验证删除结果**：检查mysql.user表中是否还有匿名用户：

   `SELECT User, Host FROM mysql.user W`

3. **测试sky1@%用户登录**：如果匿名用户已成功删除，尝试使用sky1@%用户通过socket再次登录MySQL，以确认问题是否已解决。

4. **提供sky1用户权限信息**：同时，请提供sky1用户的权限详细信息，以确保其具有正确的socket连接权限：

   `SELECT User, Host, Grant PRIVILEGES`

完成以上步骤后，根据结果反馈，我们可以确定问题是否已解决或需要进一步排查。

图 20（b）

经过检查发现用户的权限等都不存在问题，但是输入用户信息表时，ChatDBA 发现确实存在匿名用户，那么它提示直接删除用户即可解决该问题。

3.2.4 实验总结

针对该问题，ChatDBA 上述的排查步骤虽然解决了问题，但是并没有解释问题出现的根本原因。后来经过检查我们发现，当用户不输入密码的时候是可以登录的，而且登录的用户信息如下：

```
mysql> select current_user();
+------------------+
|current user      |
+------------------+
|@localhost        |
+------------------+
1 row in set (0.00 sec)
mysql> select user();
+----------------+
| user           |
+----------------+
|sky1@localhost  |
+----------------+
1 row in set (0.00 sec)
```

针对这个情况我们也问了一下 ChatDBA（见图 21）。

图 21

ChatDBA 回复说其实是因为用户匹配规则导致的：

(1)sky1@% 用户通过 socket 登录，host 被识别为 localhost。

(2) 在 mysql.user 表中，有两个匿名用户（''@localhost 和 ''@centos76）。

(3) 根据 MySQL 5.6 的匹配规则，虽然有一个名 sky1@% 的用户存在，但由于存在一个 ''@localhost 匿名用户，MySQL 优先匹配这个匿名用户。

(4) 由于匿名用户没有密码，当你输入 sky1@% 用户的密码时，MySQL 会因为匿名用户没有密码而拒绝访问，报错"Access denied"。但是，其实 MySQL 5.7 优化了用户匹配规则。当使用 socket 链接时，不会优先匹配匿名用户，因此全新安装的 MySQL 5.7 不会出现这个问题。而从 MySQL5.6 升级到 MySQL5.7 的过程中，MySQL 对用户表元数据没有更新，所以导致了本文开头出现的情况：当从 MySQL 5.6 升级到 MySQL5.7 后，新建了一个 sky1@% 用户，但是通过 socket 登录却报错 Access denied。

3.3 问问 ChatGPT-4o

我们也将相同的问题发给了 ChatGPT-4o,让我们看看效果如何(见图 22)。

> MySQL 5.6升级到5.7后,新建了sky1@%用户,但是通过socket登录报错access denied。

人类专家评论区

在 MySQL 5.6 升级到 5.7 后,如果新建了 `sky1@%` 用户但通过 socket 登录时报错 `access denied`,可能是由于 MySQL 5.7 中引入的一些新特性和安全性改进。以下是一些常见的解决方案:

1. **确认用户和权限设置**:
   确认用户 `sky1@%` 已正确创建,并且具备所需的权限。可以通过以下命令查看用户和权限:

```sql
SELECT user, host FROM mysql.user WHERE user='sky1';
SHOW GRANTS FOR 'sky1'@'%';
```

点评 1:

ChatGPT-4o 虽然也总结了这个问题的可能原因,但是遗漏了存在匿名用户导致的问题。

2. **密码插件**:
   MySQL 5.7 默认使用 `caching_sha2_password` 作为认证插件,而 MySQL 5.6 使用 `mysql_native_password`。你可以通过以下命令检查和修改用户的认证插件:

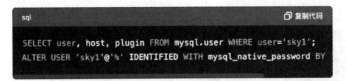

```sql
SELECT user, host, plugin FROM mysql.user WHERE user='sky1';
ALTER USER 'sky1'@'%' IDENTIFIED WITH mysql_native_password BY
```

3. **刷新权限**:
   有时候权限修改后需要刷新权限表,确保新设置生效:

```sql
FLUSH PRIVILEGES;
```

4. **检查 MySQL 配置文件**:
   确保 MySQL 配置文件(通常是 `/etc/my.cnf` 或 `/etc/mysql/my.cnf`)中没有不兼容的配置选项。特别注意 `skip-networking` 和 `bind-address` 配置项。

点评 2:

当用户输入了新的信息后,ChatGPT-4o 能够要求继续排查其他的原因。

5. **调试日志**:
   查看 MySQL 错误日志(通常在 `/var/log/mysql/` 或 `/var/log/` 目录下),查找与登录失败相关的详细信息。

6. **重新启动 MySQL 服务**:
   在修改配置文件或用户权限后,重新启动 MySQL 服务确保更改生效:

图 22(a)

如果以上步骤没有解决问题，请提供更多的错误信息和相关配置，以便进行更详细的分析。

用户 sky1@% 已正确创建，并且具备所需的权限

🌀 如果用户 `sky1@%` 已正确创建并具备所需的权限，但通过 socket 登录仍然报错 `access denied`，可以尝试以下步骤来解决问题：

1. **检查用户具体登录的来源**：
   确认 socket 登录的具体用户和主机。在 MySQL 中，`%` 代表任意主机，但对于 localhost 登录，可能需要明确的用户配置。

2. **检查 MySQL 插件**：
   确认用户的认证插件是否兼容 socket 登录：

   如果用户 `sky1` 的插件不是 `mysql_native_password`，可以通过以下命令更改：

3. **检查 MySQL 配置文件**：
   确认 MySQL 配置文件中是否有可能导致登录问题的配置。特别注意 `skip-networking` 和 `bind-address` 选项：

图 22（b）

确保没有 `skip-networking`，并且 `bind-address` 应该允许 socket 连接：

4. 清除缓存和重启 MySQL 服务：
   确保 MySQL 缓存被清除，并重启 MySQL 服务：

5. 使用正确的登录命令：
   确认使用正确的命令通过 socket 登录：

或者明确指定 socket 文件路径：

6. 检查 MySQL 错误日志：
   查看 MySQL 错误日志以获取更多信息：

这些步骤可以帮助诊断和解决通过 socket 登录 `sky1@%` 用户报错 `access denied` 的问题。如果问题仍然存在，请提供具体的错误日志内容，以便进一步分析。

图 22（c）

4 MySQL Too many open files 报错

4.1 问题

MySQL 报错"Too many open files"怎么处理？

4.2 实验

4.2.1 将问题丢给 ChatDBA

问题如图 23 所示。

图 23

我们先把这个问题丢给 ChatDBA，让它看一下具体出了什么问题（见图 24）。

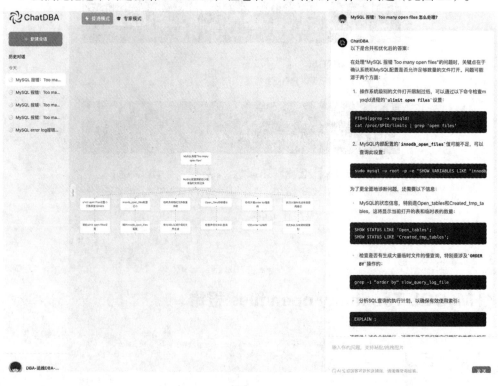

图 24

4.2.2 ChatDBA 协助问题排查

具体如图 25 所示。

人类专家评论区

MySQL 报错：Too many open files 怎么处理？

ChatDBA
以下是合并和优化后的答案：

在处理"MySQL 报错 Too many open files"的问题时，关键点在于确认系统和MySQL配置是否允许足够数量的文件打开。问题可能源于两个方面：

1. 操作系统级别的文件打开限制过低，可以通过以下命令检查mysqld进程的`ulimit open files`设置：

```
PID=$(pgrep -x mysqld)
cat /proc/$PID/limits | grep 'open files'
```

2. MySQL内部配置的`innodb_open_files`值可能不足，可以查询此设置：

```
sudo mysql -u root -p -e "SHOW VARIABLES LIKE 'innodb
```

为了更全面地诊断问题，还需要以下信息：

- MySQL的状态信息，特别是Open_tables和Created_tmp_tables，这将显示当前打开的表和临时表的数量：

```
SHOW STATUS LIKE 'Open_tables';
SHOW STATUS LIKE 'Created_tmp_tables';
```

- 检查是否有生成大量临时文件的慢查询，特别是涉及`ORDER BY`操作的：

```
grep -i "order by" slow_query_log_file
```

- 分析SQL查询的执行计划，以确保有效使用索引：

```
EXPLAIN ;
```

请提供上述命令的输出，这将有助于我们确定问题所在并提出相应的解决方案。

点评：
此处，更有效的索引和文件打开数过大之间的逻辑比较跳跃。
AI 应当简单说明两者之间的关系。

图 25

我们将问题输入 ChatDBA 后，它的反馈是先进行信息的收集。这里 ChatDBA 要求输入系统的一些基本情况，因为该报错很有可能是由于文件描述符数量引起的，要求输入的信息分别为：open_files_limit、ulimit -n、Innodb open files。

4.2.2.1 open_files_limit

具体如图 26 所示。

图 26

4.2.2.2 ulimit -n

具体如图 27 所示。

ulimit -n也是102400

ChatDBA
亲爱的用户，

感谢您提供的信息。从您描述的情况来看，`ulimit -n` 设置为 102400，`innodb_open_files` 配置为 6000，这些数值在一般情况下是合理的。为了更准确地诊断 "Too many open files" 问题，我们需要进一步分析 MySQL 的实际运行状态。以下是关键指标的检查步骤：

1. **当前打开的表数量：**

   ```
   SHOW STATUS LIKE 'Open_tables';
   ```

   这将显示 MySQL 正在管理的文件数量，如果数值较大，可能表明文件打开数接近限制。

2. **临时表的创建情况：**

   ```
   SHOW STATUS LIKE 'Created_TMP_tables';
   ```

   大量的临时表可能与临时文件生成有关，这会增加文件使用。

3. **文件打开和关闭的频率：**

   ```
   SHOW STATUS LIKE 'Open_files';
   SHOW STATUS LIKE 'Opened_files';
   ```

   这些状态将帮助我们了解文件操作的频率，如果 `Opened_files` 值持续增长，可能存在文件管理问题。

此外，如果可能，请提供以下信息以辅助分析：

- 是否存在涉及 `ORDER BY` 操作的慢查询，这可能导致临时文件生成：

  ```
  grep -i "order by" slow_query_log_file
  ```

- 具体的 SQL 查询执行计划，以评估索引使用情况。

请执行上述命令并分享结果，这将帮助我们确定是否存在临时表过度使用或文件操作频繁的问题。如果所有指标均正常，我们将进一步调查其他潜在原因。我们的目标是全面分析以找到问题的根本所在，而非直接提供未经证实的解决方案。

图 27

4.2.2.3 innodb_open_files

具体如图 28 所示。

innodb_open_files=6000

ChatDBA

通过对您提供的三个答案的整合，以下是优化后的回答：

首先，我们注意到`innodb_open_files`配置为6000，这看起来是充足的，但关键在于系统级别的文件打开限制。请提供MySQL用户和mysqld进程的`ulimit -n`设置，以评估是否达到了系统限制：

```
sudo bash -c 'ulimit -n'
sudo cat /proc/<MYSQLD_PID>/limits | grep 'open files
```

其中`<MYSQLD_PID>`是mysqld主进程的PID。

同时，我们需要关注MySQL的实际运行状况：

1. 当前打开的表数量，这反映了MySQL正在管理的文件数：

```
SHOW STATUS LIKE 'Open_tables';
```

2. 临时表的创建频率，这可能关联到临时文件的使用：

```
SHOW STATUS LIKE 'Created_TMP_tables';
```

3. 文件的打开和关闭情况，这有助于判断是否频繁超出限制：

```
SHOW STATUS LIKE 'Open_files';
SHOW STATUS LIKE 'Opened_files';
```

如果数据显示临时表或文件使用异常，我们将检查慢查询日志，特别是涉及`ORDER BY`操作的查询：

```
grep -i "order by" slow_query_log_file
```

请提供上述命令的输出，以便我们分析问题并提供相应的解决方案。如果存在系统级限制或临时文件过度使用，可能需要调整配置或优化查询以减少文件使用。

图 28

ChatDBA 发现这些配置基本上满足需求，所以要求输入打开文件的数量、临时表情况等，所以接下来就将当时系统的监控图片上传到 ChatDBA（见图 29、图 30）。

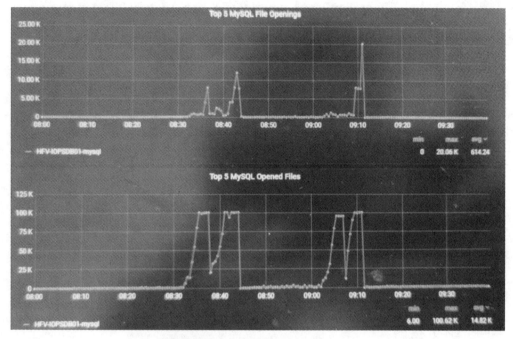

图 29

报错时打开的文件数量如图

ChatDBA

根据您提供的图表，MySQL在08:00至09:30期间，特别是在09:10-09:20，经历了大规模的文件操作，其中最高达到约20.06K次打开和100.62K次操作。这表明临时文件的过度生成是主要问题。尽管`ulimit -n`和`innodb_open_files`设置合适，但数据库操作的优化至关重要。

为了诊断并解决这个问题，我们需要以下信息：

1. **收集MySQL错误日志中涉及文件错误的部分**：

```
tail -n 100 /var/log/mysql/error.log | grep -i "o
```

2. **分析慢查询和涉及大量文件操作的SQL**：

```
# 检索一小时前的慢查询，特别是涉及排序的
grep --text -i "query_time>" 1h_before_error_slow
# 检查涉及临时表的查询
mysql -u -p -e "SHOW ENGINE INNODB STATUS \G;" |
grep -E "order by|临时表" /var/log/mysql/slow_query
```

这些步骤将帮助我们识别导致大量临时文件生成的具体查询和原因。一旦找到问题查询，我们应优化SQL，例如避免全表扫描，改进索引使用，以及调整事务处理以减少文件IO压力。同时，监控和优化数据库的内存分配以及事务并发控制也是关键。

请您执行上述命令，然后分享结果，以便我们能进一步分析并提出针对性的解决方案。如果需要帮助，随时告诉我。

图 30

首先，ChatDBA 准确识别出了监控图像的内容，同时也猜测虽然参数配置合理，但是临时文件过多是导致该问题的主要原因。通过查看监控数据我们也发现，临时文件确实在报错期间增长很快（见图 31）。

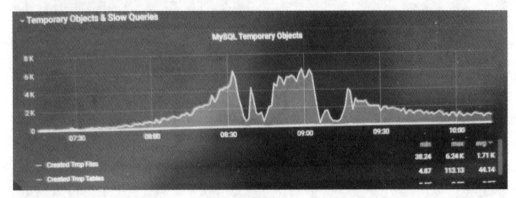

图 31

其次，ChatDBA 推测是慢查询导致的临时文件过多，所以要求我们输入对应的慢查语句（见图 32）。

slow log 在报错期间 记录有大量慢查询，慢查询语句如下：
Query_time: 80.107946 Lock_time: 0.000262 Rows_sent: 11 Rows_examined: 870
SET timestamp=1663202019;
select xxxx as xxx from xxx xx left join xxx xx on xx = xx and xx = 45041 left join xx = 0 and (xx = 0 or concat(',',xx,',') like '%,45041,%' or concat(',',xx,',') like '%,17832,%'or concat(',',xx,',') like '%,2089,%' or xx GROUP BY xxxxx order by xxx asc limit 0,11;

ChatDBA
经过整合，以下是优化后的答案：

根据之前的讨论，我们确定了问题的关键点：大量的慢查询与 `ORDER BY` 操作有关，这导致了临时文件的过度生成，从而引发"Too many open files"错误。主要问题在于执行计划未充分利用索引，特别是对于 `ORDER BY` 和 `LEFT JOIN` 操作。为了有效解决这个问题，我们需要采取以下步骤：

1. **分析查询执行计划**：对包含 `ORDER BY` 的慢查询执行 `EXPLAIN ANALYZE`，检查是否在 `ORDER BY` 列上存在有效的覆盖索引，例如：

```
EXPLAIN ANALYZE SELECT * FROM your_table
WHERE ...
ORDER BY xxx LIMIT 11;
```

替换 `your_table` 和查询条件，以评估索引使用情况。

图 32（a）

2. **审查JOIN操作**：检查`LEFT JOIN`是否必要，考虑优化连接条件、使用子查询，或调整连接顺序以减少数据扫描。

3. **优化索引**：获取受影响表的结构和索引信息，如：

```
DESC ;
SHOW INDEXES FROM ;
```

根据查询需求，创建或优化索引以提高查询效率。

4. **识别慢查询模式**：查找具有类似`ORDER BY`操作和索引使用不当的其他慢查询记录：

```
cat slow_query_log_file | grep -P "ORDER BY.*?WHER
```

这有助于识别共性问题并制定全局优化策略。

5. **分析执行时间**：了解慢查询的执行高峰时段，以便针对性地优化：

```
SELECT Hour(timeStamp) AS Hour, COUNT(*) as NumOfS
FROM slow_query_log
WHERE Query_time > 80
GROUP BY Hour(timeStamp);
```

根据上述信息，我们可以制定更精确的优化策略，如调整JOIN顺序、创建覆盖索引或考虑表分区。如果您需要进一步的帮助或有疑问，请随时提问。

图 32（b）

最后，我们将对应的 SQL 语句发给 ChatDBA，这时它根据经验推测原因是该语句未充分利用索引，所以让我们给出对应的执行计划。

4.2.3 ChatDBA 给出解决方案

我们将 EXPLAIN 的结果截图发给 ChatDBA 后，发现其根据图 33 内容推测出了问题的关键，并且给出了对应的解决方案（见图 34）。

图 33

图 34

4.2.4 实验总结

这个案例比较有趣,一般情况下"Too many open files"报错和文件描述符配置不当有关。但是在该案例中,是由于数据表没有有效利用索引导致的。后续观察表结构发现,JOIN 关联条件中等号左右两边字段的数据类型不同,一个是 VARCHAR 类型,另一个是 INT 类型,所以该条 SQL 语句没法用索引,进而创建了非常多的临时文件,导致了报错。

4.3 问问 ChatGPT-4o

我们也将相同的问题发给了 ChatGPT-4o,让我们看看效果如何(见图 35)。

> MySQL 报错：Too many open files 怎么处理？

MySQL 报错 "Too many open files" 通常是由于操作系统允许的文件描述符数量不足引起的。解决这个问题可以通过增加文件描述符的限制来实现。以下是一些步骤来解决这个问题：

1. 检查当前文件描述符限制

首先，检查 MySQL 当前使用的文件描述符限制：

```bash
mysql -u root -p -e "SHOW VARIABLES LIKE 'open_files_limit';"
```

这将显示 MySQL 当前允许打开的最大文件数量。

2. 检查操作系统的文件描述符限制

使用以下命令检查当前系统允许的文件描述符限制：

```bash
ulimit -n
```

这将显示当前会话的文件描述符限制。

3. 增加文件描述符限制

临时增加限制

你可以使用 `ulimit` 命令临时增加文件描述符限制：

```bash
ulimit -n 65535
```

此命令将当前会话的文件描述符限制增加到 65535。但这只是临时的，重启后会失效。

永久增加限制

要永久增加文件描述符限制，你需要编辑系统配置文件。

对于 CentOS/RHEL 系统

编辑 `/etc/security/limits.conf` 文件，添加以下行：

```plaintext
* soft nofile 65535
* hard nofile 65535
```

图 35（a）

编辑 `/etc/pam.d/login` 文件，添加以下行：

编辑 `/etc/sysctl.conf` 文件，添加以下行：

应用更改：

对于 Ubuntu/Debian 系统

编辑 `/etc/security/limits.conf` 文件，添加以下行：

编辑 `/etc/pam.d/common-session` 和 `/etc/pam.d/common-session-noninteractive` 文件，添加以下行：

编辑 `/etc/sysctl.conf` 文件，添加以下行：

应用更改：

4. 配置 MySQL 使用新的文件描述符限制

编辑 MySQL 配置文件（通常是 `/etc/my.cnf` 或 `/etc/mysql/my.cnf`），在 `[mysqld]` 部分添加或修改 `open_files_limit` 选项：

图 35（b）

5. 重启 MySQL 服务

应用更改后,重启 MySQL 服务:

或

完成以上步骤后,MySQL 应该能够打开更多的文件,从而解决 "Too many open files" 错误。如果问题仍然存在,请检查 MySQL 错误日志以获取更多详细信息。

> open_files_limit与ulimit -n都是102400

如果 `open_files_limit` 和 `ulimit -n` 都已经设置为 102400,但仍然遇到 "Too many open files" 错误,可能还有其他因素导致该问题。以下是一些可能的解决方案:

1. 检查 MySQL 连接数限制

确保 MySQL 的最大连接数 (`max_connections`) 也足够高:

如果需要,可以在 MySQL 配置文件(通常是 `/etc/my.cnf` 或 `/etc/mysql/my.cnf`)中增加此值:

2. 检查 MySQL 表缓存配置

MySQL 使用表缓存来管理打开的表。如果表缓存配置不足,可能会导致打开的文件过多。可以检查并调整以下参数:

图 35(c)

在 MySQL 配置文件中增加这些值：

3. 检查文件句柄泄漏

确保应用程序或 MySQL 本身没有文件句柄泄漏。你可以使用 `lsof` 命令检查 MySQL 打开的文件数量：

4. 检查系统级别文件描述符限制

确认系统级别文件描述符限制足够高：

查看当前文件描述符限制：

如果需要增加，可以在 `/etc/sysctl.conf` 中修改：

应用更改：

5. 监控和调整 MySQL 使用的文件描述符

MySQL 可以在运行时调整 `open_files_limit`：

图 35（d）

6. 检查 MySQL 版本和插件

有时，特定的 MySQL 版本或插件可能存在问题。确保使用的是最新的稳定版本，并检查是否有已知的错误或问题。

7. 检查系统日志和 MySQL 日志

查看系统日志（如 `/var/log/syslog` 或 `/var/log/messages`）和 MySQL 错误日志（如 `/var/log/mysql/error.log`）以获取更多信息。这些日志可能包含有关为什么文件描述符不足的详细信息。

8. 其他配置优化

在 MySQL 配置文件中，其他参数的优化也可能影响文件描述符的使用。例如：

```ini
[mysqld]
innodb_open_files = 4000
```

在应用任何更改后，重启 MySQL 服务：

```bash
systemctl restart mysql
```

或

```bash
service mysql restart
```

通过这些步骤，你可以更详细地检查和优化 MySQL 配置，避免 "Too many open files" 错误。

图 35（e）

我们也将相同的问题输入到 ChatGPT 中，发现其在第一步也发现是文件描述符的问题，但是当我们给出其具体参数后，它提供的操作步骤没有办法收集或定位更多有效的信息。

5 MySQL 频繁 Crash 怎么办？

5.1 问题

MySQL 一分钟内 Crash 并重启了三次，是什么原因导致的？

5.2 实验

5.2.1 将问题丢给 ChatDBA

我们先把这个问题丢给 ChatDBA，让它看一下具体出了什么问题（见图 36、图 37）。

图 36

04 一问一实验（ChatDBA）篇

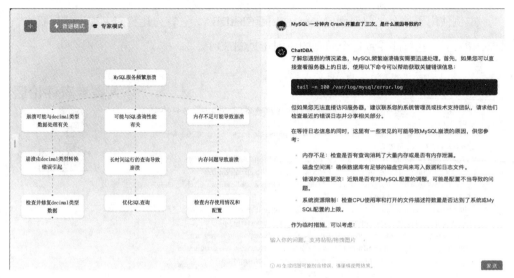

图 37

5.2.2 ChatDBA 协助问题排查

具体如图 38 所示。

图 38

本例使用了一个之前遇到过的问题来问 ChatDBA，它首先让我们去查询错误日志，同时猜测了一些可能导致这个问题的原因（见图 39）。

错误日志如下：
...
2023-11-06T14:44:05.362128+08:00 0 [Note] /mysql/mysql57/bin/mysqld: ready for connections.
Version: '5.7.36-log' socket: '/mysqldata/mysqltmp/mysqld.sock' port: 16063 MySQL Community Server (GPL)
2023-11-06T14:44:08.710518+08:00 17 [Warning] Timeout waiting for reply of binlog (file: mysqlbinlog.002386, pos: 1397), semi-sync up to file , position 0.
2023-11-06T14:44:08.710580+08:00 17 [Note] Semi-sync replication switched OFF.
06:44:29 UTC - mysqld got signal 8 ;
This could be because you hit a bug. It is also possible that this binary
or one of the libraries it was linked against is corrupt, improperly built,
or misconfigured. This error can also be caused by malfunctioning hardware.
Attempting to collect some information that could help diagnose the problem.
As this is a crash and something is definitely wrong, the information
collection process might fail.key_buffer_size=8388608
read_buffer_size=2097152
max_used_connections=49
max_threads=1000
thread_count=50
connection_count=49
It is possible that mysqld could use up to
key_buffer_size + (read_buffer_size + sort_buffer_size)*max_threads = 6165582 K bytes of memory
Hope that's ok; if not, decrease some variables in the equation.Thread pointer: 0x7f24e0000d40
Attempting backtrace. You can use the following information to find out
where mysqld died. If you see no messages after this, something went
terribly wrong...
stack_bottom = 7f2c930fce98 thread_stack 0x40000
/mysql/mysql57/bin/mysqld(my_print_stacktrace+0x35)[0xf7b545]
/mysql/mysql57/bin/mysqld(handle_fatal_signal+0x4b9)[0x7fab89]
/lib64/libpthread.so.0(+0x134c0)[0x7f332c73d4c0]
/mysql/mysql57/bin/mysqld(decimal2bin+0x1c0)[0x1487e90]
/mysql/mysql57/bin/mysqld(_Z17my_decimal2binaryjPK10my_decimalPhii+0xc5)[0xc916a5]
/mysql/mysql57/bin/mysqld(_ZN10Sort_param12make_sortkeyEPhPKh+0x33d)[0x83bf7d]
/mysql/mysql57/bin/mysqld(_Z8filesortP3THDP8FilesortbPyS3_S3_+0x1852)[0x83fdc2]
...

ChatDBA
根据您的错误日志分析，MySQL服务崩溃的根本原因在于decimal类型数据转换过程中的错误。为了解决这一问题，您可以按照以下具体步骤操作：

1. **识别问题SQL**：使用MySQL的General Query Log（一般查询日志）功能，监控并记录所有执行的SQL语句。这将帮助您定位到可能引起decimal转换错误的查询。开启日志的命令如下（在MySQL配置文件中或通过命令行临时设置）：

人类专家评论区

点评：
ChatDBA 在此处应该对"如何从日志分析出错误是和 decimal 类型有关"做出一定解释，供人类学习。

图 39（a）

```
set global general_log = 1;
set global general_log_file = '/path/to/your/logfile.log';
```

2. **审查与修复**：
   - 仔细审查General Query Log中记录的查询，特别是那些涉及decimal类型数据操作的部分。
   - 确认所有decimal类型的字段在SQL操作中正确使用，避免不匹配的数据类型转换，例如确保没有将过大的数值插入到小范围的decimal字段中。
   - 对于发现的问题SQL，修正数据类型处理逻辑，或者对数据进行预处理，确保它们符合字段定义的范围和精度。

3. **应用代码调整**：如果错误源于应用程序代码，您可能需要调整代码逻辑，增加数据验证步骤，确保在数据写入数据库之前，所有decimal值都是合法且格式正确的。

4. **测试与监控**：在应用上述更改后，进行全面的测试，确保问题得到解决，并持续监控MySQL的运行状态，以防问题再次发生。

请实施上述步骤，并在操作过程中遇到任何疑问或需要进一步帮助时，随时与我联系。

输入你的问题，支持粘贴/拖拽图片

ⓘ AI 生成回答可能包含错误，请谨慎使用结果。　　　　　　　　　　　　发送

图 39（b）

紧接着，我们将错误日志直接发给了 ChatDBA，发现 ChatDBA 识别出日志中包含的数据类型，同时也给出了临时的解决方案。接下来，我们将 SQL 语句与对应的表结构输入 ChatDBA。

5.2.3 ChatDBA 给出解决方案

解决方案如图 40 所示。

 涉及的数据表如下：
```
CREATE TABLE `xx` (
  `id` bigint(20) NOT NULL COMMENT '主键ID',
  `xx` bigint(20) NOT NULL,
  `col1` decimal(10,7) DEFAULT NULL,
  `col2` decimal(10,7) DEFAULT NULL,
  ...
  PRIMARY KEY (`id`),
) ENGINE=InnoDB DEFAULT CHARSET=utf8mb4
```

涉及的SQL语句如下：
select a, b, c, d from xx join xx on xx = xx is not null and xx is not null and xx in (12345)
ORDER BY ABS(0.000000000000000000000000000000000000000000000
00000000000000000000000000000000000000000000000000000000000
00000000000000000000000000000000000000000000000000000000000
00000000000000000000000000000000000000000000000000000000000
00000000000000000000000000000000000000000000000000000000000
0000000000000005-`xx`)+ABS(0.0000000000000000000000000000000
00000000000000000000000000000000000000000000000000000000000
00000000000000000000000000000000000000000000000000000000000
00000000000000000000000000000000000000000000000000000

 ChatDBA

经过分析，问题的根源在于您的SQL语句中复杂的decimal运算，特别是ABS函数内的表达式含有不必要的长串零及对 `xx` 列的小数值计算，这可能导致MySQL在处理时发生类型转换错误，从而引发崩溃。

优化建议：

方案一：简化SQL表达式

具体操作：考虑将ABS函数内的计算逻辑简化，例如，如果计算目标是衡量与某个值的绝对差值，可以重写为更简洁且易于MySQL处理的形式。避免使用过长的零序列和极小数值，这不仅消耗计算资源，也可能引入精度误差。简化后的示例（请注意，此为示意性修改，具体需根据业务逻辑调整）：

```sql
SELECT a, b, c, d
FROM xx
JOIN xx ON xx = xx
WHERE colX IS NOT NULL AND colY IS NOT NULL
AND xx IN (12345)
ORDER BY ABS(5 - `xx`) + ABS(`some_reasonable_constant` - `xx`);
```

其中，`some_reasonable_constant` 应替换为一个根据业务逻辑确定的合理数值。

方案二：预处理数据

· 确保在数据录入阶段，decimal类型的字段值处于合理范围内，避免在计算中因数值过小或过大导致的精度问题。

请在测试环境中先行测试上述修改，并监控其对系统稳定性的影响。如需进一步的协助或有其他疑问，请随时告知。

图 40

人类专家评论区

点评：

在这里ChatDBA SQL 转换结果中数字常量是不正确的，其想表达这只是一个示意性修改。

到这里，ChatDBA 基本上判断出问题的根本原因是 decimal 做运算时出了问题，而且识别出对应的 SQL 语句存在异常。接下来，ChatDBA 给出了对应的具体解决方案。

5.2.4 实验总结

针对该问题，其实原因有很多。ChatDBA 能够列出 decimal 字段可能导致该问题也是因为 ChatDBA 底层的知识库足够丰富，后续我们也将为其持续补充数据库领域更多的专业知识。

回到该问题，这是由于两个字段没有获取有效值，而且其数据类型为 decimal 类型，进行数值转换的时候触发了 MySQL 的 bug 导致崩溃。

最后，配合业务方针对该情况提前做数据判断，防止异常值直接入库来避免该问题的出现。

5.3 问问 ChatGPT-4o

具体如图 41 所示。

> MySQL 一分钟内 Crash 并重启了三次，是什么原因导致的?

MySQL 一分钟内 Crash 并重启了三次可能由多种原因导致，以下是一些常见的原因和解决办法：

1. **内存不足：**
   - 原因：服务器内存不足，MySQL 进程被操作系统杀死。
   - 解决办法：检查服务器的内存使用情况，增加内存或优化 MySQL 配置以减少内存使用。

2. **磁盘空间不足：**
   - 原因：磁盘空间不足，导致 MySQL 无法写入数据或日志。
   - 解决办法：检查磁盘空间，清理不必要的文件，确保有足够的可用磁盘空间。

3. **MySQL 配置问题：**
   - 原因：MySQL 配置错误或不合理的参数设置。
   - 解决办法：检查 MySQL 配置文件（如 `my.cnf`），确认配置参数的合理性，尤其是 `innodb_buffer_pool_size`、`max_connections` 等参数。

4. **数据损坏：**
   - 原因：MySQL 数据文件或表损坏。
   - 解决办法：检查 MySQL 错误日志，查找和修复损坏的表或数据文件。可以尝试使用 `CHECK TABLE` 和 `REPAIR TABLE` 命令。

图 41（a）

5. **硬件故障**：
   - **原因**：服务器硬件（如内存、硬盘）故障。
   - **解决办法**：检查硬件状态，替换故障硬件。

6. **权限问题**：
   - **原因**：MySQL 进程缺少必要的权限。
   - **解决办法**：确保 MySQL 进程拥有足够的权限访问所需的文件和目录。

7. **系统问题**：
   - **原因**：操作系统内核参数配置问题或系统自身的限制。
   - **解决办法**：检查系统日志，调整内核参数（如 `ulimit` 参数），确保操作系统配置适合 MySQL 的运行。

8. **软件 Bug**：
   - **原因**：MySQL 版本存在 Bug。
   - **解决办法**：检查 MySQL 官方发布的版本更新日志，升级到最新的稳定版本。

要具体确定原因，可以采取以下步骤：

1. **检查 MySQL 错误日志**：
   - MySQL 的错误日志通常可以在 MySQL 数据目录下找到，文件名一般为 `error.log` 或 `hostname.err`。日志中会记录详细的错误信息。

2. **检查系统日志**：
   - 系统日志文件通常位于 `/var/log` 目录下，如 `/var/log/syslog` 或 `/var/log/messages`。查看系统日志可以帮助确定是否有系统级别的问题导致 MySQL 崩溃。

3. **监控 MySQL 运行状态**：
   - 使用监控工具（如 `mysqladmin`、`top`、`htop`）实时监控 MySQL 的运行状态，观察在崩溃前是否有异常情况发生。

通过以上步骤，结合具体的错误日志和系统日志信息，应该可以找到 MySQL 崩溃和重启的具体原因，并采取相应的解决措施。

```
error log 如下：
...
2023-11-06T14:44:05.362128+08:00 0 [Note]
/mysql/mysql57/bin/mysqld: ready for connections.
Version: '5.7.36-log'  socket: '/mysqldata/mysqltmp/mysqld.sock'
port: 16063  MySQL Community Server (GPL)
2023-11-06T14:44:08.710518+08:00 17 [Warning] Timeout waiting
for reply of binlog (file: mysqlbinlog.002386, pos: 1397), semi-sync
up to file , position 0.
```

图 41（b）

```
2023-11-06T14:44:08.710580+08:00 17 [Note] Semi-sync
replication switched OFF.
06:44:29 UTC - mysqld got signal 8 ;
This could be because you hit a bug. It is also possible that this
binary
or one of the libraries it was linked against is corrupt, improperly
built,
or misconfigured. This error can also be caused by malfunctioning
hardware.
Attempting to collect some information that could help diagnose the
problem.
As this is a crash and something is definitely wrong, the information
collection process might fail.key_buffer_size=8388608
read_buffer_size=2097152
max_used_connections=49
max_threads=1000
thread_count=50
connection_count=49
It is possible that mysqld could use up to
key_buffer_size + (read_buffer_size +
sort_buffer_size)*max_threads = 6165582 K  bytes of memory
Hope that's ok; if not, decrease some variables in the
equation.Thread pointer: 0x7f24e0000d40
Attempting backtrace. You can use the following information to find
out
where mysqld died. If you see no messages after this, something
went
terribly wrong...
stack_bottom = 7f2c930fce98 thread_stack 0x40000
/mysql/mysql57/bin/mysqld(my_print_stacktrace+0x35)[0xf7b545]
/mysql/mysql57/bin/mysqld(handle_fatal_signal+0x4b9)[0x7fab89]
/lib64/libpthread.so.0(+0x134c0)[0x7f332c73d4c0]
/mysql/mysql57/bin/mysqld(decimal2bin+0x1c0)[0x1487e90]
/mysql/mysql57/bin/mysqld(_Z17my_decimal2binaryjPK10my_decim
alPhii+0xc5)[0xc916a5]
/mysql/mysql57/bin/mysqld(_ZN10Sort_param12make_sortkeyEPhP
Kh+0x33d)[0x83bf7d]
/mysql/mysql57/bin/mysqld(_Z8filesortP3THDP8FilesortbPyS3_S3_
+0x1852)[0x83fdc2]
...
```

⑤ 从错误日志来看，MySQL 崩溃可能与以下几个方面有关：

1. **Semi-sync Replication 超时**：

    - 日志中提到 `Timeout waiting for reply of binlog`，说明 MySQL 的半同步复制遇到了超时问题，可能是由于网络延迟或复制从库性能问题导致的。
    - `Semi-sync replication switched OFF` 表示 MySQL 自动关闭了半同步复制，切换到异步复制模式。这本身不应该导致崩溃，但值得注意。

2. **Signal 8 (SIGFPE) 错误**：

    - `mysqld got signal 8` 表示 MySQL 进程收到了信号 8，通常是浮点异常（例如除以零）。

图 41（c）

- 这可能是由 MySQL 内部的 Bug 引起的，或者是某些查询或操作导致了非法的浮点运算。

3. **内存分配问题**：
   - 日志中提到的 `key_buffer_size + (read_buffer_size + sort_buffer_size)*max_threads = 6165582 K bytes of memory`，表示 MySQL 可能会使用约 6GB 的内存。
   - 这对内存的需求可能超出了系统的可用内存，导致崩溃。

4. **损坏的二进制文件或库**：
   - 日志中提到 `It is also possible that this binary or one of the libraries it was linked against is corrupt, improperly built, or misconfigured`。
   - 这可能是由于 MySQL 二进制文件或其依赖库损坏、构建不当或配置错误。

5. **硬件故障**：
   - `This error can also be caused by malfunctioning hardware`，硬件故障也是一个可能的原因，例如内存或磁盘问题。

解决办法

1. **检查和修复二进制文件**：
   - 确认 MySQL 安装文件和库没有损坏。如果有备份，可以重新安装或恢复 MySQL。

2. **检查和优化 MySQL 配置**：
   - 检查 MySQL 配置文件（如 `my.cnf`），确保 `key_buffer_size`、`read_buffer_size`、`sort_buffer_size` 等参数设置合理，避免内存分配过多。
   - 例如，调整 `max_connections` 或 `max_threads` 来限制并发连接数。

3. **监控和优化复制延迟**：
   - 检查复制延迟问题，确保主从库之间的网络连接稳定，适当优化从库的性能。

4. **检查硬件健康状态**：
   - 使用系统工具（如 `memtest`、`smartctl`）检查内存和磁盘的健康状态，排除硬件故障。

5. **升级 MySQL 版本**：
   - 如果当前 MySQL 版本存在已知 Bug，考虑升级到最新的稳定版本。

6. **检查应用程序代码**：
   - 检查应用程序中是否有可能导致非法浮点运算的查询或操作，优化相关代码。

通过以上步骤，应该能够进一步诊断并解决 MySQL 崩溃和重启的问题。

到目前为止，此对话有帮助吗？

给“ChatGPT”发送消息

ChatGPT 也可能会犯错。请核查重要信息。

图 41（d）

我们将相同的内容也发给了 ChatGPT，虽然它在第一轮回复中也要求我们输入对应的 error log，但是当我们把内容发给它后，它并没有识别出什么是关键的信息，因而无法有效地指导操作。

6 为什么不建议关闭 MySQL 严格模式？

6.1 问题

处理一个 MySQL 报错：

```
ERROR 1292 (22007): Truncated incorrect DECIMAL value: ''
```

6.2 实验

6.2.1 将问题丢给 ChatDBA

我们先把这个问题丢给 ChatDBA，让它看一下具体出了什么问题（见图 42、图 43）。

图 42

图 43

6.2.2 ChatDBA 协助问题排查并给出解决方案

具体如图 44 所示。

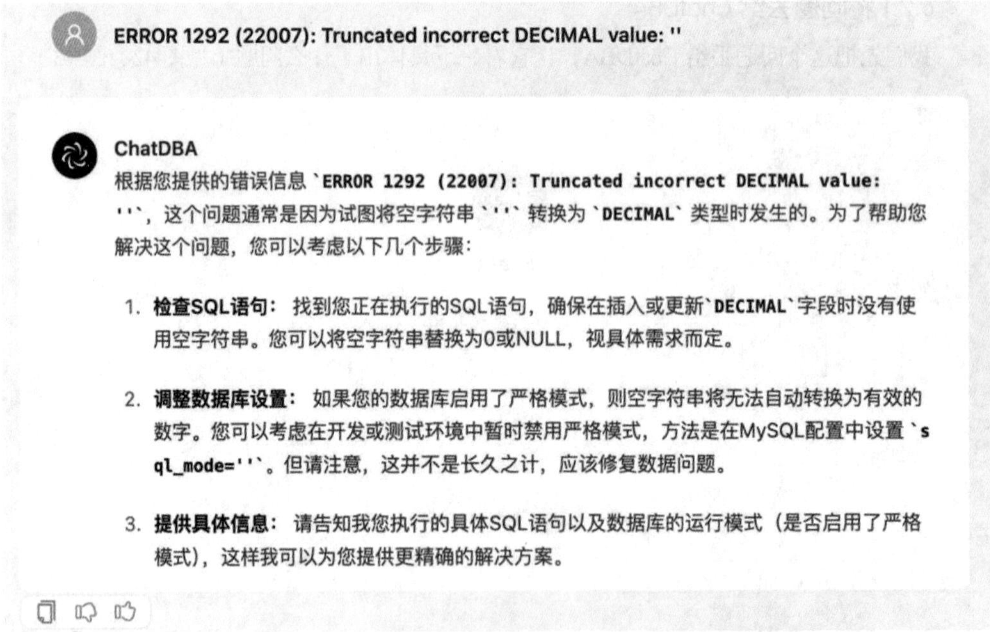

点评：
ChatDBA 意识到了人类可能需要临时的解决方案，给出了 sql_mode 的调整方案。但应当给出增量调整方案，而不是直接将 sql_mode 置空。

图 44

ChatDBA 首先对报错信息进行理解和分析，问题是由于空字符串无法被赋值给 Decimal 类型导致的。为了进一步确定原因，ChatDBA 希望确认具体的 SQL 语句以及数据库设置等信息（见图 45）。

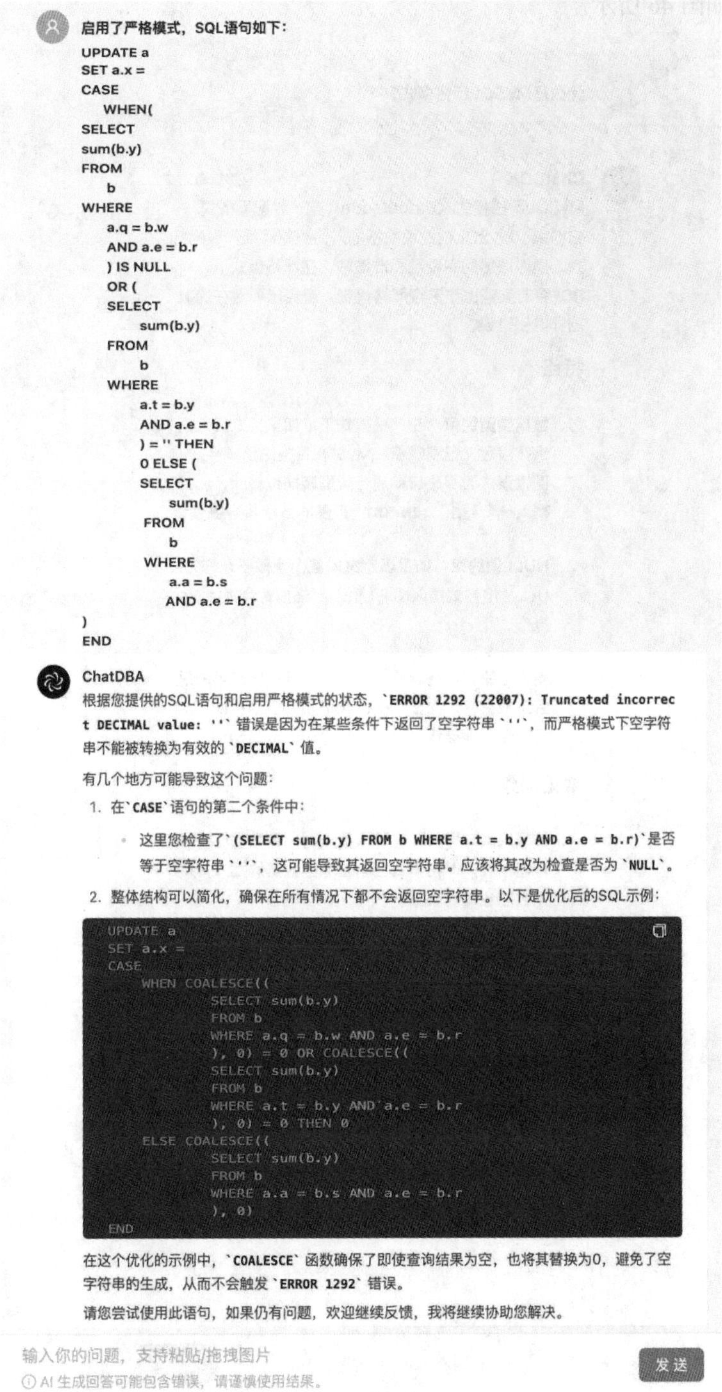
启用了严格模式，SQL语句如下：
UPDATE a
SET a.x =
CASE
WHEN(
SELECT
sum(b.y)
FROM
b
WHERE
a.q = b.w
AND a.e = b.r
) IS NULL
OR (
SELECT
sum(b.y)
FROM
b
WHERE
a.t = b.y
AND a.e = b.r
) = '' THEN
0 ELSE (
SELECT
sum(b.y)
FROM
b
WHERE
a.a = b.s
AND a.e = b.r
)
END
ChatDBA
根据您提供的SQL语句和启用严格模式的状态，`ERROR 1292 (22007): Truncated incorrect DECIMAL value: ''` 错误是因为在某些条件下返回了空字符串 `''`，而严格模式下空字符串不能被转换为有效的 `DECIMAL` 值。
有几个地方可能导致这个问题：
1. 在`CASE`语句的第二个条件中：
这里您检查了`(SELECT sum(b.y) FROM b WHERE a.t = b.y AND a.e = b.r)`是否等于空字符串 `''`，这可能导致其返回空字符串。应该将其改为检查是否为 `NULL`。
2. 整体结构可以简化，确保在所有情况下都不会返回空字符串。以下是优化后的SQL示例：
UPDATE a
SET a.x =
CASE
WHEN COALESCE((
SELECT sum(b.y)
FROM b
WHERE a.q = b.w AND a.e = b.r
), 0) = 0 OR COALESCE((
SELECT sum(b.y)
FROM b
WHERE a.t = b.y AND a.e = b.r
), 0) = 0 THEN 0
ELSE COALESCE((
SELECT sum(b.y)
FROM b
WHERE a.a = b.s AND a.e = b.r
), 0)
END
在这个优化的示例中，`COALESCE` 函数确保了即使查询结果为空，也将其替换为0，避免了空字符串的生成，从而不会触发 `ERROR 1292` 错误。
请您尝试使用此语句，如果仍有问题，欢迎继续反馈，我将继续协助您解决。
输入你的问题，支持粘贴/拖拽图片
发送
AI 生成回答可能包含错误，请谨慎使用结果。

图 45

我们将相关信息输入到 ChatDBA 后，系统指出这是在进行数据更新时导致的问题，并提出在严格模式下该类转换存在的问题，强调通过修改 SQL 语句来解决该问题。

6.2.3 什么是严格模式？

严格模式如图 46 所示。

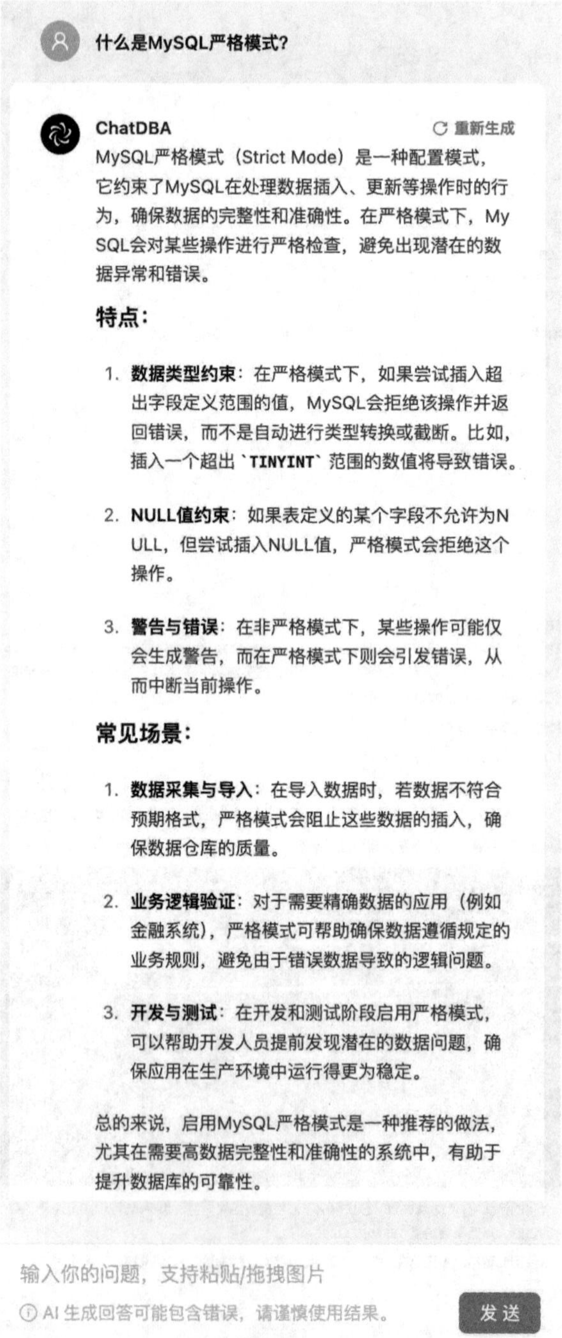

图 46

在第一次回答中，ChatDBA 提示我们可以通过暂时禁用严格模式解决该问题，但是不推荐该操作，我们也问了 ChatDBA，学习了严格模式的具体定义。

6.2.4 实验总结

MySQL 5.7 对数据检查会比较严格，在 UPDATE 或 INSERT 场景中字段为 Decimal 数据类型的话，默认严格模式下此类型是无法与""空字符串相比较的，虽然能够关闭严格模式以确保问题 SQL 的执行，但非常不建议这么做。

6.3 问问 ChatGPT-4o

具体如图 47 所示。

图 47

可以看到，ChatGPT 也能够对该问题进行初步的判断与排查，但是缺少深入追问与连续对话的能力。

7 MySQL 清理 undo log 居然用了 10 个小时？

7.1 问题

undo log 长时间处于清理状态导致备份失败。

问题现象：有客户反映有 3 套 MGR 环境全备失败，MySQL 版本为 8.0.18，Xtrabackup 版本为 8.0.9。报错信息如下：

```
xtrabackup: Generating a list of tablespaces
Directories to scan '.;./;.'
Scanning './'
Completed space ID check of 2 files.
Allocated tablespace ID 12 for zxc/a, old maximum was 0
Undo tablespace number 1 was being truncated when mysqld quit.
Cannot recover a truncated undo tablespace in read-only mode
xtrabackup: error: xb_load_tablespaces() failed with error code 57
```

7.2 实验

7.2.1 将问题丢给 ChatDBA

我们先把这个问题丢给 ChatDBA，让它看一下具体出了什么问题（见图 48）。

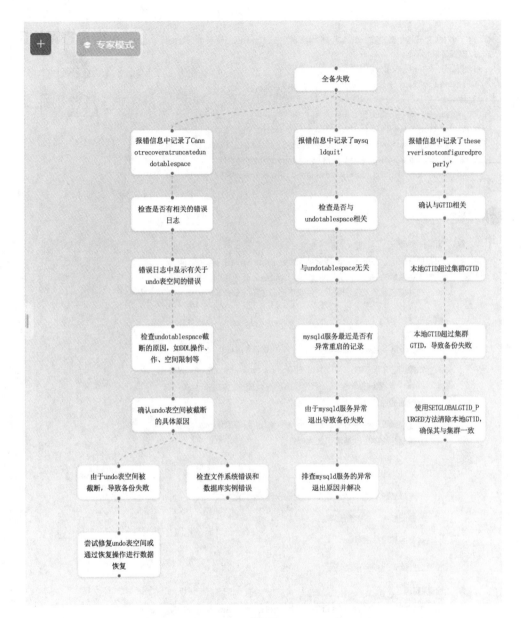

图 48

图 48 为本次问题生成的流程分析画布，展示 ChatDBA 对此问题的排查逻辑。

7.2.2 ChatDBA 问答过程

7.2.2.1 第一轮交互

首先将问题输入至 ChatDBA，ChatDBA 会根据此问题生成初步的参考结果，同时还会生成对应的检索关键词以及进行相关 Bug 的检索。针对这个问题，ChatDBA 还需要收集相关的错误日志来进行下一步分析，同时还给出了一些潜在的操作（见图 49）。

有 3 套 MGR 环境全备失败，MySQL 版本为 8.0.18，xtrabackup 版本为 8.0.9。报错信息如下：
xtrabackup: Generating a list of tablespaces
Directories to scan '.:./:.'
Scanning './'
Completed space ID check of 2 files.
Allocated tablespace ID 12 for zxc/a, old maximum was 0
Undo tablespace number 1 was being truncated when mysqld quit.
Cannot recover a truncated undo tablespace in read-only mode
xtrabackup: error: xb_load_tablespaces() failed with error code 57

人类专家评论区

ChatDBA

○ 初步参考结果

报错信息表明，在备份期间遇到的问题是 undo tablespace 被截断。在恢复这种截断的 undo tablespace 时，xtrabackup 无法以只读模式进行。这种错误通常发生在数据文件或事务日志不一致时。

解决方法如下：

1. 确保 MySQL 实例已经正常关闭，然后尝试再次运行 xtrabackup。
2. 如果问题依旧存在，可以尝试删除 undo tablespace 文件（例如 ibdata1）以强制重新生成，但是请务必在此操作之前进行全备并确保数据的完整性。
3. 另一个潜在的解决方法是升级 xtrabackup 工具和 MySQL 版本，因为较新的版本可能修复了此类问题。

升级步骤：

1. 下载最新版本的 Percona Xtrabackup 和 MySQL。
2. 卸载旧版本，安装新版本。
3. 尝试再次进行全备。

希望这些建议对你有帮助！

○ 检索关键词

检索关键词是：MySQL 8.0.18 MGR 环境，全备失败，xtrabackup 8.0.9，error code 57，Cannot recover a truncated undo tablespace.

○ 相关的已知 Bug 搜索

未搜索到相关已知 Bug

根据您提供的日志信息，我们可以初步定位到问题主要是由于 `undo tablespace` 被截断，导致 xtrabackup 无法完成全备。其报错信息包括：

- `Cannot recover a truncated undo tablespace in read-only mode`
- `Undo tablespace number 1 was being truncated when mysqld quit`

点评： ChatDBA 先尝试初步分析并在 Bug 库中查找相关 Bug，但没有找到匹配的结果。

图 49（a）

可能的解决方案:

1. 检查 MySQL 错误日志

首先,我们需要检查 MySQL 的错误日志,确认 `undo tablespace` 截断的具体原因。请执行以下命令:

```
cat /var/log/mysql/error.log | grep -i 'undo'
```

找出相关错误信息,并提供给我们进一步诊断。

2. 使用 `innodb_force_recovery` 选项启动 MySQL

如果错误日志中确实存在与 `undo tablespace` 截断相关的信息,可以尝试以下方法。注意:此方法有潜在的数据丢失风险,请务必在执行前备份所有数据。

```
innodb_force_recovery = 1
```

逐步将 `innodb_force_recovery` 的值从 1 增加到 6,尝试修复数据。如果仍未解决,请及时停止进行进一步的手动干预。

3. 关闭 undo tablespace 的 truncate 操作

为了防止将来类似的问题,可以在 `my.cnf` 配置文件中添加以下内容以禁用 `undo tablespace` 截断操作:

```
innodb_undo_log_truncate = 0
```

然后重启 MySQL 服务:

```
systemctl restart mysqld
```

进一步确认信息:

如果上述方法没有解决问题,请提供以下信息以便我们更精确地诊断问题:

- MySQL 和操作系统的详细错误日志。
- undo tablespace 截断时的具体操作记录,如是否有长时间运行的事务或DDL操作等。
- 全备失败的更详细的报错信息和步骤。

通过以上步骤和信息收集,我们希望能更精准地解决您的问题。请根据提供的信息进行操作,并将新的结果反馈给我们。

图 49(b)

7.2.2.2 第二轮交互

然后我们提供了对应的错误日志，ChatDBA 根据错误日志首先确认是由于 InnoDB 尝试访问丢失的 tablespace 导致的问题，同时给出了对应的解决方案（见图 50）。

人类专家评论区

点评：ChatDBA 在此处不再追究为什么 undo log 被截断，而是倾向于直接给出解决方案，属于治标不治本的应急措施。

图 50（a）

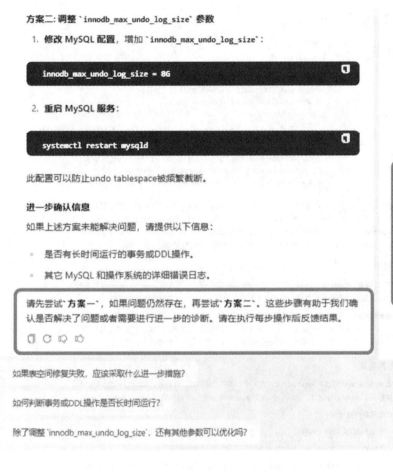

图 50（b）

7.2.3 实验总结

在该问题中，通过排查发现 undo log 过了 10 个小时依然没有清理完，正常情况下不会出现该情况。这是由于参数 super_read_only 触发的 bug 导致的。可以通过调大 innodb_max_undo_log_size 参数，使 undo log 大小达到阈值前被清理掉，这样内存空间可以重用，undo log 也很难达到阈值，所以不会触发 undo log truncate，也就不会触发 bug 而导致问题。

7.3 问问 ChatGPT-4o

具体如图 51 所示。

> 有客户反映有 3 套 MGR 环境全备失败,MySQL 版本为 8.0.18,xtrabackup 版本为 8.0.9。报错信息如下:
> xtrabackup: Generating a list of tablespaces
> Directories to scan '.;./;.'
> Scanning './'
> Completed space ID check of 2 files.
> Allocated tablespace ID 12 for zxc/a, old maximum was 0
> Undo tablespace number 1 was being truncated when mysqld quit.
> Cannot recover a truncated undo tablespace in read-only mode
> xtrabackup: error: xb_load_tablespaces() failed with error code 57

根据您提供的错误信息,问题可能与 MySQL 8.0.18 和 xtrabackup 8.0.9 之间的兼容性问题有关,特别是在备份时处理 undo tablespace(撤销表空间)时出现了问题。具体错误信息指出 "Undo tablespace number 1 was being truncated when mysqld quit. Cannot recover a truncated undo tablespace in read-only mode",这表明 xtrabackup 在处理撤销表空间时遇到了麻烦。

可能原因:

1. **MySQL 与 Xtrabackup 版本不匹配**:您使用的 MySQL 版本为 8.0.18,而 xtrabackup 版本是 8.0.9,可能是这两个版本之间存在不完全兼容的问题,尤其是对撤销表空间的处理。
2. **撤销表空间处于截断状态**:在错误日志中提到 `Undo tablespace number 1 was being truncated when mysqld quit`,表明撤销表空间在 MySQL 关闭时可能还未完成截断操作。由于 xtrabackup 处于只读模式,它无法恢复或处理处于截断状态的撤销表空间。

解决方案建议:

1. **升级 xtrabackup 版本**:Xtrabackup 8.0.9 可能存在已知问题,建议升级到与 MySQL 8.0.18 完全兼容的 xtrabackup 版本。例如,升级到 Xtrabackup 8.0.18 或更高版本来解决兼容性问题。
2. **手动清理撤销表空间**:
   - 确保 MySQL 服务已经停止。
   - 找到并手动删除未完成截断的撤销表空间。可以尝试如下步骤:
     - 在 MySQL 配置文件中,确认撤销表空间的位置(通常是 `innodb_undo_directory` 配置项)。
     - 检查这些文件的状态,尝试通过 MySQL 提供的工具来修复或删除损坏的撤销表空间。
     - 重启 MySQL 并确保它可以正常启动。
3. **跳过撤销表空间的恢复**:如果问题仅限于撤销表空间,并且这些撤销表空间对恢复过程不是必需的,可以尝试配置 xtrabackup 以跳过这些表空间的恢复。您可以在备份命令中添加选项,以忽略撤销表空间的处理(如使用 `--innodb-undo-tablespaces` 选项)。
4. **从新备份开始**:如果撤销表空间的问题持续发生,可能需要通过新的全备来替代旧的备份。确保在新的备份过程中,MySQL 服务没有未完成的撤销表空间操作。

后续操作:

1. 首先,建议您检查是否有可能升级 xtrabackup 到与 MySQL 8.0.18 完全兼容的版本。
2. 确保 MySQL 的撤销表空间处于健康状态。
3. 进行完整备份之前,检查 MySQL 日志,确保没有其他数据库问题干扰备份过程。

如果升级 xtrabackup 和手动修复撤销表空间之后问题仍然存在,可以进一步检查数据库文件系统权限、xtrabackup 日志以及 MySQL 的错误日志以获取更多线索。

图 51

8 MySQL Slave 异常掉电后主从同步出现异常怎么排查？

8.1 问题

MySQL 主从同步出现异常。

问题现象：客户的 MySQL 高可用集群中某台 Slave 机异常掉电后，主从同步出现了异常。

下面我们将带着问题，与 ChatDBA 进行五轮交互，精彩内容即将开始！

8.2 实验

8.2.1 ChatDBA 专家模式

专家模式的 ChatDBA 在第一轮对话开始后，会根据问题生成"根因分析树"，展示 ChatDBA 对问题的排查逻辑，启发 DBA 快速定位问题。

8.2.2 ChatDBA 问答过程

8.2.2.1 第一轮交互

将故障问题的描述输入给 ChatDBA。

(1) ChatDBA 会根据问题生成初步参考结果（见图 52）。

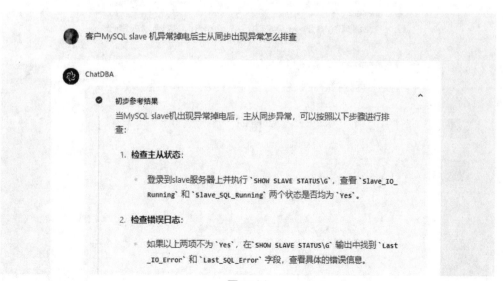

图 52（a）

3. 重启主从复制:

- 尝试停止并重启复制:

```
STOP SLAVE;
START SLAVE;
```

然后重新执行 `SHOW SLAVE STATUS\G`，查看状态是否恢复。

4. 检查relay log和master log:

- 查看`slae`的Relay Log以及`master`的Binary Log状态。确保没有日志损坏或者缺失。
- 如果日志受损，可以尝试跳过错误，执行如下命令:

```
SET GLOBAL sql_slave_skip_counter = 1;
START SLAVE;
```

5. 数据一致性检查:

- 如果发现数据不一致，可以使用工具如`pt-table-checksum`和`pt-table-sync`进行数据修复。

6. 全量恢复:

- 如果问题依然没有解决，可能需要重新搭建`slave`，包括全量备份主服务器上的数据，并重新进行数据同步。

这些步骤通常可以帮助你诊断和解决MySQL主从同步出现的问题。

图 52（b）

(2) 生成对应的检索关键词并检索相关资料（Bug）（见图 53）。

- 检索关键词

  检索关键词是:MySQL slave 异常掉电，主从同步异常，问题排查，数据库同步故障。

- 相关资料搜索

  - 未搜索到相关已知Bug

图 53

(3)ChatDBA给出了排查步骤,提醒需要收集MySQL Slave的具体状态来进一步分析,同时提供了具体的查询命令以及需要重点观察的内容(见图54)。

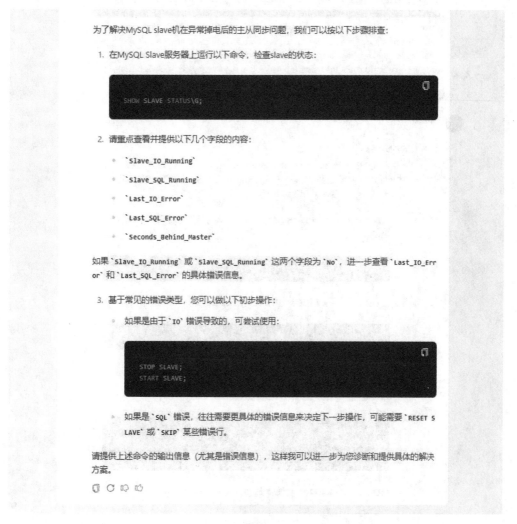

图54

8.2.2.2 第二轮交互

根据上一轮ChatDBA给出的排查步骤,查询Slave状态并输入给ChatDBA。

(1)ChatDBA根据MySQL Slave的具体状态以及报错信息,初步判断问题是由MySQL Slave的中继日志或Master的二进制日志损坏所导致的(见图55)。

在 slave mysql 上采用show slave status\G查看 slave 状态：

发现 IO 线程正常，但是 SQL 线程 NO，错误如下：Last_SQL_Errno: 1594
Last_SQL_Error: Relay log read failure: Could not parse relay log event entry. The possible reasons are: the master's binary log is corrupted (you can check this by running 'mysqlbinlog' on the binary log), the slave's relay log is corrupted (you can check this by running 'mysqlbinlog' on the relay log), a network problem, or a bug in the master's or slave's MySQL code. If you want to check the master's binary log or slave's relay log, you will be able to know their names by issuing 'SHOW SLAVE STATUS' on this slave.

 ChatDBA

- 初步参考结果

  根据您提供的 `SHOW SLAVE STATUS\G` 输出，错误码 `1594` 表示在读取relay log 时发生故障，有可能是 relay log 损坏、master的 binary log 损坏、网络问题或 MySQL 代码中的问题。以下是解决步骤：

  1. 检查 relay log 和 master 的 binary log： 在 slave 上运行以下命令查看 relay log：

     ```
     mysqlbinlog /path/to/relay-log-file
     ```

     在 master 上运行以下命令检查 master 的 binary log：

     ```
     mysqlbinlog /path/to/binlog-file
     ```

     确认 log 文件是否有损坏。

  2. 如果确认是 relay log 损坏，可以尝试重置 relay log：

     ```
     STOP SLAVE;
     RESET SLAVE;
     START SLAVE;
     ```

     注意，`RESET SLAVE` 会清除当前的 relay logs 和 master info。

  3. 如果问题仍然存在，建议：

     - 确保网络连接稳定无误。
     - 尝试从最新的 master 数据库备份重建 slave 机器。

  这可以帮助明确问题的根本原因，并采取相应的措施解决主从同步故障。

- 检索关键词

  检索关键词是:MySQL主从同步异常，slave SQL线程 NO，Last_SQL_Errno 1594，Relay log read failure，binary log corrupted，mysqlbinlog。

- 相关资料搜索

  - 未搜索到相关已知Bug

图 55

(2)为了进一步确定具体原因,ChatDBA 还需要收集错误日志来进一步分析(见图 56)。

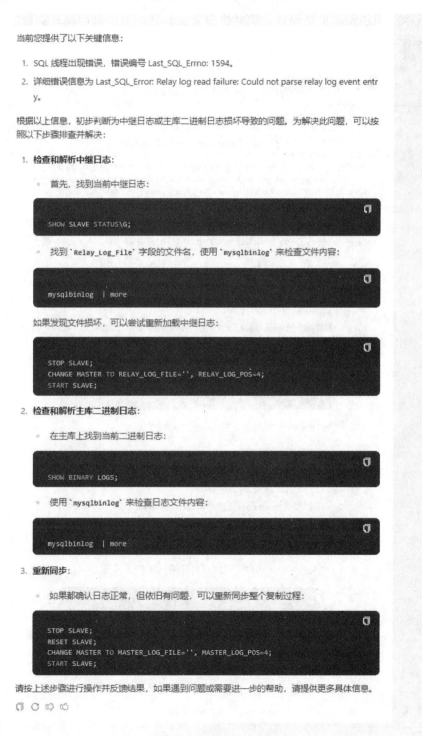

图 56

8.2.2.3 第三轮交互

根据上一轮 ChatDBA 给出的排查步骤，提供相应的错误日志给 ChatDBA。

(1)ChatDBA 根据错误日志信息确定主从异常的原因是中继日志读取失败（见图 57）。

查看 mysql 日志报错如下:
160817 18:01:14 [ERROR] Error in Log_event::read_log_event(): 'read error', data_len: 1011, event_type: 23
160817 18:01:14 [ERROR] Error reading relay log event: slave SQL thread aborted because of I/O error
160817 18:01:14 [ERROR] Slave SQL: Relay log read failure: Could not parse relay log event entry. The possible reasons are: the master's binary log is corrupted (you can check this by running 'mysqlbinlog' on the binary log), the slave's relay log is corrupted (you can check this by running 'mysqlbinlog' on the relay log), a network problem, or a bug in the master's or slave's MySQL code. If you want to check the master's binary log or slave's relay log, you will be able to know their names by issuing 'SHOW SLAVE STATUS' on this slave. Error_code: 1594
160817 18:01:14 [ERROR] Error running query, slave SQL thread aborted. Fix the problem, and restart the slave SQL thread with "SLAVE START". We stopped at log 'mysql-bin.000022' position 1068142925

ChatDBA

初步参考结果

根据您提供的错误日志信息，问题出在中继日志（relay log）读取错误上，并提示可能的原因是中继日志或主库二进制日志的损坏。具体错误如下:

```
[ERROR] Error in Log_event::read_log_event(): 'read error', data_len: 1
[ERROR] Error reading relay log event: slave SQL thread aborted because
[ERROR] Slave SQL: Relay log read failure: Could not parse relay log ev
```

为了解决这个问题，可以按以下步骤进行排查和修复:

1. **检查并解析中继日志**:
    - 找到当前的中继日志文件名，可以通过 `SHOW SLAVE STATUS\G;` 查看并找到 `Relay_Log_File` 字段。
    - 使用 `mysqlbinlog` 工具解析中继日志文件内容:

    ```
    mysqlbinlog /path/to/relay-log-file | more
    ```

    如果发现中继日志确实损坏，可以重新加载中继日志:

    ```
    STOP SLAVE;
    CHANGE MASTER TO RELAY_LOG_FILE='relay-log-file', RELAY_LOG_POS=4;
    START SLAVE;
    ```

2. **检查主库的二进制日志**:
    - 在主库上找到当前二进制日志文件名，通过 `SHOW BINARY LOGS;` 查看。
    - 使用 `mysqlbinlog` 工具解析日志文件内容:

图 57(a)

```
mysqlbinlog /path/to/binlog-file | more
```

3. 重新同步：
- 如果中继日志和二进制日志没问题，但是主从依然同步失败，可以考虑重新同步：

```
STOP SLAVE;
RESET SLAVE;
CHANGE MASTER TO MASTER_LOG_FILE='master-binlog-file', MASTER_LOG_F
START SLAVE;
```

请针对上述步骤进行排查。如果问题依然存在，请提供更多的日志信息进行深度分析。

- 检索关键词

  检索关键词是:MySQL主从同步，Relay log read failure，Error_code: 1594, 中继日志损坏，二进制日志损坏，slave SQL 线程错误。

- 相关资料搜索

  相关的已知Bug: Bug #75746 set log_bin_use_v1_row_events lead to corrupted row event 在MySQL的主从复制环境中，当使用row格式的binlog时，频繁的写/更新/删除操作以及切换全局变量log_bin_use_v1_row_events可能导致主库二进制日志损坏，从而引发从库的IO和SQL线程停止，并报错1236和1594。错误日志中还可能显示内存不足和无效事件的问题。根本原因是log_bin_use_v1_row_events变量的不一致使用，导致在解析行事件时数据大小计算错误。解决方法是修改Rows_log_event::write_data_header()的代码逻辑以避免不一致性。

图 57（b）

(2) 生成对应的检索关键词以及检索相关资料（Bug）。发现该问题相关的已知 Bug 为 75746 号 Bug（见图 58）。

- 检索关键词

  检索关键词是:MySQL主从同步，Relay log read failure，Error_code: 1594, 中继日志损坏，二进制日志损坏，slave SQL 线程错误。

- 相关资料搜索

  相关的已知Bug: Bug #75746 set log_bin_use_v1_row_events lead to corrupted row event 在MySQL的主从复制环境中，当使用row格式的binlog时，频繁的写/更新/删除操作以及切换全局变量log_bin_use_v1_row_events可能导致主库二进制日志损坏，从而引发从库的IO和SQL线程停止，并报错1236和1594。错误日志中还可能显示内存不足和无效事件的问题。根本原因是log_bin_use_v1_row_events变量的不一致使用，导致在解析行事件时数据大小计算错误。解决方法是修改Rows_log_event::write_data_header()的代码逻辑以避免不一致性。

图 58

(3)ChatDBA 分析出线程停止的具体位置并给出了对应的解决方案（见图 59）。

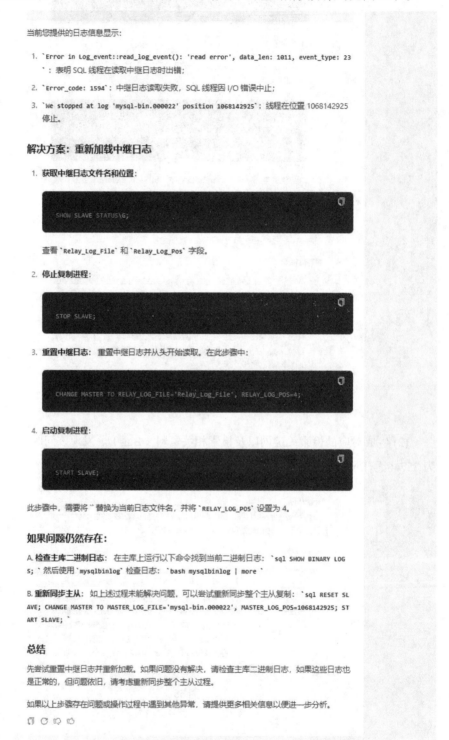

图 59

8.2.2.4 第四轮交互

根据上一轮 ChatDBA 给出的解决方案,客户顺利解决了问题!

ChatDBA 根据解决情况提醒用户需要继续观察一段时间,确保问题不再发生(见图 60)。

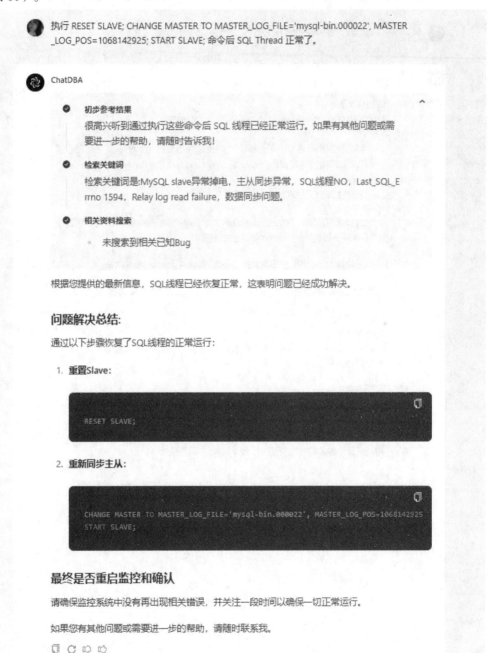

图 60

8.2.2.5 第五轮交互

根据上一轮中 ChatDBA 给出的提醒，对数据库继续观察一段时间，发现数据库一切运行正常。

(1) 将观察到的信息提供给 ChatDBA 并询问问题产生的原因（见图 61）。

> 观察查一段时间后，数据库一切运行正常。这个错误的原因是什么?

ChatDBA

✅ **初步参考结果**
很高兴得知数据库已经恢复正常。关于这个错误的原因：

1. **中继日志损坏**：在slave服务器异常掉电后，中继日志（relay log）可能会损坏。在您执行SQL时出现 `Relay log read failure: Could not parse relay log event entry` 和 `Error in Log_event::read_log_event()` 就表明中继日志出现了无法解析的错误。

2. **二进制日志问题**：主库的二进制日志文件可能也存在损坏或异常，但您的最终解决步骤表明更可能是中继日志的问题。

3. **I/O问题**：掉电后可能导致磁盘I/O错误，影响日志的完整性和可读性。

您通过重置slave并重新设置主从同步位置的操作有效绕过或修复了损坏的日志，从而解决了问题。

✅ **检索关键词**
检索关键词是:MySQL slave，主从同步异常，SQL线程错误，1594错误，中继日志损坏，重置Slave。

✅ **相关资料搜索**
- 未搜索到相关已知Bug

观察了一段时间后，数据库一切运行正常，说明之前的问题已经解决。

图 61

(2)ChatDBA 总结了导致问题的原因、最终解决方法，并给出避免产生类似情况的建议（见图 62）。

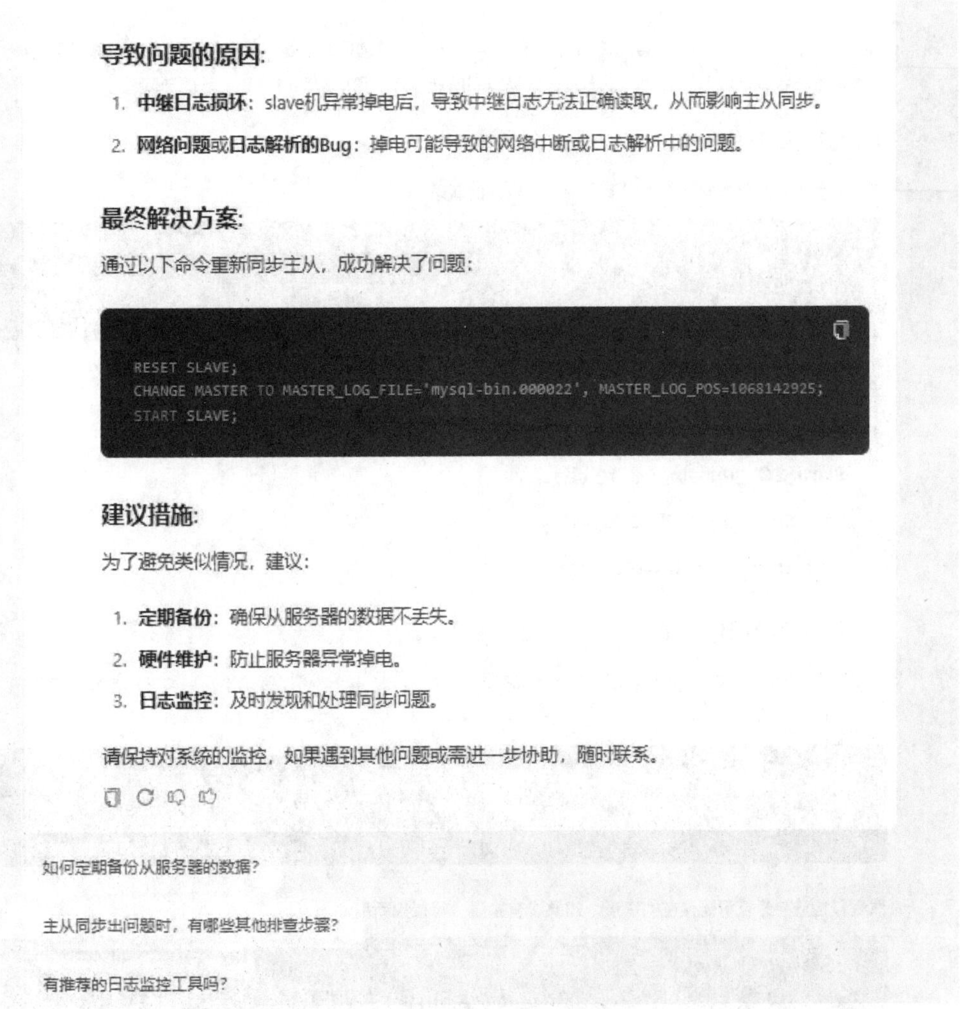

导致问题的原因：
1. **中继日志损坏**：slave 机异常掉电后，导致中继日志无法正确读取，从而影响主从同步。
2. **网络问题**或**日志解析的 Bug**：掉电可能导致的网络中断或日志解析中的问题。

最终解决方案：
通过以下命令重新同步主从，成功解决了问题：

```
RESET SLAVE;
CHANGE MASTER TO MASTER_LOG_FILE='mysql-bin.000022', MASTER_LOG_POS=1068142925;
START SLAVE;
```

建议措施：
为了避免类似情况，建议：
1. **定期备份**：确保从服务器的数据不丢失。
2. **硬件维护**：防止服务器异常掉电。
3. **日志监控**：及时发现和处理同步问题。

请保持对系统的监控，如果遇到其他问题或需进一步协助，随时联系。

如何定期备份从服务器的数据？

主从同步出问题时，有哪些其他排查步骤？

有推荐的日志监控工具吗？

图 62

8.2.3 实验总结

MySQL 主从同步异常通常可能由多种原因引起，如中继日志损坏、二进制日志错误、网络问题或 MySQL Bug 等。在此次的问题中，由于 MySQL Slave 机异常掉电，中继日志（relay log）受损，导致 SQL 线程无法正确解析日志事件，进而引发同步中断。

8.3 问问 ChatGPT-4o

具体如图 63 所示。

客户 MySQL slave 机异常掉电后主从同步出现异常

MySQL 主从同步在从服务器异常掉电后出现异常是一个比较常见的问题。通常情况下，这种情况可能导致数据不一致或从库无法启动。下面是一些排查和解决步骤，可以帮助你修复主从同步问题：

1. 检查从服务器状态

首先，登录到从服务器，使用以下命令检查当前的状态：

注意检查以下字段：

- **Slave_IO_Running**: 是否在运行。
- **Slave_SQL_Running**: 是否在运行。
- **Last_Errno**: 上一个错误的错误码。
- **Last_Error**: 上一个错误的具体信息。

2. 启动从服务器

如果发现 `Slave_IO_Running` 或 `Slave_SQL_Running` 为 `No`，可以尝试重新启动主从同步。首先执行：

然后，检查并修复可能存在的错误。如果没有错误，可以重新启动同步：

3. 检查错误日志

查看 MySQL 的错误日志文件，以获取更详细的信息，可能会有关于异常掉电的相关信息。

4. 数据一致性检查

由于掉电可能导致数据不一致，可以考虑使用 `pt-table-checksum` 工具来检查主从之间的数据一致性。如果发现不一致，可以使用 `pt-table-sync` 来进行数据同步。

5. 恢复备份

如果问题较为严重，且无法通过上述步骤解决，可以考虑从主库重新备份数据到从库。确保在备份过程中，主库处于一致性状态。

图 63（a）

6. 配置检查

确保在 `my.cnf` 配置文件中，主从服务器的配置正确，包括 `server-id`、`log_bin`、`relay_log` 等参数。

7. 查看网络连接

如果从服务器无法连接到主服务器，也会导致同步失败。检查网络连接是否正常。

8. 版本兼容性

确认主从服务器的 MySQL 版本兼容。如果主库和从库版本不一致，可能会引起问题。

通过以上步骤，通常可以找出并解决 MySQL 主从同步异常的问题。如果仍然存在问题，建议进一步查看官方文档或寻求专业的技术支持。

图 63（b）

更多内容见图 64。

爱可生开源社区

什么是 ChatDBA？

ChatDBA

ChatDBA 是一款基于大模型的智能数据库助手（**Database Copilot**）。通过对话交互的方式进行数据库故障诊断、数据库专业知识学习、SQL 生成与 SQL 优化等功能，大幅提升数据库从业者工作效率。

无论中小企业还是大型组织，ChatDBA 都能提供强大支持，确保业务数据库平稳高效运行。

加入社区

预约试用

联系我们
（请备注 ChatDBA）

图 64